SYNCHROTRON RADIATION INSTRUMENTATION

SYNCHROTRON RADIATION INSTRUMENTATION

Tenth US National Conference

Ithaca, New York   June 1997

EDITOR
Ernest Fontes
*Cornell High Energy Synchrotron Source,
Cornell University*

American Institute of Physics

AIP CONFERENCE PROCEEDINGS 417

Woodbury, New York

Authorization to photocopy items for internal or personal use, beyond the free copying permitted under the 1978 U.S. Copyright Law (see statement below), is granted by the American Institute of Physics for users registered with the Copyright Clearance Center (CCC) Transactional Reporting Service, provided that the base fee of $10.00 per copy is paid directly to CCC, 222 Rosewood Drive, Danvers, MA 01923. For those organizations that have been granted a photocopy license by CCC, a separate system of payment has been arranged. The fee code for users of the Transactional Reporting Service is: 1-56396-742-1/ 97/$10.00.

© 1997 American Institute of Physics

Individual readers of this volume and nonprofit libraries, acting for them, are permitted to make fair use of the material in it, such as copying an article for use in teaching or research. Permission is granted to quote from this volume in scientific work with the customary acknowledgment of the source. To reprint a figure, table, or other excerpt requires the consent of one of the original authors and notification to AIP. Republication or systematic or multiple reproduction of any material in this volume is permitted only under license from AIP. Address inquiries to Office of Rights and Permissions, 500 Sunnyside Boulevard, Woodbury, NY 11797-2999; phone: 516-576-2268; fax: 516-576-2499; e-mail: rights@aip.org.

L.C. Catalog Card No. 97-77402
ISBN 1-56396-742-1
ISSN 0094-243X
DOE CONF- 9706157

Printed in the United States of America

Contents

Preface ix

Facilities: History, Status, and Upgrades

The NIST SURF II Storage Ring is Upgraded 3
R. P. Madden

APS Insertion Devices: Magnetic Performance and Radiation Characteristics 4
E. Gluskin

The Structural Biology Center at the APS: An Integrated User Facility for Macromolecular Crystallography 5
G. Rosenbaum and E. M. Westbrook

Optical Design and Performance of the Phase II Inelastic Scattering Beamline at the National Synchrotron Light Source 6
W. A. Caliebe, C.-C. Kao, M. Krisch, T. Oversluizen, P. Montanez, and J. B. Hastings

UHV Surface-Analysis Endstation with X-ray Scattering and Spectroscopic Capabilities 10
P. F. Lyman, D. T. Keane, and M. J. Bedzyk

Commission of the PNC-CAT Insertion Device Beamline at the APS 15
D. T. Jiang, S. M. Heald, E. A. Stern, E. D. Crozier, B. Barg, K. H. Kim, R. A. Gordon, D. Brewe, and F. Brown

BioCAT Undulator Beamline at APS 16
G. B. Bunker, T. Irving, E. Black, K. Zhang, R. Fischetti, S. Wang, and S. Stepanov

Expanded Capabilities from SRC's High Energy Resolution Variable Line Density Grating Monochromator Beamline: HERMON 17
M. Bissen, M. Fisher, R. Reininger, G. Rogers, and H. Höchst

A New Time-of-Flight Mass Spectrometer for Electron-Ion and Ion-Ion Coincidence Experiments (PEPICO, PIPICO and PEPIPICO) 22
J. B. Maciel, E. Morikawa, and G. G. B. de Souza

Sources of X-ray and IR: Insertion Device, Laser Excited, and Other

Design of a X-ray Free Electron Laser Undulator 29
R. Carr

Long-Wavelength Edge Radiation in an Electron Storage Ring 35
R. A. Bosch and O. V. Chubar

Control System for Insertion Devices at the Advanced Photon Source 43
O. A. Makarov, P. D. Hartog, E. R. Moog, and M. L. Smith

Characterization of the Coherent Microwave Emission from the SURF II Synchrotron Storage Ring 48
G. Fraser, A. R. Hight Walker, U. Arp, T. Lucatorto, and K. K. Lehmann

Characterization of the Elliptical Multipole Wiggler at the Advanced Photon Source 49
P. Ilinski, C. T. Venkataraman, J. C. Lang, and G. Srajer

Computation of Undulator Tuning Curves 55
R. J. Dejus

Magnetic Field Characterization of the NIST Undulator 56
L. E. Johnson, G. Denbeaux, J. M. J. Madey, and K. D. Straub

Soft X-Ray Sources on the Duke Storage Ring 57
L. E. Johnson, G. Denbeaux, N. Hower, J. M. J. Madey, and K. D. Straub

Resorting the NIST Undulator Using Simulated Annealing for Field Error Reduction 58
G. Denbeaux, L. E. Johnson, J. M. J. Madey, and K. D. Straub

Making Beams: Novel Optics Design, Fabrication, Testing, and Use

Cryogenic High-Heat-Load Optics at the Advanced Photon Source 61
    C. S. Rogers

Silver Bonded, Internally Water-Cooled Monochromators for CHESS Wiggler Beamlines 66
    K. W. Smolenski, Q. Shen, and P. Doing

Performance of the Double Multilayer Monochromator on the NSLS Wiggler Beam Line X25 71
    L. E. Berman, Z. Yin, S. B. Dierker, E. Dufresne, S. G. J. Mochrie, O. K. C. Tsui, S. K. Burley, F. Shu,
    X. Xie, M. S. Capel, and R. M. Sweet

Inelastic X-Ray Scattering at Modest Energy Resolution 80
    K. D. Finkelstein, J. Z. Tischler, and B. C. Larson

X-Ray Monochromators with Sub-MeV Resolution 88
    T. S. Toellner, M. Y. Hu, W. Sturhahn, P. Hession, E. Alp, and J. Sutter

Test Results of a Diamond Double-Crystal Monochromator at the Advanced Photon Source 89
    P. B. Fernandez, T. Graber, S. Krasnicki, W.-K. Lee, D. M. Mills, C. S. Rogers, and L. Assoufid

A Tunable Laue Bent-Laue Monochromator with Fixed Second Crystal for Synchrotron Radiation 95
    Z. Zhong, G. Le Duc, D. Chapman, and W. Thomlinson

Beam Size Measurement of the Stored Electron Beam at the APS Storage Ring Using Zone Plate Optics and Undulator Radiation 101
    Z. Cai, B. Lai, W. Yun, E. Gluskin, D. Legnini, P. Ilinski, E. Trakhtenberg, S. Xu, W. Rodrigues,
    and H. R. Lee

Beam-Smiling in Bent-Laue Monochromators 106
    B. Ren, F. A. Dilmanian, L. D. Chapman, X. Y. Wu, Z. Zhong, I. Ivanov, W. C. Thomlinson, and X. Huang

The Performance of a Wide Band X-Ray Bragg Polarizer Grown by Molecular Beam Epitaxy 117
    J. O. Cross, B. R. Bennett, M. I. Bell, and K. J. Kuhn

Measurement of Diffraction Gratings with a Long Trace Profiler with Application for Synchrotron Beamline Gratings 118
    S. C. Irick and W. R. McKinney

Design and Performance of Two New Double Crystal Monochromators 124
    D. J. Holly, W. P. Mason, F. H. Middleton, T. Sailor, and R. E. Smith

New Actively Bent Plane Mirror at SRC 130
    T. Kubala, M. Fisher, R. Reininger, and M. Severson

Preliminary Results From A New Plane Grating Monochromator at SRC 135
    M. Severson, M. Bissen, R. Reininger, M. V. Fisher, G. Rogers, D. Eisert, T. Kubala, and W. Wood

Variation of Q with Energy in Mosaic Analyzers for Inelastic X-Ray Measurements 140
    J. Z. Tischler, B. C. Larson, and P. Zschack

Small Stuff: Making and Imaging Microstructures

Glass Capillary Optics for Making X-Ray Beams of 0.1 to 50 Microns Diameter 147
    D. H. Bilderback and E. Fontes

Beyond Sunshine: Hard X-Rays for Precision Microfabrication 156
    E. D. Johnson, D. P. Siddons, J. C. Milne, H. Gückel, and J. L. Klein

X-Ray Lithography at CXrL–3D Nanostructures 157
    Y. Vladimirsky

X-Ray Fluorescence Correlation Spectroscopy for Studying Particle Dynamics in Condensed Matter ... 158
    J. Wang, A. K. Sood, P. Satyam, Y. Feng, X. Wu, Z. Cai, W. Yun, and S. K. Sinha

A Beamline for Micromachining and Micro-Characterization at the APS 159
    B. Lai, W. Yun, D. C. Mancini, F. DeCarlo, D. Shu, and J. Chang

X-Ray Microdiffraction Studies of an Integrated Laser-Modulator System 161
    W. Rodrigues, Z. Cai, W. Yun, H.-R. Lee, P. Ilinski, E. Isaacs, and J. Grenko

Exposure Station with Precision Scanning Stage for Deep X-Ray Lithography 165
    D. C. Mancini, F. DeCarlo, and B. Lai

X-Ray Microdiffraction Studies to Measure Strain Fields in a Metal Matrix Composite 166
    H.-R. Lee, W. Yun, Z. Cai, W. Rodrigues, and D. S. Kupperman

High-Tech: Beam Stability, Beamline Hardware, and Software Control

Progress of the APS High Heat Load X-Ray Beam Position Monitor Development 173
  D. Shu, J. Barraza, H. Ding, T. M. Kuzay, and M. Ramanathan

High Heat Load Fixed Primary Aperture for an Undulator Beamline with Integral Beam Position Monitors 178
  G. Rosenbaum and T. Fornek

Mirror Mounts Designed for the Advanced Photon Source SRI-CAT 179
  D. Shu, C. Benson, J. Chang, J. Barraza, T. M. Kuzay, E. E. Alp, W. Sturhahn,
  B. Lai, I. McNulty, K. Randall, G. Srajer, Z. Xu, and W. Yun

Miniaturized Kappa Goniometer for Macromolecular Crystallography 186
  G. Rosenbaum and E. M. Westbrook

User-Friendly Interfaces for Control of Crystallographic Experiments at CHESS 187
  D. M. E. Szebenyi, A. Deacon, S. E. Ealick, J. M. LaIuppa, and D. J. Thiel

Synchrotron Beam Stabilization Techniques at CHESS 192
  J. White, E. Fontes, and S. Peck

Real-Time: *In-situ* Measurements and Materials Characterization

Real-Time X-Ray Scattering Studies of Thin-Film Growth and Processing 195
  R. L. Headrick

X-Ray Studies of Annealing in Thin-Film Semiconductors 196
  R. Clarke

Low Energy X-Ray Dosimetry Studies (6–16 keV) at SSRL Beamline 1–5 197
  N. Ipe, S. Chatterji, A. Fassò, K. Kase, R. Seefred, P. Olko, P. Bilski, and C. Soares

Ultrafast X-Ray Diffraction of Laser-Irradiated Crystals 204
  P. A. Heimann, J. Larsson, Z. Chang, A. Lindenberg, P. J. Schuck, E. Judd, H. A. Padmore,
  P. H. Bucksbaum, R. W. Lee, M. Murnane, H. Kapteyn, J. S. Wark, and R. W. Falcone

Information Stored in High-Q Space: Role of High Energy Scattering 209
  T. Egami, S. J. L. Billinge, S. Kycia, W. Dmowski, and A. S. Eberhardt

Measurements of EUV Optical Constants of *in-situ* Deposited Films 214
  C. Tarrio, R. N. Watts, T. B. Lucatorto, J. M. Slaughter, and C. M. Falco

Real-Time X-Ray Diffraction Measurements of GaN Growth on Sapphire (0001) 218
  A. R. Woll, R. L. Headrick, S. Kycia, and J. D. Brock

Radiation Doses to Insertion Devices at the Advanced Photon Source 219
  E. R. Moog, P. K. Den Hartog, E. J. Semones, and P. K. Job

Application of Electroreflectance to Stark Spectroscopy 224
  A. K. Gaigalas, T. Ruzgas, and G. Niaura

High-Energy X-Ray Experiments at the APS Sector 1 Beamlines 230
  D. R. Haeffner, S. D. Shastri, and D. M. Mills

Detectors: Large, Small, Fast, Energy-Resolving

X-Ray Imaging Characteristics of a Direct Conversion Detector Using Selenium and Thin Film Transistor Array 233
  B. Rodricks, D. L. Lee, L. K. Cheung, L. S. Jeromin, and E. F. Palecki

Photoemission from Silicon Photodiodes and Induced Changes in the Detection Efficiency in the Far Ultraviolet 234
  R. E. Vest and L. R. Canfield

X-Ray Imaging with Amorphous Silicon Active Matrix Flat-Panel Imagers (AMFPIs) 241
  Y. El-Mohri, L. E. Antonuk, K. W. Jee, M. Maolinbay, X. Rong, J. H. Siewerdsen,
  M. Verma, and Q. Zhao

A New Large Area X-Ray Image Sensor 243
  D. Ouimette

Actual Conference Program	245
Attendee Address and E-mail Directory	249
List of Attending Vendors	259
Author Index	265

Preface

The 10th United States National Conference on Synchrotron Radiation Instrumentation was held June 17-20, 1997 at Cornell University in Ithaca, New York. The SRI National meeting is held every two years, and this year was proudly hosted by CHESS, the Cornell High Energy Synchrotron Source. As a national meeting, it serves as a venue for all light sources in the United States to discuss the latest, most advanced capabilities in the fields of X-ray and VUV synchrotron radiation. Topics for this year's meeting included:

Facilities: history, status, and upgrades
Sources: insertion devices, laser excited and other
High-tech: beam stabilization, beamline hardware and control
Optics: novel design, fabrication, testing and use
Small stuff: making and imaging microstructures
Real-time: *in-situ* measurements and materials characterization
Detectors: large, small, fast, and energy-resolving.

In addition to experts in the field of synchrotron radiation, the detector session included speakers from other disciplines whose work on detector technology may someday impact synchrotron sources.

During the meeting, posters and vendor exhibits provided a backdrop for discussions and refreshment breaks. At the Thursday evening conference banquet, Denis McWan (BNL) provided an honorable and humorous tribute for Professor Boris Batterman, who has served as the Director of CHESS for 19 years and retired from that post in 1997. Adding to the week were several social events, including a picnic outing at Taughannock State Park and a tour of Cayuga Lake wineries.

In all honesty, this meeting and conference could not have been organized without the constant and cheerful assistance of Virginia Bizzell and Karl Smolenski. Virginia managed all business aspects of the week, and Karl single-handedly organized the vendor show. The CHESS staff provided tremendous support; John Kopsa, Dave Corridon and Walt Protas helped get the facility set and rigged the vendor exhibits, and CHESS Staff Scientists Bob Batterman, Don Bilderback, Keith Brister, Ken Finkelstein, Randy Headrick Stefan Kycia, and Qun Shen lent scientific guidance. I am also indebted to the Program Committee members who helped contact speakers from their home institutions.

As far as this Proceedings volume is concerned, my goal was to document the full scope of the meeting, thereby providing memories to those who attended and insightful reading to those who couldn't. Never an easy task, 1997 proved an especially strenuous exercise because members of the synchrotron community were significantly overextended. Nearly a dozen conferences were held in 1997 on topics related to synchrotron radiation sources, X-ray optics, hardware, and scientific applications. In addition to the two SRI meetings (this meeting as well as the 6th International SRI held in Hyogo Japan), some of the other conferences included a Gordon Conference on X-ray Physics, a SPIE meeting section on Synchrotron Radiation Optics, the annual American Crystallographic Association meeting, and additional topical conferences covering X-ray lithography, Micromachining, etc. Of course these meetings added to the docket of the synchrotron radiation scientist, who had also to consider the half dozen or so annual User Meetings and Workshops hosted by each US light source.

This activity is welcomed, of course, since it shows how successful and applicable synchrotron radiation has become. Most of the attendees of the SRI meetings are scientists who work to build and support synchrotron facilities, so they routinely measure success by record demand for their experimental stations and services. I am sorry to report, though, that this Proceedings must use "over-extension" as an excuse for not being able to include a full paper summarizing each talk and poster presented at the 1997 National Conference. Where papers could not be provided in a timely manner, an abstract from the meeting program captures the essence of the presentations.

Ernie Fontes
SRI'97 National Conference Chair

FACILITIES:
HISTORY, STATUS, AND UPGRADES

The NIST SURF II Storage Ring is Upgraded

R. P. Madden

*National Institute of Standards and Technology
Gaithersburg, MD 20899*

The Synchrotron Ultraviolet Radiation Facility (SURF II) at the National Institutes of Standards and Technology (NIST) rose from the ashes of SURF I (a synchrotron light source) in 1974. It has served NIST well for 23 years and proved itself a valuable radiometric calibration source for the UV/VUV spectral region. During the next year, the SURF II storage ring magnet system will be dismantled and a new one installed. The SURF electron orbit is a circle embedded in a single large electromagnet, and the uniformity of the magnetic field therein determines the accuracy of the trajectory and the radiometric predictability of the radiated photon flux. The new system, designed by the Physical Sciences Laboratory in consultation with the Synchrotron Radiation Center of the University of Wisconsin, has been designed to have a magnetic field accuracy of 2 parts in 10^4. The predictability of the radiation distribution will be improved by an order of magnitude, making the SURF III system a cornerstone of the NIST arsenal of radiometric standards throughout the VUV/UV/visible and infrared. Furthermore, the new magnet system will support an increase in the maximum electron energy of the machine to be increased from 300 MeV to over 400 MeV. Thus SURF III will also be able to serve as a source for soft x-ray physics, chemistry and radiometry. The new capability should be on-line in about one year.

APS Insertion Devices: Magnetic Performance and Radiation Characteristics

E. Gluskin

Argonne National Laboratory, Advanced Photon Source
9700 S. Cass Ave., Argonne, IL 60439

There are 23 insertion devices (IDs) at the APS now, and 17 of them have been installed and commissioned. Most of the IDs are 2.4-meter-long hybrid-type undulators with a 3.3 cm period. Two wigglers are used at the APS, one of them is an elliptical multipole wiggler with an electromagnet that produces alternate horizontal magnetic fields.

All APS IDs have undergone fine magnetic tuning in order to minimize various integrals of the field through the ID for the whole range of IDs gaps, including the higher-order integrated multipole moments. Measurements of the particle-beam motion as the field strength of each ID was changed throughout its range found that the closed-orbit distortion agreed well with predictions based on magnetic measurement results. Radiation from the APS undulator has been characterized in terms of absolute spectral flux. Direct measurements of the positron beam emittance have been performed. In addition, the experimentally measured brilliance tuning curves for the APS undulator have been obtained.

This research was supported by the U.S. Department of Energy, BES Materials Sciences, under contract No. W-31-109-ENG-38.

The Structural Biology Center at the APS: an integrated user facility for macromolecular crystallography

G. Rosenbaum and E. M. Westbrook

Structural Biology Center
Argonne National Laboratory
Advanced Photon Source
9700 S. Cass Ave., Argonne, IL 60439

The Structural Biology Center (SBC) has developed and operates a sector (undulator and bending magnet) of the APS as a user facility for macromolecular crystallography. Crystallographically determined structures of proteins, nucleic acids and their complexes with proteins, viruses, and complexes between macromolecules and small ligands have become of central importance in molecular and cellular biology.

Major design goals were to make the extremely high brilliance of the APS available for brilliance limited studies, and to achieve a high throughput of less demanding studies, as well as optimization for MAD-phasing. Crystal samples will include extremely small crystals, crystals with large unit cells (viruses, ribosomes, etc.) and ensembles of closely similar crystal structures for drug design, protein engineering, etc. Data are recorded on a 3000x3000 pixel CCD-area detector (optionally on image plates).

The x-ray optics of both beamlines has been designed to produce a highly demagnified image of the source in order to match the focal size with the sizes of the sample and the resolution element of the detector. Vertical focusing is achieved by a flat, cylindrically bent mirror. Horizontal focusing is achieved by sagitally bending the second crystal of the double crystal monochromator. Monochromatic fluxes of $1.3*10^{13}$ ph/s into focal sizes of 0.08 mm (horizontal) x 0.04 mm (vertical) FWHM (flux density $3.5*10^{15}$ ph/s/mm^2) have been recorded.

This work was supported by the US Department of Energy, Office of Health and Environmental Research, under Contract W31-109-ENG-38.

Optical Design and Performance of the Phase II Inelastic Scattering Beamline at the National Synchrotron Light Source

W.A. Caliebe†, C.-C. Kao, M. Krisch‡, T. Oversluizen, P. Montanez, J.B. Hastings

National Synchrotron Light Source, Brookhaven National Laboratory, Upton NY, 11973
†*present address: Hamburger Synchrotronstrahlungslabor HASYLAB, Deutsches Elektronen-Synchrotron DESY, 22603 Hamburg, Germany*
‡*European Synchrotron Radiation Facility, F-38043 Grenoble Cedex, France*

Abstract We report the optical design and performance of the phase II inelastic scattering beamline at the National Synchrotron Light Source. The new beamline consists of a four-crystal Si(220) monochromator followed by a bent cylinder mirror. The monochromator is tunable from 5 to 10 keV with about 0.2 eV energy resolution throughout the tuning range. The size of the focused beam is about 0.5 mm (H) × 0.3 mm (V).

INTRODUCTION

The main research program at the X21 beamline at the National Synchrotron Light Source (NSLS) is inelastic x-ray scattering studies of electronic excitations in condensed matter. A horizontally focusing monochromator was installed in 1993 to allow initial experiments to be performed with total energy resolution of 1 eV, and to provide a facility for the testing of high energy resolution crystal analyzers[1, 2]. Experiments ranging from collective excitations of simple metals[3], single particle excitations in semiconductors and insulators[4], high resolution Compton scattering [5], and high resolution x-ray resonant Raman scattering[6, 7] were performed. Spherically and cylindrically bent Si and Ge analyzers developed at the NSLS, European Synchrotron Radiation Facility, and Advanced Photon Source were also tested[8, 9].

However, the tuning range of the horizontally focusing monochromator is rather small, which severely limits the number of systems that can be studied by high resolution x-ray resonant Raman scattering. To improve the total energy resolution and to accommodate the increasing demand for resonant Raman scattering, a new beamline, Phase II, was designed and commissioned during spring 1996. The new beamline extends the tunability of the monochromator to cover the K edges of most of the transition elements and the L edges of most of the rare earth elements. The extended energy range also allows the use of either Si(333), Si(444) or Si(555) analyzer in non-resonant Raman scattering experiments. Furthermore, the new beamline improves the energy resolution of the monochromator to 0.2 eV, and provides a doubly focused monochromatic beam. In this paper, the optical design and the performance of the new monochromator is reported.

BEAMLINE DESIGN

The radiation source of the beamline is a 27-pole hybrid wiggler located at one of the straight sections of the 2.584 GeV X-ray storage ring. The horizontal opening angle of the wiggler is 2.5 mrad, and the full vertical opening is about 0.25 mrad. The critical energy at the center of the horizontal fan is 4.885 keV. Detailed information and parameters of the wiggler have been reported previously[10]. The design criteria of the beamline are: (1) high monochromator energy resolution of the order of 0.1 eV in order to match that of the spherically bent analyzers; (2) large monochromator tunability; (3) constant energy resolution throughout the monochromator tuning range; (4) large angular acceptance in both horizontal and vertical direction; (5) doubly focused monochromatic beam.

The new beamline consists of the front end, the monochromator, and the focusing mirror. The front end was not changed during this beamline upgrade, and has been described elsewhere[1]. A schematic of the monochromator and the mirorr is shown in Fig. 1. The monochromator is a dispersive Si(220) four-crystal arrangement with miscut angle of -16°, 0°, 0°, +16°, respectively. The particular choice of crystal arrangement, Si reflection, and miscut angles are results of optimization among the following conflicting factors: large vertical acceptance, wide energy tunability, and high energy resolution. With these parameters, the designed tuning range of the monochromator is from 5 keV to 10 keV; the vertical acceptance is about 0.06 mrad and the energy resolution is about 0.2 eV throughout the whole tuning range. The horizontal acceptance is about 1 mrad, which is determined mainly by the size of the monochromator crystals. It is interesting to note that this type of dispersive monochromator was first proposed by DuMond[11], though only symmetric reflections were used. The use of asymmetric reflections was proposed by Nakayama[12], and has been widely adopted in nuclear resonant scattering experiments[13].

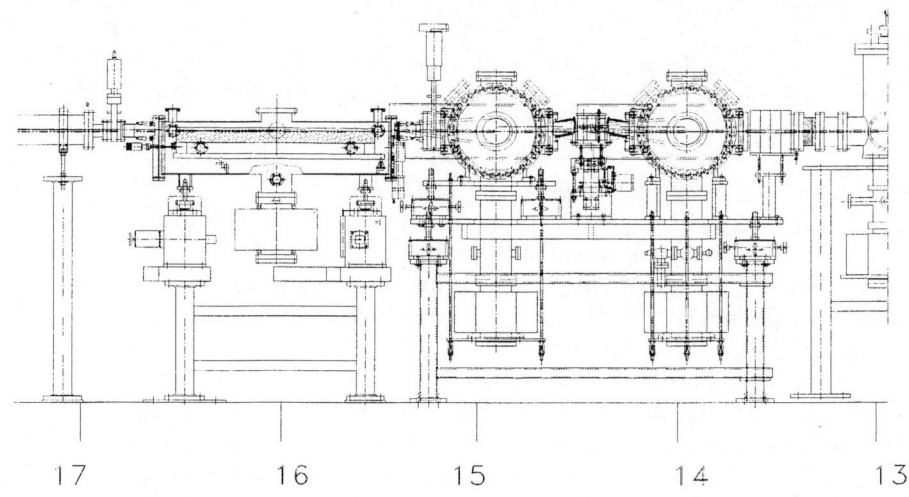

FIGURE 1. Schematic of the phase II beamline X21.

To realize the four-crystal arrangement, two channel-cut crystals are used. The crystals are located at 14 m and 15 m away from the source, respectively, as shown in Fig. 1. The details of the channel-cut crystals are shown in Fig. 2. The second and the third crystals can be detuned by pushing the stainless steel springs using two piezoelectric micrometers. It should be noted also that the purpose of the second and third symmetric Si(220) reflections is to make the monochromator an in-line monochromator and to keep the exit beam height fixed. The resolution of this monochromator is determined by the first and fourth reflections.

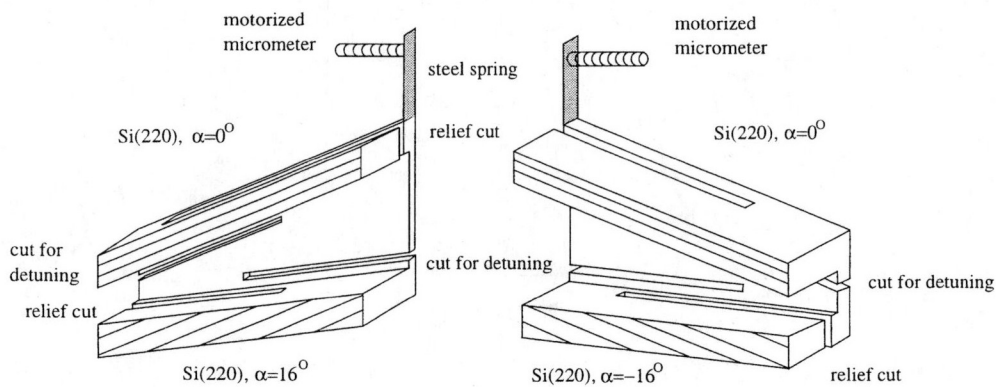

FIGURE 2. Detailed drawing of the channel-cut crystals.

Since the monochromator is the first optical element, the heat load on the first crystal is a major concern. Because of the large miscut of the first crystal, the power density on the first crystal is reduced significantly. The power density is about $0.2\,\text{W/mm}^2$ at 10 keV and increases with decreasing photon energy to about $1.6\,\text{W/mm}^2$ at 5 keV [14]. The heat is removed by coupling the first crystal to a water cooled copper block via GaIn–eutectic.

Downstream from the monochromator is a one meter long bendable cylindrical mirror. The center of the mirror is located at 16 m from the source, and the incident angle is 4 mrad. The fixed radius of the cylinder is 42.7 mm, and the bendable meridional radius is set to 26.7 m. The focus of the mirror is located at 24 m from the source which results in a 2:1 demagnification. The mirror is made of ultra-low expansion glass and coated with 500 Å of platinum. Since the mirror is only exposed to monochromatic radiation, no cooling is provided. The mirror position and angle are adjusted externally by a set of motorized kinematic mounts.

BEAMLINE PERFORMANCE

The energy resolution of the monochromator is measured at the photon energies corresponding to the backscattering energies of Si(333), Si(444), and Si(555) reflections by using a flat Si(111) analyzer crystal operated at exact backscattering angle. A single ionization chamber filled with air was used to measure the photon flux for both the incident and the reflected x-rays as the monochromator tuned through the backscattering energy of the particular reflection. The energy resolution of the monochromator is then deduced from the measured width of the reflected beam and the intrinsic width of Si analyzer. For example, for Si(444) reflection, the backscattering energy is 7908.4 eV and the intrinsic resolution is 37.5 meV. The measured width of the reflected beam is 220 meV (FWHM) as shown in Fig. 3. After removing the width of the analyzer, the monochromator resolution is 217 meV which is in very good agreement with the theoretical value of 209 meV.

The photon flux from the monochromator is measured with an air-filled ionization chamber. At 10 keV and 300 mA ring current, the measured flux is about $5 \cdot 10^{10}$ photons/s. The flux is roughly constant between 6 and 10 keV. However, at the lower end of the tuning range, a much larger variation of photon flux with changing ring current is observed. In comparison with the calculated photon flux, the measured flux is lower by roughly a factor of 4. This difference is mainly due to the remaining heat-load problem of the first monochromator crystal.

The improved energy resolution and tunability of the monochromator is demonstrated in the following example. It is well known that in many transition metals and their compounds, a pre-edge peak is observed in the K edge absorption spectrum. The origin of the pre-edge peak is usually assigned to the $1s \rightarrow 3d$ quadrupolar transition[15]. In Fig. 4, the pre-edge region of the Fe K edge absorption spectrum of Hematite is shown. The crystal-field splitting of the 3d level is clearly observed. The separation between the peaks is about 1.4 eV. With 0.2 eV incident energy resolution, the incident energy can be selectively tuned to either t_{2g} or e_g state in resonant Raman scattering experiments. By reducing the number of intermediate states excited in the resonant process, more detailed spectral features can be revealed in the resonant Raman spectra. For example, Fig. 5. shows resonant Raman spectra near the Fe K-alpha fluorescent energy with the incident energy tuned to the pre-edge peaks. Note that the sharp features in these spectra would be smeared if the incident resolution was not good enough to separate the crystal-field split pre-edge peaks. Detailed discussion of these results will be published elsewhere[16].

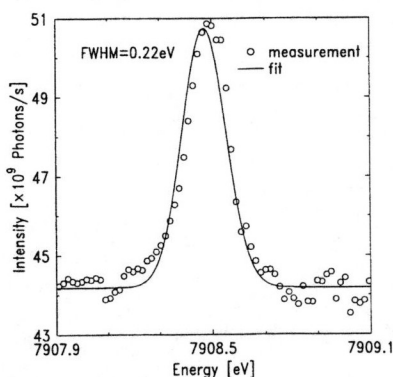

FIGURE 3. Resolution of the monochromator arrangement. The incident beam is scattered back into the ion-chamber by a flat perfect Si(111) crystal.

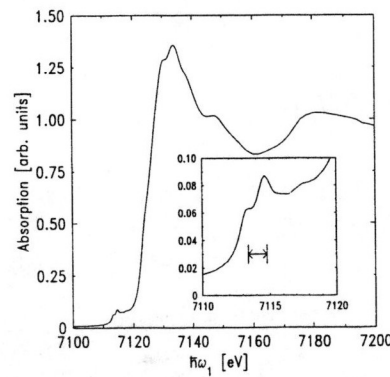

FIGURE 4. Fe K-absorption edge in Hematite. The inset shows the pre-edge structure in more detail. The arrow indicates the crystal field splitting.

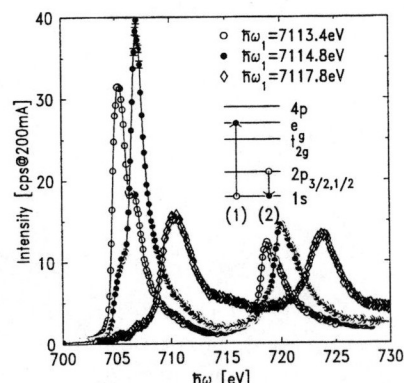

FIGURE 5. Resonant Raman scattering of Hematite. The emission of Fe is measured as the incident energy is tuned to different energies in the pre-edge peak region. The diagram shows the different electronic states which are involved in the scattering process.

SUMMARY

In summary, a high energy resolution and widely tunable monochromator was designed and commissioned for the inelastic x-ray scattering beamline at the NSLS. The new beamline improves the total energy resolution for non-resonant Raman experiments to be better than 0.5 eV, and expands the systems that can be studied by resonant Raman scattering to include most of the 3d transition elements, 4f rare earth elements, and their compounds.

ACKNOWLEDGMENTS

We would like to thank Rick Greene and Walter Stoeber for their technical assistance and valuable help toward design, construction and comissioning of the beamline. This work has been supported by the Division of Materials Science, U.S. Departement of Energy under Contract No. DE-AC02-76CH00016.

REFERENCES

[1] Kao, C.-C., Hamalainen, K., Krisch, M., Siddons, D.P., Oversluizen, T. *Rev. Sci. Instrum.* **66**, 1699-1702 (1995).

[2] Kao, C.-C., Caliebe, W.A., Hamalainen, K., Krisch, M., Hastings, J.B., *Rev. Sci. Instrum.* **67**, CDROM (1996).

[3] Hill, J., Kao, C.-C., Caliebe, W.A., Gibbs, D., Hastings, J.B., *Phys. Rev. Lett.* **77**, 3665-3668 (1996).

[4] Macrander, A.T., Montano, P.A., Price, D.L., Kushnir, V.A., Blasdell, R.C., Kao, C.-C., Cooper, B.R., *Phys. Rev. B* **54**, 305-312 (1996).

[5] Hamalainen, K., Manninen, S., Kao, C.-C., Caliebe, W.A., Hastings, J.B., Bansil, A., Kaprzyk, S., Platzman, P.M., *Phys. Rev. B* **54**, 5453-5459 (1996).

[6] Krisch, M., Kao, C.-C., Sette, F., Caliebe, W.A., Hamalainen, K., Hastings, J.B., *Phys. Rev. Lett.* **74**, 4931-4934 (1995).

[7] Kao, C.-C., Caliebe, W.A., Hastings, J.B., Gillet, J.-M., *Phys. Rev. B* **54**, 16361-16364 (1996).

[8] Hamalainen, K., Krisch, M., Sette, F., Kao, C.-C., Caliebe, W.A., Hastings, J.B., *Rev. Sci. Instrum.* **66**, 1525-1527 (1995).

[9] Macrander, A.T., Kushnir, V.I., Blasdell, R.C., *Rev. Sci. Instrum.* **66**, 1546-1548 (1995).

[10] Decker, G., Galayada, J., Solomon, L., and Kitamura, M., *Rev. Sci. Instrum.* **60**, 1845-1848 (1989).

[11] DuMond, J.W.M., *Phys. Rev.* **52**, 872-883 (1937).

[12] Nakayama, K., Hashizume, H., Migoshi, A., Kikuta, S., Kohra, K., *Z. Naturforsch.* **28a**, 632-638 (1973).

[13] Bergmann, U., *Resonant Nuclear Scattering Using Synchrotron Radiation*, thesis, State University of New York at Stony Brook (1994).

[14] Berman, L.E., Hastings, J.B., Oversluizen, T., and Woodle, M., *Rev. Sci. Instrum.* **63**, 428-432 (1992).

[15] Dräger, G., Frahm, R., Materlik, G., and Brümmer, O., *phys. stat. sol. (b)* **146**, 287-294 (1988).

[16] Kao, C.-C., Caliebe, W.A., Hastings, J.B., Kotanti, A., DeGroot, F.M.F., *in preparation*.

UHV Surface-Analysis Endstation with X-ray Scattering and Spectroscopic Capabilities

Paul F. Lyman [a], Denis T. Keane [b], and Michael J. Bedzyk [a,b,c]

(a) Department of Materials Science and Engineering,
Northwestern University, Evanston, IL 60208

(b) DuPont-Northwestern-Dow CAT,
Northwestern University, Evanston, IL 60208

(c) Materials Science Division
Argonne National Laboratory, Argonne, IL 60439

Abstract: The design of a versatile ultrahigh vacuum (UHV) endstation for use at the Advanced Photon Source is described. The capabilities of the endstation include x-ray scattering *and* x-ray spectroscopic techniques for the investigation of surfaces, interfaces, and thin films. The UHV analytical chamber also includes facilities for surface preparation, thin film growth, and standard (non-x-ray) surface analyses. The endstation, which is inspired by previous successful implementations for surface scattering, incorporates several novel design features to facilitate the use of both scattering and spectroscopic techniques, and also allows the examination of small samples. Its capabilities include x-ray reflectivity and crystal truncation rod studies, grazing-incidence x-ray diffraction, x-ray standing waves, surface extended x-ray absorption fine structure, x-ray holography, and x-ray photoelectron spectroscopy.

INTRODUCTION

A component of the research mission of the DND-CAT at the Advanced Photon Source (APS) will focus on surfaces, interfaces, and thin films. This research thrust will exploit extremely bright undulator radiation, utilizing a dedicated experimental enclosure (5-ID-C). This paper will describe endstation instrumentation capable of x-ray scattering *and* spectroscopic investigations of these systems in ultrahigh vacuum (UHV). The endstation will reside at the undulator, but will be transportable to a bending magnet beamline enclosure (5-BM-D). The purpose of the endstation instrumentation is to provide the full complement of x-ray analytical techniques useful for investigating surfaces, interfaces, and thin films. Additionally, the full suite of conventional (non-x-ray) techniques for surface preparation and analysis was desired. These requirements present a design challenge, since the conventional implementations of many of the desired x-ray techniques nearly preclude the use of any other technique. For example, UHV x-ray scattering requires that the incoming and outgoing rays will enter and exit through a Be window, so these windows generally subtend a large fraction of the solid angle visible to the surface. However, spectroscopic techniques also require access to a large solid angle. This paper will outline aspects of the design that overcome these conflicts, and will describe other novel features of the instrument.

DESIGN REQUIREMENTS

Desired capabilities of the instrument include x-ray scattering studies [including x-ray reflectivity, grazing-incidence x-ray diffraction (GIXD) studies of crystal truncation rods (CTR) and 2D crystallography], spectroscopies [including surface extended x-ray absorption fine structure (SEXAFS) and x-ray photoelectron spectroscopy (XPS)], and hybrid techniques [x-ray standing waves (XSW) and x-ray holography]. These techniques require a wide range of sample orientations with respect to the x-ray beam, as well as access to several detectors. For example, in-plane surface diffraction is typically carried out with the sample normal oriented horizontally, reflectivity is typically studied with the sample normal in the vertical plane, and XSW may require that the sample normal take on an arbitrary orientation. It is worth noting that some of these constraints are imposed in practice by the characteristics of radiation from a bending magnet; generally, there is a large horizontal angular divergence, leading to the above choices of orientation for surface diffraction and reflectivity. Although this instrument will be primarily deployed on an undulator beamline, it is prudent to honor the conventional constraints so that the endstation can be used expeditiously on a bending magnet beamline.

Desired conventional surface analytical techniques and preparation facilities include: low energy electron diffraction (LEED), Auger electron spectroscopy (AES), temperature programmed desorption (TPD), ion sputtering, molecular beam epitaxy (MBE), and sample heating and cooling. Excellent pumping is required, as the expected base pressure will be $\sim 5 \times 10^{-11}$ torr. Moreover, at a third generation source where beamtime is limited, a sample transfer system is imperative.

The overall system, then, requires a precision x-ray diffractometer that can be used with the sample oriented in any direction, a scattered x-ray detector with analyzing crystal, a fast energy-dispersive x-ray detector for fluorescent x-rays, and a fast electron energy detector. Also useful would be an x-ray detector with a large angular acceptance for XSW, such as an *in situ* PIN diode. Moreover, the extremely narrow (sub-arc-second) rocking curve widths encountered in XSW place extreme demands on the sample stability.

SPECIFIC DESIGN FEATURES

A. Diffractometer Geometry

Perhaps the most fundamental choice to made in designing the endstation is that of diffractometer geometry. Over the years, several approaches have been implemented for UHV diffractometers. (See Refs. [1,2].) Earlier schemes featured small- to moderate-sized UHV chambers mounted directly on a standard four-circle [3,4] or z-axis [5,6] diffractometer. More recent implementations have favored using feedthroughs and/or bellows to couple some of the precision motions into a stationary vacuum envelope [7-9]. Since it is not practical to carry a large, multiple-use endstation completely on a precision diffractometer, only the latter will now be considered.

A remaining fundamental choice is how many degrees of freedom (DOF) will be carried on the sample, and how many will be carried on the detector. Although it is possible in principle to specify the momentum transfer with a total of 3 DOF, a total of 4 DOF are almost universally employed at the minimum; this overdetermination then allows the experimenter to choose, for example, the angle between the sample surface and the incoming x-ray (α), or outgoing x-ray (β) [10]. The standard four-circle geometry features 3 DOF on the sample and one on the detector. The z-axis geometry [5] places 2 DOF on the sample, and 2 independent DOF on the detector. (Note, however, that in the z-axis geometry the two detector goniometers are mounted on, and rotate with, one of the sample rotations.)

Later, more complex implementations contain more than 4 DOF. For example, a rotation of a standard four-circle about a vertical axis has been used to create a five-circle (which has 4 DOF on the sample, and 1 independent DOF on the detector) [11,12]. Finally, the full combination of the motions allowed by the four-circle and z-axis geometries has also been implemented, resulting in a six-circle with 4 DOF on the sample, and 2 independent DOF on the detector [13-15].

A refinement of the basic z-axis layout has been proposed [16] and independently constructed [17] that completely mechanically decouples the DOF of the sample and detector. This diffractometer, having 2 DOF on the sample and 2 *fully independent* DOF on the detector, is termed the "S2D2 geometry." This arrangement is simple and has many practical advantages. It is particularly well suited to the requirements of this endstation, as discussed below, and is presently under construction [18].

The S2D2 geometry is depicted in Fig. 1. (The angular nomenclature differs from Ref. [17], and was chosen to most closely integrate the nomenclature used for all the principal diffractometer geometries.) The sample is rotated about a horizontal axis ω, which in turns rotates about a vertical axis μ. (The UHV chamber also rotates with μ.) The detector is rotated about a horizontal axis δ, which in turns rotates about a vertical axis γ. The other angles depicted, φ and χ, are not precision motions used for scanning the diffractometer, and will be explained later. For surface diffraction, χ will be set to zero, placing the surface normal in the horizontal plane.

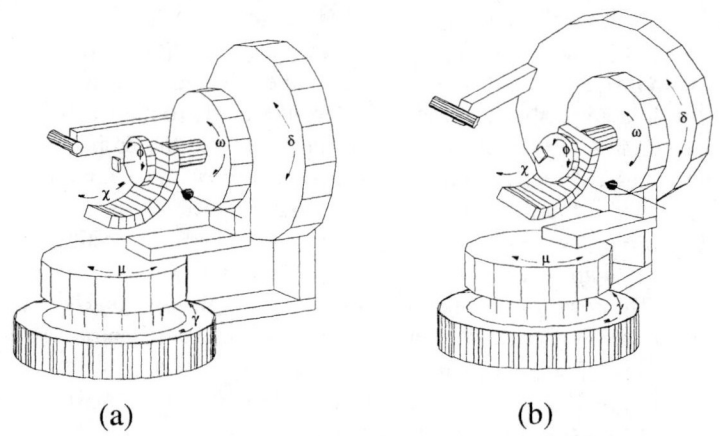

Fig. 1 Schematic of the S2D2 geometry (a) with all diffractometer angles at zero and (b) with all angles offset to a small positive value.

With this arrangement, only one precision DOF needs to be coupled into the UHV chamber, greatly simplifying the vacuum interface. In the present design, as in many others, the principal axial rotation is coupled by a differentially pumped rotary seal [19]. Unlike geometries that encompass the standard four-circle arrangement [7,8,13,14], however, it is possible to mount the rotary seal so that no torque is placed on the expensive bellows attached to the rotary seal.

As pointed out in Ref. [17], the S2D2 geometry allows the greatest possible reciprocal-space access for a given size of Be window. This is important in the present application, because the instrumentation necessary for spectroscopy and surface science limit the size of allowable Be windows. Also, in the S2D2 setup, the sample normal and the UHV chamber rotate as a unit during x-ray diffraction scans. This feature facilitates the use of techniques that combine scattering and spectroscopy (*e.g.*, x-ray holography). Finally, the sample is not rotated about an external χ arc, and so the sample may be physically far from the ω-circle without limiting reciprocal-space access. As discussed in the next section, this ability to incorporate a lengthy manipulator arm to hold the sample is advantageous for the design.

To provide positioning of the diffractometer with respect to the x-ray beam, the entire assembly will reside on a heavy-duty, motorized translation table [20]. The table will allow precise translations both vertically and transverse to the incident beam direction. The diffractometer will have a mass of 1700 kg, and the table is capable of supporting over 3000 kg.

B. UHV Chamber Layout

There are two critical issues to consider for the UHV chamber layout. The first consideration is how the sample will be moved from a position where the Be windows are accessible (the "x-ray spot") to a position where the conventional surface analyses are accessible (the "surface analysis spot"). If the sample is moved with respect to the diffractometer, the sample alignment will be lost. Although providing an internal kinematic mounting for the sample can help to recover the orientation quickly [7], a better approach is to hold the sample fixed with respect to the diffractometer and to translate the *chamber* instead [12].

The next consideration is whether to locate the surface analytical instruments closer to the goniometer ("inboard"), or farther away ("outboard"), than the x-ray spot. In any design that uses an external χ arc, the distance from the x-ray spot to the goniometer should be minimized to maximize the range of χ available. Almost universally, then, the x-ray spot is placed inboard, and the surface science equipment is located farther out, away from the goniometer. To access these, either the entire chamber or the individual instruments are translated inward so that the sample is at the focal point of the analytical apparatus. This arrangement is inappropriate for this case, however, because the spectroscopic detectors require unrestricted access to the sample, and will occupy the space opposite the sample support arm. Thus, in the chosen layout, the large fluorescence detector and hemispherical electron energy analyzer are located at the most-outboard location, the Be windows defining the x-ray spot are adjacent, and the surface analytical instrumentation is in the most-inboard position.

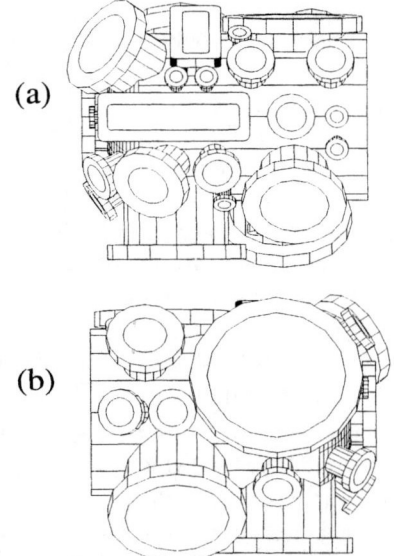

Most surface diffraction chambers utilize a semi-cylindrical strip of Be brazed or welded into a section of the chamber wall [7]. This gives the incoming and outgoing x-rays access to 180° angular range in the traditional scattering plane, as well as a much smaller range (~30° typical) in the out-of-plane direction. As appealing as that layout is, it could not be incorporated for several reasons. Firstly, the hemispherical electron detector is heavy and needs to be close to the sample; a traditional Be strip window would not allow the port for that detector to be located close enough to the sample position. Secondly, the replacement of a large section of the stainless steel chamber wall with thin Be would substantially weaken the chamber; it was feared that the torque that the detector could (and inevitably would) apply to the chamber would likely rupture the Be window or open a leak at the weld. Finally, the desired inclusion of a full array of analytical and preparation instruments inboard of the x-ray spot forced the sample to reside far from the diffractometer; it is highly desirable to minimize this distance for better sample stability, and this goal was achieved by "nesting" the surface analytical instruments with several discrete Be windows.

The resulting chamber layout is visible in Fig 2. There are three Be windows [21] mounted on Conflat®-type flanges. The horizontal entrance window, visible in the center of Fig. 2(a), is mounted on a 222 mm × 76 mm rectangular flange; with inner dimensions of 193 mm × 41 mm, this window will accept x-rays at incidence angles of $\mu < 46°$. The principal exit window is formed by a large (203 mm diameter) Be window mounted on a 254 mm OD round flange, which is prominent in Fig. 2(b). The port is offset so that the straight-through beam would exit near the periphery of the window.

Fig. 2 Views of the UHV chamber from the (a) beam-entrance side and (b) beam exit side.

This arrangement allows x-rays to exit at up to 45° in the vertical plane (δ), and up to 35° out-of-plane ($\gamma - \mu$). Lastly, there is a secondary, rectangular exit window, visible near the top of Fig. 2(a), which spans angles of 90° < δ < 143° for out-of-plane angles of 8° to 20°. The S2D2 geometry allows these two compact exit windows, coupled with a wide entrance window, to provide access to a sufficient amount of reciprocal space to conduct surface diffraction studies.

Other notable features of the UHV chamber include two ports for laser-alignment of the sample; these ports are located 180° apart from one another, with a shared focus at the x-ray spot. Another port, aimed at the focus of the hemispherical analyzer, will contain an electron gun for exciting the sample for AES analysis. For MBE growth, up to three effusion cells or e-beam evaporators can be placed in ports with water-cooled jackets and shutters, all mounted on a demountable 203 mm flange. Lastly, spare ports that can contain deposition sources are focused on the x-ray spot to allow studies of real time, *in situ* growth.

C. Sample Manipulator

A key feature of the endstation design is the introduction of a sample manipulator, having several DOF, interposed between the sample and the ω-circle. The manipulator encompasses an X-Y-Z stage [22] and two angular DOF [23], corresponding to φ and χ in a conventional four-circle diffractometer. These angular motions have precision of $\approx 0.01°$, and thus do not represent precision motions that can be used to scan a sample during diffraction measurements (although, as noted by Robinson [1], the requirements on the precision of χ are comparatively modest). In any event, the angular DOF provided by this setup provide two very important functions. Firstly, the χ rotation allows the sample to be mounted with its surface normal oriented horizontally (for surface diffraction), vertically (for reflectivity), and anywhere in between (for XSW and magic-angle SEXAFS). Secondly, the combination of (even relatively imprecise) φ and χ rotations overcomes a shortcoming in the S2D2 and z-axis geometries: if the sample normal is not perfectly aligned with the ω axis, the incident angle α cannot be held at a fixed value during a diffraction scan [17]. In essence, one can use the internal φ and χ rotations in the same way as a "goniometer head" is used on a conventional, non-vacuum diffractometer, namely, to align the plane of interest with a rotational DOF. Moreover, aside from diffraction considerations, these rotations allow the sample to be oriented to face the conventional surface analytical instruments, and to facilitate sample transfer.

The analogy to the goniometer head extends to the X-Y-Z stages of the manipulator as well. As on a conventional diffractometer, translations of the goniometer head allow an arbitrary point on the sample to be placed at the center of rotation of the diffractometer. Such an ability is vital for conducting diffraction studies of small samples with a focused beam. X-ray-based analyses of surfaces have, up until now, been almost exclusively conducted on large, homogeneous samples, and it mattered little which point lay at the center of rotation. With the advent of third generation sources, however, the exciting possibility of employing microbeams for surface and interface work can be realized. There has been a rising interest in microdiffraction; to the authors' knowledge, this instrument will be the first surface diffractometer equipped with the sample translations necessary for microdiffraction.

The sample manipulator also incorporates facilities to heat the sample (by irradiation) to 1300 K, and to cool to 125 K. Cooling is accomplished by flexible Cu braids attached to an *in vacuo* LN_2 reservoir. Moreover, the sample sits on a transferable platen, allowing for quick exchanges of samples. This is vital in a new facility where beamtime will be limited.

There are practical difficulties in constructing a sample manipulator with all these features. The foremost concern has been with stability and rigidity of the sample support, and many steps were taken to stabilize the sample. The rotational gears are preloaded by springs to limit sample vibrations. The manipulator probe assembly is supported by a thick-walled, 64 mm OD tube for rigidity. The X-Y stages slide on highly preloaded, crossed-roller-bearing slides with hardened sways. The Z stage slides on preloaded, full-area-contact linear bushings.

D. Detectors

To effectively conduct spectroscopic studies at a third-generation insertion device, fast detectors are required. The hemispherical electron energy analyzer [24] incorporates a multielement detector: a multichannel plate amplifies each detected photoelectron, and the resultant electron shower is directed onto a 16-wire anode. Sixteen separate sets of pulse-counting electronics can provide a collective count rate of over 1 MHz. The x-ray fluorescence from the sample will be detected by a multielement solid state detector with fast electronics, allowing a composite count rate of several hundred kHz.

For x-ray diffraction, a standard 2-circle crystal analyzer will be mounted on the end of the detector arm; a fast scintillator detector accommodating count rates up to 1.2 MHz will be used. For XSW, precise knowledge of the scattering angle is less important, and it is convenient to use a detector with a large solid angle, which should have a 2θ range of 0° to close to 180°. This will be accomplished by placing a PIN photodiode on a moveable track inside the UHV chamber.

CONCLUSIONS

With little compromise to any of the techniques desired, a versatile UHV endstation with capabilities for surface diffraction and spectroscopy has been designed. This diffractometer incorporates the most desirable elements from several other successful surface scattering endstations (*e.g.*, S2D2 geometry, movable UHV chamber), as well as novel features (spectroscopic detectors, outboard sample position, highly adjustable sample manipulator) to achieve these goals. Experiments are planned to commence with this system in the winter of 1997-98 at the APS.

ACKNOWLEDGEMENTS

We gratefully acknowledge discussions with Kenneth Evans-Lutterodt and help from Yonglin Qian. The endstation construction is supported by the DOE under contract numbers DE-FG02-94ER45527 and DE-FG02-96ER45588 and by the MRSEC program of the NSF through the MRC at Northwestern University under contract number DMR-9632472.

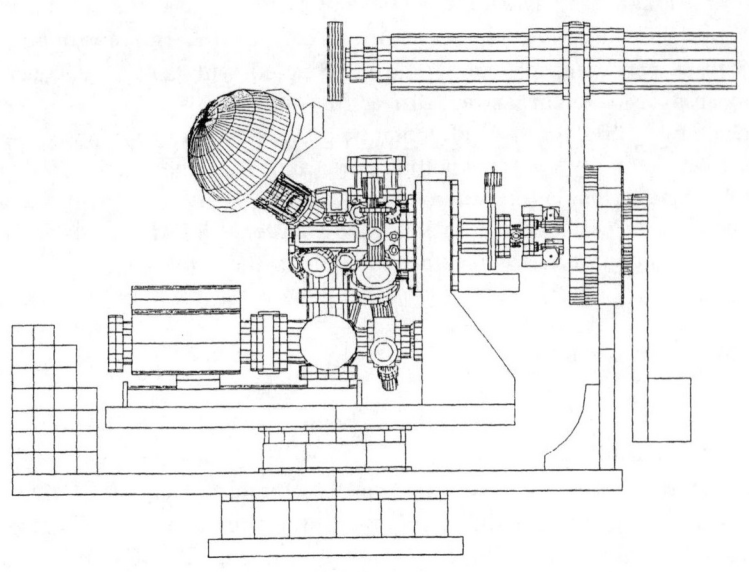

Fig. 3 View of complete diffractometer assembly along beam direction.

REFERENCES

[1] I.K. Robinson, "Surface Crystallography," in *Handbook on Synchrotron Radiation*, Vol. 3, p. 221, eds. G. Brown and D.E. Moncton (Elsevier Science Publishers, Amsterdam, 1991).
[2] R. Feidenhans'l, *Surf. Sci. Rep.* **10**, 105 (1989).
[3] R.L. Johnson, et al., in *The Structure of Surfaces*, p. 313, eds. M.A. van Hove and S.Y. Tong (Springer, Berlin, 1985).
[4] G.A. Held, J.L. Jordan-Sweet, P.M. Horn, A. Mak, and R.J. Birgeneau, *J. Phys. (Paris)* **50**, C7-245 (1989).
[5] J.M. Bloch, *J. Appl. Cryst.* **18**, 33 (1985).
[6] S. Brennan and P. Eisenberger, *Nucl. Instrum. Methods* **222**, 164 (1984).
[7] P.H. Fuoss and I.K. Robinson, *Nucl. Instrum. Methods* **222**, 171 (1984).
[8] D. Gibbs, B.M. Ocko, D.M. Zehner, and S.G.J. Mochrie, *Phys. Rev. B* **38**, 7303 (1988).
[9] G.A. Held, D.M. Goodstein, and J.D. Brock, *Phys. Rev. B* **51**, 7269 (1995).
[10] S.G.J. Mochrie, *J. Appl. Cryst.* **21**, 1 (1988).
[11] D. Gibbs, B.M. Ocko, D.M. Zehner, and S.G.J. Mochrie, *Phys. Rev. B* **42**, 7330 (1990).
[12] E. Vlieg, A. van't Ent, A.P. Jongh, H. Neerings, and J.F. van der Veen, *Nucl. Instrum. Methods* **262**, 522 (1987); E. Vlieg, J.F. Van der Veen, J.E. Macdonald, and M. Miller, *J. Appl. Cryst.* **20**, 330 (1987).
[13] S. Ferrer and F. Comin, *Rev. Sci. Instrum.* **66**, 1674 (1995).
[14] M. Takahasi, S. Nakatani, Y. Ito, T. Takahashi, X.W. Zhang, and M. Ando, *Surf. Sci.* **357-358**, 78 (1996).
[15] M. Lohmeier and E. Vlieg, *J. Appl. Cryst.* **26**, 706 (1993).
[16] H.-H. Hung, *J. Appl. Cryst.* **25**, 761 (1992).
[17] K.W. Evans-Lutterodt and M. Tang, *J. Appl. Cryst.* **28**, 318 (1995).
[18] Blake Industries, Scotch Plains, NJ.
[19] Model DPRF-400, McAllister Technical Services, Coeur d'Alene, ID.
[20] T6-22 Goniometer Support Table, Dial Machine Corp., Rockford, IL.
[21] Brush-Wellman Corp., Electrofusion Products, Fremont, CA.
[22] Model MB2002 X-Y-Z stage, McAllister Technical Services, Coeur d'Alene, ID. The X and Y stages have a range of ± 13 mm, and the Z stage has a total stroke of 38 mm.
[23] Model GB16, Thermionics Northwest Laboratories, Port Townsend, WA. The azimuthal (φ) and flip (χ) have ranges of $\pm 90°$ and $-5°/+95°$, respectively.
[24] Model 10-360 SCA, Physical Electronics, Inc., Eden Prairie, MN.

Commission of the PNC-CAT insertion device beamline at the APS

D. T. Jiang,* S. M. Heald,** E. A. Stern,*** E. D. Crozier,* B. Barg,*** K. H. Kim,*** R. A. Gordon,* D. Brewe,*** and F. Brown***

*Simon Fraser University, Burnaby BC, Canada
**PNNL, Richland, WA 98058
***University of Washington, Seattle, WA 98101

The PNC-CAT has commissioned the front optics for its insertion device beamline at the APS, which enables experiments to be performed in the front optics station with monochromatic x-rays. The beamline consists of a liquid nitrogen cooled monochromator followed by a variety of focusing optics and is optimized for spectroscopic and microbeam applications. Two downstream experimental stations are under construction and will be equipped with microfocusing optics (Kirkpatrick-Baez mirrors and/or tapered capillary optics) to produce beams ranging from 0.1 to 20 microns with expected intensity of 10^{10}-10^{11} ph/sec/micron^2; and with two UHV systems capable of MBE sample growth and combined in-situ structural characterization by glancing-incidence XAFS, x-ray reflectivity, SEXAFS, XSW, and surface x-ray diffraction. The initial performance of the beamline and experimental activities will be reported.

Project support by the US DOE (BES and EM), the Canadian NSERC, Univ. of Washington, and Simon Fraser University.

BioCAT Undulator Beamline at APS

G. B. Bunker, T. Irving, E. Black, K. Zhang,
R. Fischetti, S. Wang, S. Stepanov

BioCAT
Argonne National Laboratory
Advanced Photon Source
9700 S. Cass Ave., Argonne, IL 60439

The Biophysics Collaborative Access Team (BioCAT) project will be an outstanding research facility for biological small angle scattering, non-crystalline diffraction, and X-ray absorption spectroscopy at the Advanced Photon Source, Argonne National Labs. BioCAT operates as an NIH Research Resource under a cooperative agreement with NIH. BioCAT has an aggressive program of core and collaborative research, service, and training. Central to the facility is the undulator beam line (designed by G. Rosenbaum, Argonne National Labs) capable of delivering ca. 10^{13} - 10^{14} ph/s to the sample. Focusing optics will allow focal spot ranges from 1.5 x 3.5 mm to 30 x 80 micron, independently adjustable in the vertical and horizontal direction. Up to 8 m camera lengths can be accommodated in the 12m experimental enclosure. The accessible beam energy range will be from 3.5-13 keV using the undulator fundamental and 10-40 keV using the third harmonic. Energy resolution will exceed 2×10^{-4} $\Delta E/E$. Detectors will include image plates, CCD detectors and some novel detectors designed to accommodate the high count rates expected at the APS. The multi-element detector will be a very high count-rate (up to 10^9 ph/s global), one dimensional detector optimized for scattering applications. We are also developing a multilayer analyser detector which maximizes solid angle of collection with high background rejection for biological spectroscopy applications.

Expanded Capabilities from SRC's High Energy Resolution Variable Line Density Grating Monochromator Beamline: HERMON

M. Bissen, M. Fisher, R. Reininger, G. Rogers, and H. Höchst

University of Wisconsin Synchrotron Radiation Center, 3731 Schneider Drive, Stoughton, WI 53589

Abstract. The capabilities and performance of the variable groove density spherical grating monochromator have been extended through the installation of new gratings. Two new varied line spacing (VLS) gratings have been installed in the SRC High Energy Resolution Monochromator (HERMON). The specified polynomial variation in groove density was successfully fabricated on a blazed grating with a conventional ruling engine and on a laminar profile grating by ion beam etching. The theoretical resolving power of 5000 to 10000 is achieved in the energy ranges of 250 to 550 eV for the low energy laminar grating and 500 to 1100 eV for the high energy blazed grating. Experiments in magnetic circular dichroism are now possible with the implementation of a motorized water cooled aperture. The performance throughout the energy range is characterized by a variety of solid state and gas phase measurements.

1. INTRODUCTION

The high resolution beamline for the soft-x-ray region implemented at SRC is based on a monochromator with a single variable line density spherical grating between fixed entrance and exit slits [1-5]. The second order polynomial variation of the line density along the grating length allows the defocus and coma aberration to be zeroed at two wavelengths [6]. Furthermore, by optimizing the line density variation and implementing a scan mechanism which also involves the translation of the grating, the defocus aberration is zeroed over the scan range of the grating, and the coma aberration is minimized [2].

The HERMON beamline based on the above concepts has been operational at SRC since 1995, and its very high resolution at the oxygen and neon K edges has already been documented [4]. The performance of this bending magnet beamline was, however, impaired by a relatively low efficiency grating. Two new gratings were installed in the monochromator and the present beamline performance, in terms of flux and energy resolution, achieves the design values [7]. In addition, the beamline has been equipped with a motorized aperture to allow experiments which require circularly polarized radiation. The present contribution summarizes some of the data obtained with the new high energy grating, which has been already used by several user groups, as well as first results obtained with the low energy grating during its initial tests.

2. FLUX AND RESOLUTION

The monochromator covers the energy range 245-1100 eV with two spherical gratings having radii of 73.0 m. The 200 mm long grating substrates were manufactured from a single block by Hyperfine (Boulder, CO) with an RMS slope error of less than 0.2 arcseconds. The line density, ρ, as a function of the distance, w, from the center of the gratings is given by

$$\rho = \rho_0\left(1 + 2b_1 w + 3b_2 w^2 + 4b_3 w^3\right). \tag{1}$$

The parameters in Eq. 1 for the two gratings and their tuning range are listed in Table 1. The laminar profile low energy grating was holographically recorded with the specified groove density variations and ion etched by Zeiss (Oberkochen, Germany). The blazed high energy grating was mechanically ruled on gold by Hyperfine.

Fig. 1a shows the measured photon flux through the monochromator for the two gratings with 200 mA stored beam current in Aladdin operating at an energy of 800 MeV. These values were measured with a calibrated XUV-100 photodiode and

TABLE 1. Energy range and parameters (Eq. 1) for the line density variations of the low and high energy gratings.

	LEG	HEG
Energy Range (eV)	245-550	490-1100
ρ_0 (mm^{-1})	703.6	1407.3
b_1 (mm^{-2})	8.25x10^{-5}	8.25x10^{-5}
b_2 (mm^{-3})	3.51x10^{-8}	3.54x10^{-8}
b_3 (mm^{-4})	-3.8x10^{-12}	0

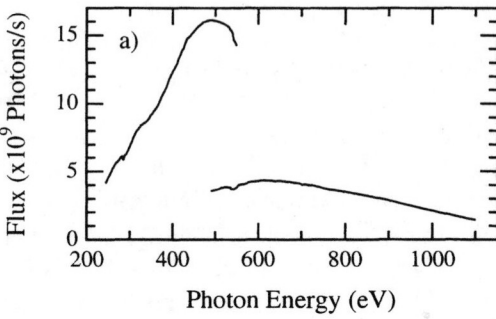
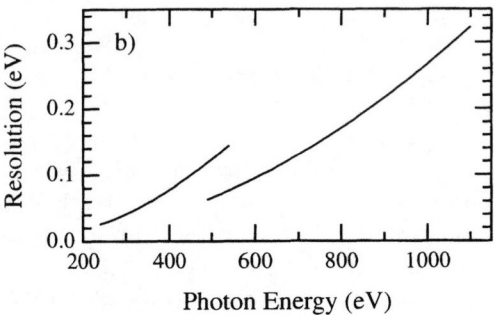

FIGURE 1: a) Measured photon flux for 200 mA using 20 µm slits. b) Calculated resolution for these slits settings.

using entrance and exit slits set to 20 µm. The calculated resolution for these slits settings is displayed in Fig. 1b. It should be noted that the flux increases linearly with the entrance slit up to a slit value of approximately 100 µm.

3. SPECTRA

Resonant Auger Emission in Ne

The new high energy grating ruled by Hyperfine not only provides excellent resolution but also the intensity to record the Auger decay of Ne after selective excitation of the 1s to 3p, 4p and 5p [8]. The decay spectra shown in Fig. 2 were obtained using a photon energy resolution of 0.4 eV. The Auger electrons were retarded by 700 VDC before they entered a spherical electron analyzer working at a resolution of 0.6 eV and equipped with a position sensitive detector. Similar spectra for the resonant Auger decay after 1s → 3p and 4p excitation were recorded by Neeb *et al.* [9] at the soft-x-ray undulator beamline BW3 at HASYLAB at DESY, Hamburg [10, 11]. Their assignment [9] for the peaks in these two spectra are given below. The peaks at 811.5 and 807.7 eV in the lowest spectrum correspond to the $2p^{-2}(^1D)3p$ and $2p^{-2}(^1S)3p$ configurations. The lower energy peaks at 806.3 and 802.5 eV are due to the spectator electron shake-up $2p^{-2}(^1D, ^1S)4p$. The Auger decay following 1s → 4p excitation shows lines at 807.8 and 804.1 eV which correspond to the $2p^{-2}(^1D, ^1S)4p$ configuration as well as the spectator electron shake-down $2p^{-2}(^1D)3p$ at 813.2 eV. The spectator electron shake-up $2p^{-2}(^1D, ^1S)5p$ peaks are observed at 806.0 and 802.2 eV. The good photon energy resolution allowed us to record the Auger decay following 1s → 5p without any 4p contamination. The peaks at 806.5 and 802.7 eV in this spectrum are assigned to the $2p^{-2}(^1D, ^1S)5p$ configurations. The spectator electron shake-down $2p^{-2}(^1D)4p$ is observed at 808.5 eV. The two additional peaks in this spectrum at 806.5 eV and 801.9 eV are most probably due to spectator electron shake-ups to n>5 Rydberg states. The topmost spectrum in the graph was obtained after exciting with photon energies slightly above the 1s ionization threshold. The strong asymmetry of the most prominent peak in the spectrum is related to post collision interactions [12].

MCD Results

A motorized water-cooled aperture was installed downstream the front end of the beamline to allow for experiments which require circularly polarized radiation. The upper part of Fig. 3 shows the total electron yield obtained from two cobalt samples magnetized in opposite directions. These spectra were obtained by aperturing 80% of the bending magnet radiation and using

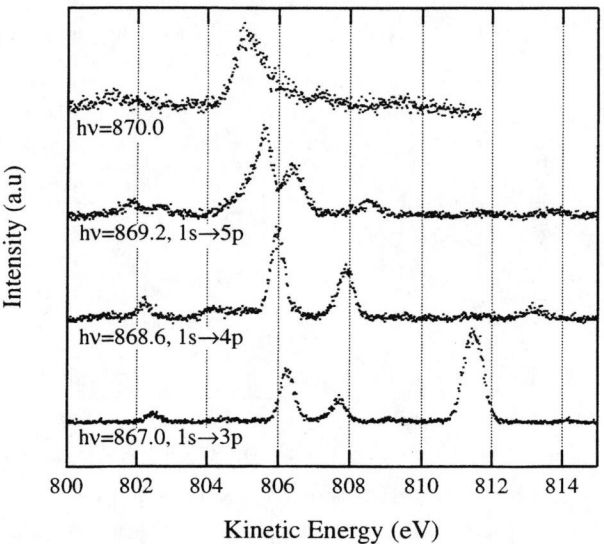

FIGURE 2. Auger decay of Ne after selective excitation of the $1s$ to $3p$ (867.0 eV), $4p$ (868.6 eV), $5p$ (869.2 eV), and above threshold (870.0 eV).

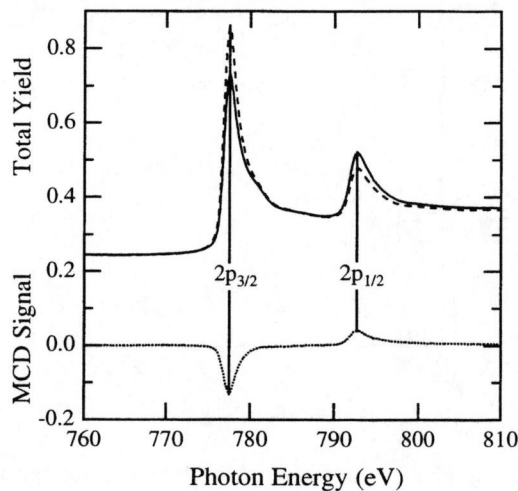

FIGURE 3. Solid and dashed lines: Total yield of Co for opposite magnetizations recorded with circularly polarized radiation. Dotted line: magnetic circular dichroism signal.

100 μm entrance and exit slits. The energy resolution with these slit settings is approximately 0.9 eV, far less than the width of the peaks seen in the graph. The asymmetry in the region of the $2p_{3/2}$ and $2p_{1/2}$ edges for the two magnetizations is clearly seen in the figure. The magnetic circular dichroism signal, obtained from the difference between the two total yield spectra, is displayed in the lower part of the figure.

$3d$ Absorption of Pr in $Y_{0.6}Pr_{0.4}Ba_2Cu_3O_{6.9}$

The overlap between the ionization edges of the multi-element high Tc superconductor $Y_{0.6}Pr_{0.4}Ba_2Cu_3O_{6.9}$ hinders the determination of the ionization state of Pr. Using X-ray absorption spectroscopy (XAS), a confirmation that Pr is mostly in its triple ionized state in this material can be obtained by comparing the measured spectrum with theoretical calculations. The total electron yield spectrum of the high Tc superconductor $Y_{0.6}Pr_{0.4}Ba_2Cu_3O_{6.9}$ in the energy region of the Pr $3d_{5/2}$ and $3d_{3/2}$ thresholds is displayed in the lower trace of Fig. 4 [13]. This energy region is not free from absorption features due to other elements in the material. The sensitivity achieved with 80 micron slits is sufficient to resolve the various components in the

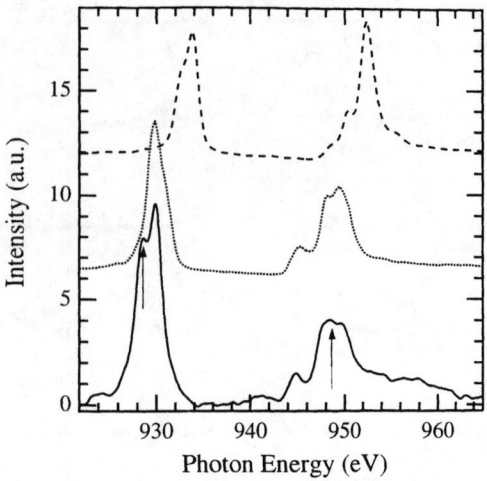

FIGURE 4. Dotted and dashed lines: Calculated soft x-ray absorption spectra [14] of the Pr^{+3}: $3d^{10}4f^2 \rightarrow 3d^94f^3$ and Pr^{+4}: $3d^{10}4f^1 \rightarrow 3d^94f^2$ transitions. Solid line: Total yield spectrum of $Y_{0.6}Pr_{0.4}Ba_2Cu_3O_{6.9}$. The arrows mark the energy positions of the Cu $2p_{3/2}$ and $2p_{1/2}$ edges.

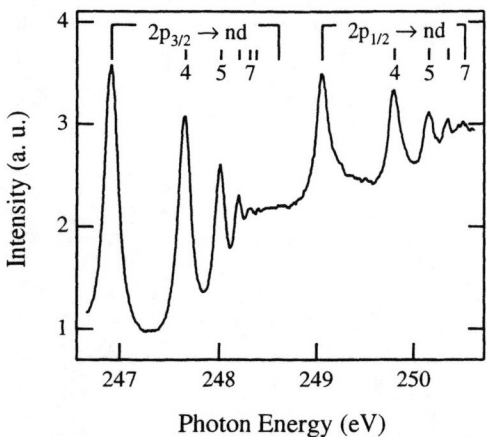

FIGURE 5. Photoionization of Ar in the $L_{2,3}$ thresholds region.

compound. The arrows in the figure mark the energy positions of the Cu $2p_{3/2}$ and $2p_{1/2}$ edges. The dotted and dashed lines in the figure correspond to the calculated soft x-ray absorption spectra [14] of the $3d^{10}4f^2 \rightarrow 3d^94f^3$ transition of Pr^{+3} and $3d^{10}4f^1 \rightarrow 3d^94f^2$ transition of Pr^{+4}, respectively. A comparison between the measured and the theoretical spectra clearly shows that most of the Pr in $Y_{0.6}Pr_{0.4}Ba_2Cu_3O_{6.9}$ is in the +3 valence state.

Argon $L_{2,3}$ Thresholds

The photoionization of argon in the region of the $L_{2,3}$ thresholds was measured during the commissioning of the low energy grating. The data were taken with a gas cell separated from the UHV of the beamline by an aluminum window. The spectrum and its assignments are presented in Fig. 5. The assignments for n<6 were taken from the work by King *et al.* [15]. For n ≥ 6 we used the quantum defect of the $5d$ Rydberg and the ionization thresholds (248.628 and 250.776 eV) given by the same authors [15]. The calculated energy positions of the 6 and $7d$ states of both series agree with the peaks in the spectrum to within the energy step, 5 meV. The width of the $2p \rightarrow 4s$ transition has been frequently used to determine the resolution of various monochromators in this energy region [16-18]. Unfortunately, this transition is below the tuning energy range of HERMON, thus preventing a direct comparison with the performance of other monochromators.

5. CONCLUSION

The new variable line density gratings have significantly improved the performance of the HERMON beamline. Equipped with these gratings, HERMON fulfills its design parameters not only in terms of high resolution but also in terms of flux. Currently the beamline is being used for high-resolution photoemission at the carbon K edge. The additional capability of delivering circularly polarized radiation further extends the types of experiments that can now be performed in this soft-x ray beamline.

ACKNOWLEDGMENTS

This work is based upon research conducted at the Synchrotron Radiation Center, Univ. of Wisconsin, which is supported by the NSF under award No. DMR-9531009.

REFERENCES

1. This high energy resolution soft x ray SRC user beamline is based on an In-Focus Monochromator (IFM) invented by Michael C. Hettrick of Hettrick Scientific, Inc. and covered by US. patent No 4,991,934 and foreign patents pending. HERMON was jointly designed and built by SRC and the Physical Laboratory (PSL) under a licensing agreement between the University of Wisconsin-Madison and Hettrick Scientific, Inc.
2. Höchst H., Bissen M., Engelhardt M.E., and Crossley D., Nucl. Instr. Methods A **319**, 121 (1992).
3. Fisher M.V., Steinhauser N., Eisert D., Winter B., Mason B., Middelton F., and Höchst H., Nucl. Instr. Methods A **347**, 264 (1994).
4. Bissen M., Fisher M., Rogers G., Eisert D., Kleman K., Nelson T., Mason B., Middleton F., and Höchst H., Rev. Sci. Instrum. **66**, 2072 (1995).
5. Fisher M.V., Bissen M., Bourgeois F., Eisert D., Kubala T., Reininger R., and Severson M., SPIE **2856**, 212 (1996).
6. Namioka T., J. Opt. Soc. Am. **49**, 446 (1959).
7. Bissen M., Fisher M.V., Höchst H., Reininger R., and Rogers G., SPIE, submitted.
8. Mossessian D., Jürgenson A., and Cavell R., unpublished.
9. Neeb M., Rubensson J.-E., Biermann M., Bringer A., and Eberhardt W., in *Atomic and Molecular Photoinization*, edited by Yagishita A. and Sasaki T., (Universal Academy Press Inc., Tokyo, Japan, 1996), pp. 109.
10. Reininger R. and Saile V., Nucl. Instr. Methods A **288**, 343 (1990).
11. Larsson C.U.S., Beutler A., Federmann F., Rieck A., Verbin S., and Möller T., Nucl. Instr. Methods A **337**, 603 (1993).
12. Armen G. B., Tulkki J., Åberg T., and Crasemann B., Phys. Rev. A **36**, 5606 (1987).
13. Bissen M., Olson C.G., and Höchst H., unpublished.
14. Thole B.T., van der Laan G., Fuggle J.C., Sawatzky G.A., Karnatak R.C., and Esteva J.-M., Phys. Rev. B , 5107 (1985).
15. King G.C., Tronc M., Read F.H., and Bradford R.C., J. Phys. B **10**, 2479 (1977).
16. Domke M., Mandel T., Puschmann A., Xue C., Shirley D.A., Kaindl G., Petersen H., and Kuske P., Rev. Sci. Instrum. 63, 80 (1992).
17. Abrami A., Jark W., et al., Rev. Sci. Instrum. **66**, 1618 (1995).
18. Aksela S., Kivimäki A., Sairanen O.-P., Naves de Brito A., Nömmiste E., and Svensson S., Rev. Sci. Instrum. **66**, 1621 (1995).

A New Time-of-Flight Mass Spectrometer for Electron-Ion and Ion-Ion Coincidence Experiments (PEPICO, PIPICO and PEPIPICO).

Joselito B. Maciel
Instituto de Química, Universidade Federal do Rio de Janeiro, Cidade Universitária, Rio de Janeiro, RJ, Brazil 21949-900

Eizi Morikawa
Center for Advanced Microstructures and Devices, Louisiana State University, Baton Rouge, Louisiana 70803

G.G.B. de Souza
Center for Advanced Microstructures and Devices, Louisiana State University, Baton Rouge, Louisiana 70803
and
Instituto de Química, Universidade Federal do Rio de Janeiro Cidade Universitária, Rio de Janeiro, RJ, Brazil 21949-900 (permanent address)

A time-of-flight mass spectrometer (TOFMS), has been designed and built at the Federal University of Rio de Janeiro (Brazil), with the aim of studying the ionic fragmentation of gas-phase compounds, using synchrotron radiation. Our main research interest consists on the observation of single, double and multiple ionization as a function of the incident photon energy, with particular emphasis on core- excited molecules. These processes are usually associated with the formation of very fast ions. The spectrometer employs, accordingly, a collimating lens which allows for the collection of ions having kinetic energies up to 40 eV with almost 100 % efficiency. By changing the lens voltage, the spectrometer can also be operated in a highly discriminating condition, which provides an easy and accurate method for the determination of the kinetic energy release.

In order to be tested, the spectrometer was adapted to a brazilian beamline installed at CAMD (Center for Advanced Microstructures and Devices). Mass and coincidence (PEPICO, PIPICO and PEPIPICO) spectra were obtained, in the 12 to 310 eV photon energy range, for several compounds. A mass resolution (M/ΔM) of 450 was attained at 132 a.m.u., with xenon.

I. INTRODUCTION:

The study of molecular core ionization as a function of the incident photon energy using synchrotron radiation has revealed a wealth of interesting phenomena, such as simple and double resonant Auger, and the associated molecular breakdown mechanisms[1]. Double (or even multiple) ionization is usually associated with these core ionized molecules. The resulting doubly (or multiply) charged species are usually quite unstable and hard to detect through standard mass spectrometric techniques. More sophisticated techniques, based on the coincident detection of electrons and ions, must be used and are therefore essential to the characterization of the ionic fragmentation of highly ionized molecules[2].

In the present paper, we describe a recently developed time-of-flight mass spectrometer, whose design has been specifically optimized towards the realization of electron-ion, ion-ion and electron-ion-ion measurements.

II. THE TIME-OF-FLIGHT MASS SPECTROMETER (TOFMS):

The spectrometer, schematically shown in Figure 1, is similar in concept to some previously described time-of-flight mass spectrometers[3], utilizing two sets of microchannel plates, one for the detection of photoelectrons and the other for the photoions. The ions and electrons are extracted in opposite directions through the aplication of a strong extracting field (typically, 500 V/cm). The drift tube for the ions is 25cm long and extraction gap is 2cm long. An important feature of the present TOFMS consists in the use of a collimating electrostatic lens, which allows for the collection of ions having kinetic energies of up to 40 eV with 100% collecting efficiency. The trajectories of Ar+ ions with 40 eV kinetic energy, simulated with use of the SIMION program, are presented in Figure 2. The mass resolution of the spectrometer was measured with Xenon gas, figure 3, and a resolution of M/ΔM= 450 at 132 a.m.u.

(Xe^+) was achieved (for thermal ions with extraction field of 500 V/cm).

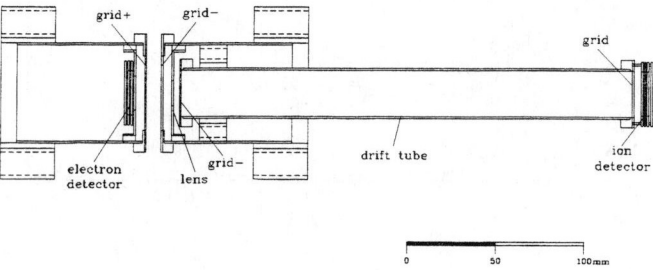

Figure 1: Time of Flight Mass Spectrometer

Figure 2: Simulated trajectory for Ar^+ ions with = 40eV using the lens with focusing potential (extraction field= 500v/cm).

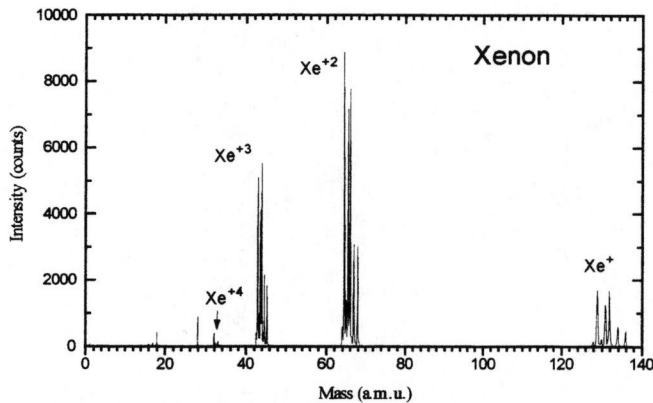

Figure 3: Photoelectron-photoion coincidence spectrum of Xenon. Resolution (M/ΔM)= 450 at 132 a.m.u. (Xe^+).

III. PERFORMANCE OF THE TOFMS

The present TOFMS was basically designed for the study of the ionic fragmentation of gaseous compounds either valence or core-excited, using synchrotron radiation. In order to be tested, the TOFMS was attached to the brazilian beamline[4] installed at the CAMD ring (this beam line was designed and built at the brazilian national synchrotron radiation laboratory, LNLS, and transported to Baton Rouge for testing before its definite installation at the brazilian facility).

The time-of-flight acquisition system is based on a single board time-to-digital converter (TDC), with 1ns resolution[5]. It has four main operating modes:

1) Photoelectron-Photoion Coincidence (PEPICO) spectrum.

The ejected photoelectron triggers the TDC, providing the start signal and the arriving ion provides the stop, generating the time of flight for the latter. The experimental arrangement is shown in Figure 4. The resulting spectrum is similar to a mass spectrum, with the important restriction that multi-ionization processes do not contribute to the spectrum. The PEPICO spectrum can be consequently considered as a single event mass spectrum. By using a defocusing potentials on the lens, the spectrometer may work in a highly discriminating condition, what means that the spectrometer will in this case detect only ions ejected along the spectrometer axis. Figures 5 and 6 show, respectively, mass spectra for the CF_4 molecule obtained with a 100 % collection efficiency and with the defocusing potential applied.

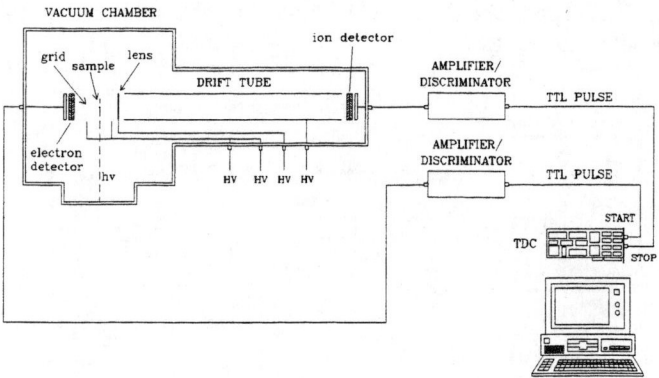

Figure 4: Experimental setup for photoelectron-photoion coincidence acquisition.

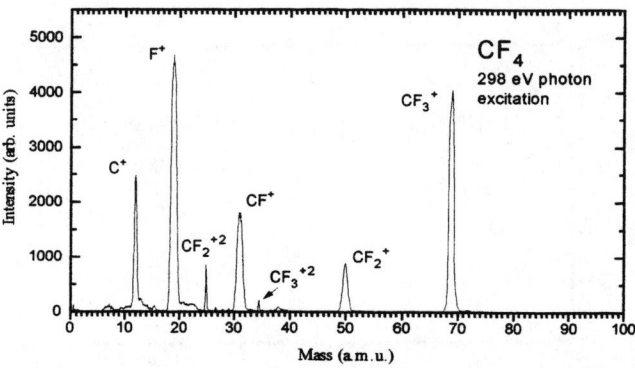

Figure 5: Photoelectron-photoion coincidence spectrum of CF_4 acquired with an applied focusing potential on the lens.

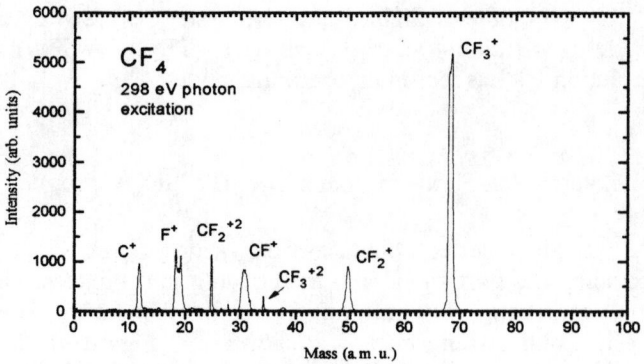

Figure 6: Photoelectron-photoion coincidence spectrum of CF_4 acquired with defocusing potential on the lens.

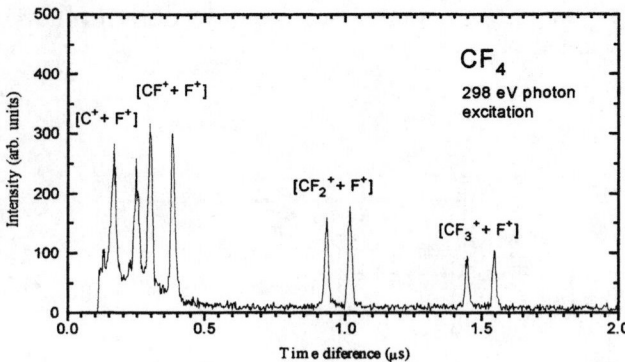

Figure 9: Photoion-photoion coincidence spectrum of CF_4 acquired with a defocusing potential on the lens.

2) Photoion-Photoion Coincidence (PIPICO) spectrum;

Both the start and stop signals are generated by the photoions associated with a double (or multiple) ionization event. Only the ion detector is used for data acquisition. The experimental arrangement for PIPICO measurements is shown in Figure 7. The resulting spectrum may again be obtained with a 100 % efficiency or in a discriminating mode, as exemplified in Figures 8 and 9, which show the PIPICO spectrum for the CF_4 molecule, obtained at 298 eV excitation energy.

3) Photoelectron-Photoion-Photoion Coincidence EPIPICO) spectrum

A photoelectron triggers the TDC, which registers the time of arrival for two correlated ions (provenient of the same ionization event). If X and Y are the arrival times for two ions formed, the PEPIPICO spectrum will consist in a XYZ graph, where Z is the number of total coincidences. Figure 10 shows the PEPIPICO spectrum for the methyl methacrylate, MMA, molecule, obtained at 284 eV.

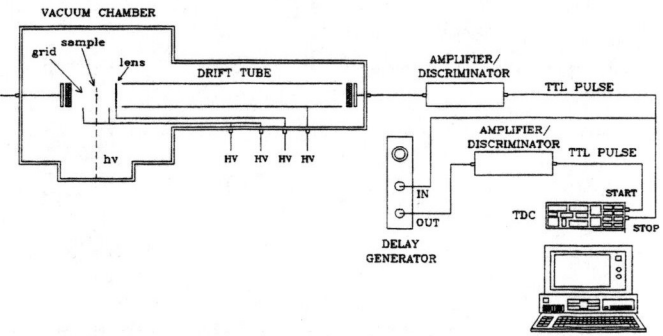

Figure 7: Experimental setup for photoion-photoion coincidence acquisition.

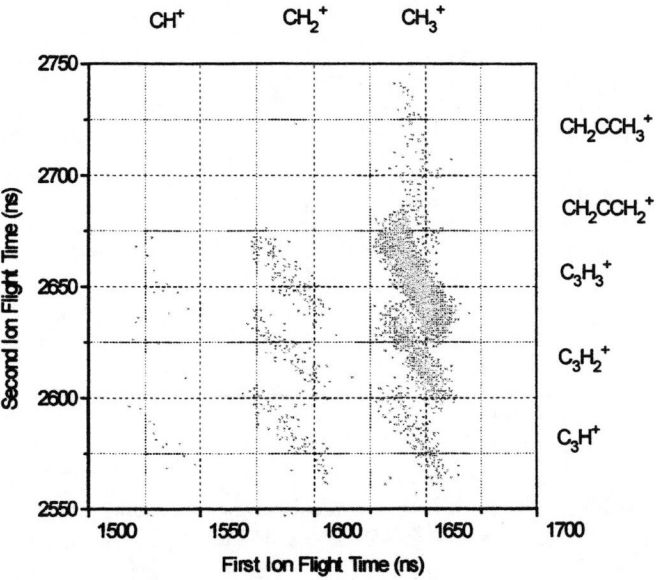

Figure 10: Photoelectron-photoion-photoion coincidence spectrum of MMA, 284eV photon excitation.

4) Total Ion Yield (TIY) spectrum

The total ion yield is measured as a function of the incident photon energy. For high photon energies, particularly in the range associated with core electron excitation, this spectrum is entirely equivalent to a photoabsorption spectrum. Figure 11 shows the TIY spectrum for MMA, measured around the C1s edge.

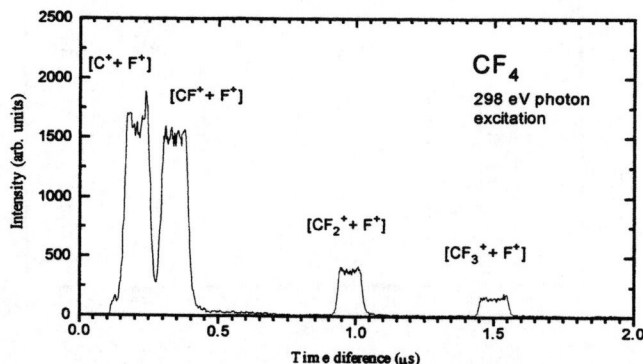

Figure 8: Photoion-photoion coincidence spectrum of CF_4 acquired with an applied focusing potential on the lens.

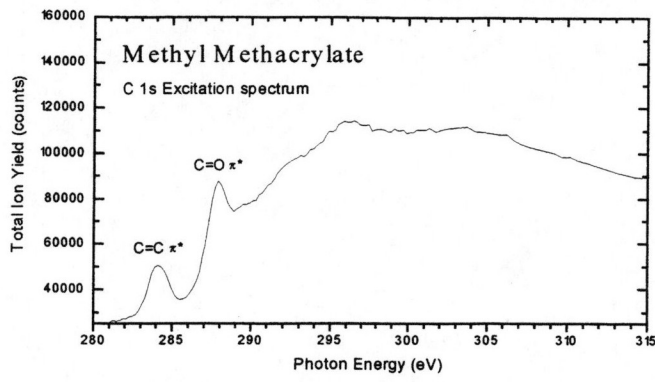

Figure 11: C1s Total Ion Yield Spectrum for Methyl Methacrylate.

Detailed analysis of the main experimental results highlighted in the present paper will be the focus of forthcoming publications.

IV. ACKNOWLEDGEMENTS

The authors acknowledge the CAMD staff for operating the synchrotron storage ring, and LNLS (Campinas, Brazil) for use of the TGM beamline. Financial assistance from CNPq (Brazil), LNLS and CAMD are also gratefully acknowledged.

V. REFERENCES

1) For example, (a)P. Morin, G.G.B.de Souza, I. Nenner and P. Lablanquie, Phys. Rev. Lett., 56,131 (1986); (b) P. Morin and I. Nenner, 56,1913(1986); (c) M.Simon, P. Morin, P. Lablanquie, M.Lavollee, K.Ueda and N.Kosugi, Chem. Phys. Lett., 238,42 (1995)

2) For example, (a) J.H.D.Eland, F.S.Wort and R.N.Royds, J. El. Spectr. Rel. Phen., 41, 297 (1986); (b) M.Simon, T. Lebrun, R.Martins, G.G.B.de Souza, I. Nenner, M.Lavollee and P. Morin, J.Phys. Chem., 97, 5228 (1993)

3) For example, (a) K. Ueda, E. Shigemasa, Y. Sato, A. Yagashita, T. Sasaki and T. Hayaishi, Rev.Sci. Instrum., 60, 2193 (1989); (b) B.P.Tsai and J.H.D.Eland, Int. J. Mass Spectrom. and Ion Phys., 36, 143 (1980)

4) P.T. Fonseca, J.G. Pacheco, E. d'A Samogin and A.R.B. de Castro, Rev. Sci. Instr. 63, 1256-9 (1992)

5) FLYTDC card for Macintosh, MIPSYS; TOF2 time-of-flight data acquisition board, Schimdt Instruments.

SOURCES OF X-RAY AND IR: INSERTION DEVICE, LASER EXCITED, AND OTHER

Design of an X-ray Free Electron Laser Undulator

Roger Carr

Stanford Synchrotron Radiation Laboratory
Stanford Linear Accelerator Center
Stanford, CA 94309 USA

Abstract. An undulator designed to be used for an x-ray free electron laser has to meet a set of stringent requirements. With no optical cavity, an x-ray FEL operates in the single pass Self Amplfied Spontaneous Emission (SASE) mode; an electron macropulse is microbunched by an undulator and the radiation it creates. The microbunched pulse emits spontaneous radiation and coherent FEL radiation, whose power may reach saturation in a sufficiently long and perfect undulator. The pulse must have low emittance and high current, and its trajectory in the undulator must keep the radiation and the pulse together with a very high degree of overlap. We shall consider the case of the Linear Coherent Light Source (LCLS) FEL project at SLAC, which is intended to create 1.5 Å x-rays using an electron beam with 15 GeV energy, 1.5π mm-mrad normalized emittance, 3400 A peak current, and 280 fsec FWHM bunch duration. We find that this 65 µm rms diameter beam must overlap its radiation with a walkoff of no more than 5 µm for efficient gain. This places severe limitations on the magnetic field errors and other mechanical tolerances. The following is a discussion of the undulator design, specifications, alignment, engineering, and beam position monitoring we plan to implement for the LCLS X-ray FEL.

INTRODUCTION

In a single pass free electron laser operating in the Self Amplified Spontaneous Emission (SASE) mode, exponential gain of coherent radiation intensity is predicted by theory, and power saturation is achieved in about 10 field gain lengths, assuming no tapering of the undulator. [1] It is desirable to build an FEL undulator to a full saturation length, so that the output is more stable than it would be on the exponential part of the gain growth curve. Given the desired output radiation wavelength and the energy of the available electron beam, saturation implies a certain output power, desired or not. If one reduces the beam current or enlarges the emittance in order to lower the output power, saturation will not be achieved.

At SLAC, we are studying the design of a 1.5 Å x-ray FEL based on a laser photocathode electron gun capable of generating a 1 mm-mrad electron pulse with 8 psec duration. [2] The bunch is accelerated in the last kilometer of the SLAC linac to E_e = 15 GeV, and passes through two bunch compressors that reduce the bunch duration to 280 fsec FWHM and raise the peak current to 3400 A. The expected normalized emittance is $\varepsilon_n = \gamma \varepsilon = 1.5\pi$ mm-mrad at the entrance to the undulator, with a bunch diameter of $2\sigma_r$ = 65 mm rms. The energy spread of the beam is σ_E = 0.0002 E_e, and the field gain length is 11.7 m. We shall discuss x-ray FEL undulators with this example in mind. The task of the undulator design is to achieve saturation with an undulator that is not excessively longer than the saturation length based on ideal undulator parameters. In the case of the LCLS, a perfect undulator is predicted to saturate in 94 m, based on both GINGER simulations [3] and semi-analytic models [4]. We hope that building a 100-110 meter device will allow for imperfections and reach saturation.

CHOICE OF UNDULATOR TYPE

After studying several undulator options, we chose a plane polarizing design using hybrid permanent magnet technology. We examined harmonic generation strategies where longer period bunching would feed radiation to successive shorter period devices, but no overall length savings were predicted [5] We looked at putting optical klystron style dispersive sections in the undulator to allow the energy modulated bunch to become spatially modulated, but found that energy spread ruled out this approach. [6] We also considered superconducting bifilar helix and other helical devices, but chose the planar hybrid on grounds of simplicity of construction, focusing, and correction strategy.

Simulations of the LCLS show that strong focusing must be added to the undulator lattice in order to maintain a small beam size. Natural focusing would give a beta function length of more than 50 m, but optimal focusing occurs with a beta function of less than 20 m. We intend to operate the LCLS from 5 GeV to 15 GeV, and if we choose an optimum beta function for the 15 GeV limit, we are not far off optimum at 5 GeV. Our models do not presently consider the decrease in energy of the electron beam due to emission of spontaneous radiation; we do not expect this to necessitate tapering the undulator when it is accounted for.

In a hybrid structure, one could add strong focusing with quadrupole magnets between segments of the undulator, or distribute strong focusing along the undulator lattice. To keep the beta function modulation ('sausaging') small, we would chose distributed focusing, for which we reviewed several strategies. They included putting permanent magnet quadrupole magnets between the poles of a hybrid undulator [7] a canted and wedged pole design [8], permanent magnets placed alongside the beampipe [9], transversely staggered poles, canted poles, and combined function pole pieces. We also considered a design with lumped quadrupole focusing and no quadrupole moment in the undulator segments. We chose this design because it affords much easier tolerance control, and a separated lattice is more flexible. The focusing and defocusing quadrupole magnets are moveable transversely, so they can act as trajectory correctors, with feedback from beam position monitors. Working with the Tesla-TTF group at DESY, we have determined that the resulting large beta function modulation does not have a deleterious effect on FEL performance [10]

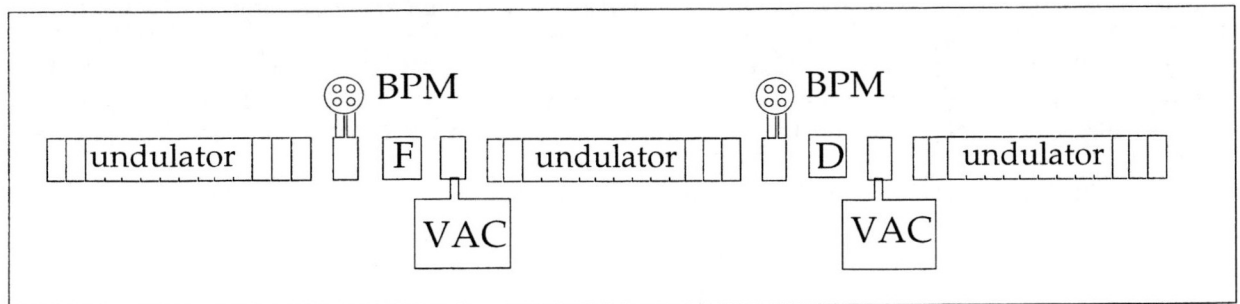

FIGURE 1. A schematic side view of the undulator structure, showing the FODO lattice with separations between 2m undulator sections for diagnostics, focusing (F and D) correctors, and vacuum ports.

The present undulator design has fifty 1.92 meter segments, mounted rigidly to high precision on aluminum girders. The undulator period is $\lambda_u = 30$ mm and the gap is 6 mm. Aluminum was chosen for its high ratio of thermal conductivity/thermal expansion. Each end of the girder will be supported by a pier, and a separation between segments of $\lambda_u (1 + K^2/2) = 23.75$ cm will be allowed between undulators, to allow the electrons to slip behind the photons by one period. The separation will be used for diagnostics, pumping, and focusing corrector magnets.

SPECIFICATIONS AND TOLERANCES

The resonance condition for the energy of the radiation first harmonic of an undulator is:

$$E_1 \text{ (eV)} = \frac{950 \, E_e^2 \text{ (GeV)}}{\lambda_u \text{(cm)} \left(1 + K^2/2\right)} \qquad (1)$$

where E_e is the energy of the electron beam, and $K = 0.934 \, B_{max} \text{(T)} \, \lambda_u \text{(cm)}$. For an NdFeB hybrid, the maximum value of the magnetic field is: [11]

$$B_{max} \text{(T)} = 3.44 \, e^{-\text{gap} / \lambda_u} \left(5.08 - 1.54 \, \text{gap} / \lambda_u\right) \qquad (2)$$

An energy difference of $\sigma_E = 0.0002 \, E_e$ would cause a resonance shift of $0.0004 \, E_1$, as would a gap change of 0.77 µm, a period change of 1.4 µm, or a change of field strength of 15 Gauss. These tolerances would be extremely severe. However, FRED-3D simulations show that FEL efficiency drops off only if random $\Delta B_{max}/B_{max}$ is greater than 0.1%. [12] This tolerance is reached when we have a period change of 70 µm, or a gap change of 13.5 µm. These are merely tight tolerances, which we would meet with machining, assembly, and magnetic shimming techniques.

Even with precision machining, sorting of magnets, tight specifications, and precise assembly, we expect measurable field errors. A standard Hall probe can, in principle, detect fields in the Tesla range with 20 μTesla resolution. [13] It is our strategy to shim each pole, to minimize the trajectory and photon resonance errors. We have a coordinate measuring machine with 0.8 μm absolute positioning accuracy over a volume of 1.2 m x 1 m x 0.6 m; it can be fitted with a temperature compensated Hall probe (calibration drift = 80 ppm/K [14] We have modeled the effect of placing small shims on the ends of the pole pieces away from the gap as shown in figure 1. The effect is to lower the field in the gap by a few Gauss, or a few 10's of Gauss. It is possible to order magnets with $\Delta B/B_{rms} < 0.5\%$, and an angular error of 0.5° from the nominal easy axis, but 1.0%, 1° magnets are much less costly. We have found in the case of pure permanent magnet devices that we can use computerized sorting techniques to improve undulator errors by a factor of about 5 over the case of a randomly assembled undulator; we would hope to be able to do as well with a hybrid structure. [15] If we purchase magnets with 1% errors, and lower the errors to 0.2% with sorting, we are well within the range of this shimming strategy.

GEOPHYSICAL, THERMAL, AND MECHANICAL DESIGN

The LCLS undulator will be placed in the existing Final Focus Test Beam (FFTB) facility at SLAC. The first third of the tunnel for the unduator is underground; the last two thirds are in an above-ground heavily shielded concrete structure. The substrate throughout is miocene sandstone. We propose to place the undulator on piers made of a proprietary sand and expoxy material [16], which is much more damping than metal piers, and much more stable against water related swelling and shrinking than ordinary concrete. In the range below 100 Hz, there is cultural and ocean generated vibration, but at the top of existing sand/epoxy FFTB piers, the measured amplitude of these vibrations is less than 50 nm.

Diurnal thermal distortions of the tunnel outside the hillside can be as much as 100 μm /day, but they can be reduced by isolating the tunnel with trenches cut into the substrate alongside it. The slow diffusive ground motion can separate points 100 m apart by about 100 μm /year. [17] These measurements indicate that our feedback system needs to have a dynamic range of some fraction of a millimeter. Presently the tunnel air temperature is stable to about 1K, but we plan to stabilize the undulator structure with flowing water that can be stabilized to 0.1K.

We propose to detect and correct diurnal, thermal, and ground motions with a system that was successfully implemented on the FFTB. [18] Two wires will be suspended inside hollow chambers that run the length of the undulator. The chambers are fixed to the magnets, and one end of the wire runs over a pulley and is tensioned by a fixed hanging weight. The exact shape of the curve described by the wire is not a catenary, but is distorted by imperfections in the wire at the micron level. However, a wire suspended this way is very stable. We will monitor the position of the wire with optical LED and split photodiode position sensors that are capable of resolving motions of the wire with submicron resolution.

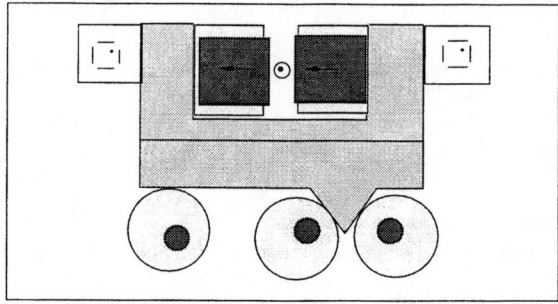

FIGURE 2. End view of the undulator magnet, showing three of the cams on one end of the girder (there are two on the other end) and the wire position monitors (upper left and right). The beampipe is 6 mm OD stainless steel tube.

The undulator magnets will be mounted on 5 eccentric cams, which are driven by stepping motors and harmonic drive gear reduction units. [19] The stepping motors might have 200 steps/turn, the harmonic drives might have 100:1 gear ratios, and the cam might have an eccentricity of 1 mm, for a net resolution of 10 steps/μm. In the

FFTB, these units have been shown to have submicron resolution and repeatability, which matches that of the wire position monitoring system. Rotation of the 5 cams controls pitch, yaw, roll, x, and y motion; an kinematic algorithm is used to control the motors; all 5 may have to move to correct an error in any of these 5 degrees of freedom. It is our intention to energize the stepping motors only when they move, so as not to create unwanted heat. Each motor dissipates about 10 W and only while it is in motion.

ALIGNMENT, STABILIZATION, AND BEAM POSITION MONITORING

Once the undulator is aligned, the wire position monitoring and cam system should be able to maintain its alignment at the micron level, with a response time on the order of seconds. However, the initial alignment of the undulator is a major challenge. The basic specification for alignment is that the beam trajectory must be straight to within about 5 μm over each 10 m field gain length. High performance laser tracker survey instruments could be used to position absolute beam position monitors (such as carbon wires) to within about 50 μm over this distance, perhaps somewhat less with great care and thermal stability. A hydrostatic level is capable of absolute micron tolerances, but only in the vertical. [20] Simulations show us that with the 50 μm tolerances over 10 meters that we expect from conventional surveying, we should achieve SASE laser action at 5 GeV, but not necessarily at 15 GeV.

We used trajectory modeling codes to find out how to monitor and correct the beam position. We find that beam position monitors and correctors with micron resolution should be placed at 4-5 meter intervals for optimum correction. If such correctors are installed, we calculated that initial trajectory errors of 100-200 μm could be reduced to 5 μm. Motions less than 200 μm are required of the quadrupole movers.

Beam based alignment is generally preferred to external mechanical methods. Our primary beam based alignment approach is to use the sensitivity of beam dispersion to quadrupole misalignment. We can run the LCLS from 5 to 15 GeV, with a corresponding FEL fundamental wavelength of 15 to 1.5 Å. With moveable quadrupole magnet corrections, we should be able to find a trajectory that does not change with energy, and such a trajectory must be a straight line. The concept here is that we do not measure straightness directly, which is hard, but subtsitute a measurement of a change in position with energy, which is much easier.

Another strategy for beam based alignment is to use a beam position monitor that is sensitive both to photons and electrons. We considered allowing the beam to strike a 30 μm carbon wire, as shown in figure 3. Experience at the SLC shows that such a wire can survive indefinitely when struck by 1 nC, 15 GeV electron pulses, and damage by x-radiation is less than that from electrons. [21] When the wire is struck by the electron beam, it will yield bremsstrahlung in a flat power spectrum from 15-0 GeV. Sweeping a 30 μm wire through a 65 μm rms diameter electron beam should allow us to locate the center of the electron beam to a few microns, using a downstream detector.

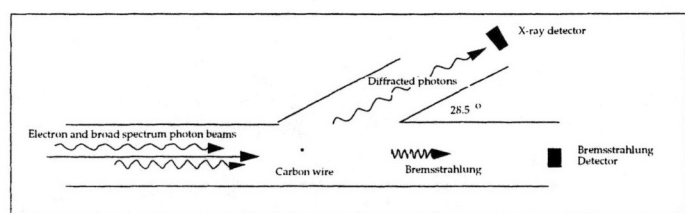

FIGURE 3. The combined photon and electron beam position monitor system. The x-ray detector could be a solid state photodiode; the bremsstrahlung detector is a radiation length of lead that converts bremsstrahlung into electron positron pairs, followed by a block of silica in which the pairs emit Cerenkov radiation that is detected by a photomultiplier tube.

We have examined the x-ray diffraction pattern from 7 μm carbon wires, and saw a peak at an angle of about 28 degrees with a FWHM of about 5 degrees, and sufficient intensity for BPM use. The spontaneous radiation at the 1.5 Å fundamental diverges by $\sigma_{r'} = \sqrt{\lambda_{rad}/L_u} = 1.2$ μrad, or 120 μm at 100 meters. The coherent FEL beam diverges by $\varepsilon / \sigma_r = 0.74$ μrad, or 74 μm at 100 meters. Therefore, we should be able to detect spontaneous radiation in the two-thirds of the undulator, where it is stronger than the FEL radiation, and use FEL radiation downstream.

After the initial alignment, we propose to turn on a feedback system to stabilize the trajectory. Electron beam position monitors will sense changes in the beam position, and correctors will null those changes. We examined many beam position monitor technolgies, including intercepting carbon wires, Compton scattering from laser 'wires', striplines and wall current monitors, diffraction radiation monitors, RF cavity monitors, fluorescent materials such as YAG crystals, and transition radiation monitors. We required simplicity, stability, low drift, low impedence, high resolution. and fabricability. The candidate technologies we are presently entertaining are carbon wires for initial beam finding, and striplines or RF cavity monitors for feedback control. We have advanced the design of a stripline for this particular application with MAFIA modeling, and feel that RF cavity technology and carbon wire technology are in hand at SLAC and CERN. [22]

RADIATION EFFECTS AND WAKEFIELDS

Our electron beam is calculated not to endanger the beampipe, but a single electron beam strike in the magnet material could fracture it, so we will protect the undulator with collimators, which will also scrape the halo off the beam. We calculate that 2-3 radiation lengths of Ti, followed by 20-30 radiation lengths of Cu is an appropriate collimator, with a possible sacrificial stopper at the entry to the undulator, for safety. [23] Radiation induced demagnetization is also a problem; it is modeled like a thermal demagnetization process, and thermal stabilization is apparently a defense against radiation induced demagnetization. [24] High coercivity NdFeB [25] used in the KEK in-vacuum undulator has been exposed to circulating beam for 7 years at a 12 mm gap with no apparent demagnetization. [26] The material was thermally stabilized to 125C, with a loss of about 2% of B_r. If NdFeB is not robust enough against radiation damage, we could use Sm_2Co_{17}.

We need not have very good vacuum in the LCLS beampipe; 100 nTorr should be adequate. The main disadvantage of higher pressure is the creation of gas bremsstahlung. However, we did need to examine the 5 mm ID beampipe for the effects of wakefields. For stainless steel, resistive wall losses would generate an energy spread in the electron beam of $\Delta E/E = 0.0016$ and an emittance growth of $\Delta \varepsilon/\varepsilon = 100\%$, which is unacceptable, but it is simple enough to plate the inner wall with a skin depth of copper, which reduces $\Delta E/E$ to 0.0003 and $\Delta \varepsilon/\varepsilon$ to 3%. [27] The total heat generated by resistive wall losses is on the order of a watt for 120 Hz operation, so heat dissipation will not be a problem. We have also calculated the expected effects of pumping ports, flange joints, etc and find them to be much smaller than the wall effects. Beampipe roughness may lead to longitudinal wakefield effects, so a highly polished beampipe is desirable.

SUMMARY

We have briefly described above some of the problems that we addressed in studying the design of an x-ray FEL at SLAC. Among the problems that any such design must address are: 1) tight mechanical tolerances, 2) geophysical and thermal environmental problems 3) beam position monitoring 4) initial alignment strategy 5) stability of alignmnent 6) radiation dose managment and 7) wakefield effects. Of these, we consider the initial alignment the most challenging, because we have no precedent for it. The other issues have been addressed in existing SLC and FFTB machines at SLAC, and in work at other laboratories.

ACKNOWLEDGEMENTS

The author is pleased to thank the many contributors to the LCLS undulator design study, including Heinz-Dieter Nuhn, Max Cornacchia, Robert Hettel, Don Martin, Roman Tatchyn, John Arthur, Jim Sebek, Richard Boyce, Jeff Corbett, (SSRL), Gordon Bowden, Dieter Walz, Robert Ruland, Clive Field, Cho Ng, Karl Bane, John Sheppard, Paul Emma, Vinod Bharadwaj, Pantaleo Raimondi, Vaclav Vylet, Alberto Fasso, Sayed Rokni, (SLAC), Klaus Halbach, Ross Schlueter, Steve Lidia, Kwang-je Kim, Ming Xie (LBNL), Lou Bertolini, Marcus Libkind, Lee Griffith (LLNL), Ilan Ben-Zvi (BNL), and Claudio Pellegrini (UCLA). This work was supported by the US Department of Energy, Office of Basic Energy Sciences under contract number DE-AC03-76SF00515.

REFERENCES

1. R. Bonifacio, C. Pellegrini, L.M. Narducci, *Opt. Commun.* 50 (1985)
2. H. Winick, et. al. *Nucl. Inst & Methods* A347 (1994) p. 199
3. H-D Nuhn, to be published
4. M. Xie, Proceedings of the US Particle Accelerator Conference 1995
5. H-D Nuhn, private communication
6. H-D Nuhn and C. Pellegrini, to be published
7. Yu.M. Nikiyina and J. Pflueger, *Nucl. Inst. & Methods* A375 (1996) p. 325
8. R. Schlueter, *Nucl. Inst. & Methods* A358 (1995) p. 44
9. A.A. Varfolomeev, et. al. *Nucl. Inst. & Methods* A341 (1994) p. 341
10. B. Faatz, K. Floettmann, private communication
11. K. Halbach, *Nucl. Inst. & Methods* 187 (1981) p. 109
12. H-D Nuhn, to be published
13. Group 3 Technology Ltd. model DTM 130 Teslameter
14. Leitz Enhanced Accuracy Model PMM 12-10-6 Coordinate Measuring Machine
15. S. Lidia and R. Carr, *Rev. Sci. Inst.* 66 (1995) p. 1865
16. Anocast Division of Anorad Corportation, Chagrin Falls, Ohio
17. Zero Order Design Report for the Next Linear Collider, Appendix C. SLAC report 474
18. R. Ruland - private communication
19. G. Bowden, P. Holik, S.R. Wagner, G. Heimlinger, R. Settles *Nucl. Instr.&Methods* A368 (1996) p. 579
20. Fogale Nanotech 190 Parc Georges Besse, 30000 Nimes, France
21. C. Field, *Nucl. Instr.& Methods.* A 360 (1995) 467.
22. W. Schnell, J.P.H. Sladen, I Wilson, and W. Wuensch, CLIC note 170, CERN SL/92-33
23. D. Walz, private communication
24. O.P. Kahkonen, M. Talvitie, E. Kauutto and M. Mannigen, *Phys. Rev.* B49 (1994) p. 6052
25. Sumitomo Special Metals Co. Neomax 28SH NdFeB material (1993)
26. S. Yamamoto, private communication
27. K. Bane, private communication

Long-wavelength edge radiation in an electron storage ring

R. A. Bosch

Synchrotron Radiation Center, University of Wisconsin-Madison
3731 Schneider Dr., Stoughton, WI 53589

O. V. Chubar*

Russian Research Center "Kurchatov Institute," 123182, Moscow, Russia

Abstract. The edge radiation emitted from a bending magnet edge in an electron storage ring is considered for wavelengths sufficiently long that the spatial extent of the magnet edge may be neglected. For a typical electron storage ring with energy ≥ 800 MeV, this "zero edge length" model applies to the infrared region. The far-field edge radiation is radially polarized, directed in a cylindrically-symmetric hollow cone along the straight-section axis with peak intensity at an opening angle of $1/\gamma$, where γ is the relativistic mass factor. At the wavelength, λ, the near field occurs at observer distance, R, less than $\lambda\gamma^2$. The near-field edge radiation is also radially polarized, forming a hollow cylindrical distribution with peak intensity at a distance $(R\lambda)^{1/2}$ from the straight-section axis.

I. INTRODUCTION

The bending magnets of an ultrarelativistic electron storage ring are increasingly utilized as a source of infrared radiation (IR) (1), whose wavelength is typically much longer than the critical wavelength in the bending magnets. At such long wavelengths, the radiation from electrons entering and exiting the bending magnets (edge radiation (2,3)) may be brighter than standard synchrotron radiation from the uniform magnetic field region (4). Consequently, the edge radiation may be used as a bright IR source; a beamline for this purpose has recently been constructed at the Synchrotron Radiation Center.

For sufficiently-long wavelengths, the edge radiation may be approximated by neglecting the spatial extent of the magnet edge. In this "zero edge length" model, the far-field radiation is radially polarized, directed in a cylindrically-symmetric hollow cone along the straight-section axis with peak intensity at an opening angle of $1/\gamma$, where γ is the relativistic mass factor (5). The far-field zero edge length model applies for $R > \lambda\gamma^2 \gg d$, where R is the observer distance, λ is the wavelength, and d is the distance traveled by an electron in the bending magnet edge field while being deflected an angle of $1/\gamma$. For the 800 MeV electron storage ring, Aladdin (6), with $d \sim 10$ cm, the zero edge length model applies for the infrared region ($\lambda \gg d/\gamma^2 = 40$ nm), while for the 7 GeV Advanced Photon Source with $d \sim 1$ cm, the zero edge length model is expected to apply for infrared, visible, ultraviolet, and soft xray wavelengths ($\lambda \gg d/\gamma^2 = 0.05$ nm).

For observer distance, R, less than $\lambda\gamma^2$, the observer is in the near field. This region is important in edge-radiation IR beam lines, where apertures and optical elements may intercept the edge radiation before its far-field characteristics are attained. In the zero edge length model, the near-field edge radiation is radially polarized, forming a hollow, cylindrically-symmetric distribution with peak intensity at a distance, $r_{peak} \approx (R\lambda)^{1/2}$, and angle, $\theta_{peak} \approx (\lambda/R)^{1/2}$, from the straight-section axis.

II. FAR-FIELD EDGE RADIATION

Consider the far-field radiation emitted along a straight-section axis from an electron passing through the downstream bending magnet. We assume that the observed radiation is produced during the time interval $0 < t < t_f$, where $t = 0$ describes

---

* Present address: *European Synchrotron Radiation Facility, Boite Postale 220, F-38043 Grenoble CEDEX, France*

the entrance, and $t = t_f$ is sufficiently large that the electron has been deflected by an angle $\gg \theta_{opening}$, the opening angle of the edge radiation. In the far field, the electric field results from the acceleration experienced by the electron. In SI units, this "acceleration" field, $\vec{E}_a$, obeys (7)

$$\vec{E}_a(\vec{x},\omega) = \int_{-\infty}^{\infty} \vec{E}_a(\vec{x},t_o) e^{i\omega t_o} dt_o = \frac{e}{4\pi\varepsilon_o cR} \int_0^{t_f} \frac{d}{dt}\left[\frac{\hat{n} \times (\hat{n} \times \vec{\beta})}{(1-\hat{n}\cdot\vec{\beta})}\right] e^{i\omega t_o(t)} dt. \qquad (1)$$

Here, $\vec{x}$ is the observer location, ω is angular frequency, c is the speed of light, $\vec{\beta}(t)$ is the electron velocity divided by c, e is the electron charge, ε_o is the permittivity of free space, R is the distance from the electron to the observer, and $\hat{n}(t)$ is the unit vector pointing from the electron location to the observer. In the far field of radiation emitted near the origin, $R \approx |\vec{x}|$ and $\hat{n} \approx \vec{x}/R$ are constant in time, while the observation time, t_o, is related to the time, t, of radiation emission by

$$\frac{dt_o}{dt} = 1 - \hat{n}\cdot\vec{\beta}(t). \qquad (2)$$

For observer angles (measured relative to the straight section axis) $\theta \sim \theta_{opening}$, the variation in the phase term of eq. (1), $\omega[t_o(t_f) - t_o(0)] \sim (2\pi c/\lambda)(1/2\gamma^2 + \theta^2/2)t_f$, is small compared to π provided that

$$d \ll \lambda/\theta_{opening}^2, \qquad (3)$$

where $d = \beta c t_f$, the "edge length," is the distance traversed while being deflected through an angle $\theta_{opening}$. For wavelengths obeying eq. (3), we consider the zero edge length ($d \to 0$) model in which $\exp(i\omega t_o(t))$ is considered constant in eq. (1). Integrating an exact differential yields, for $t_o(0) \equiv 0$,

$$\vec{E}_a(\vec{x},\omega) = \frac{e}{4\pi\varepsilon_o cR} \left.\frac{\hat{n} \times (\hat{n} \times \vec{\beta})}{1-\hat{n}\cdot\vec{\beta}}\right|_0^{t_f}. \qquad (4)$$

The RHS of eq. (4) evaluated at time $t = t_f$ is negligible for observer locations with $\theta \sim \theta_{opening}$, by definition of t_f, so that

$$\vec{E}_a(\vec{x},\omega) = -\frac{e}{4\pi\varepsilon_o cR} \left.\frac{\hat{n} \times (\hat{n} \times \vec{\beta})}{1-\hat{n}\cdot\vec{\beta}}\right|_0. \qquad (5)$$

The far-field edge radiation described by eq. (5) is independent of wavelength, forming a cylindrically-symmetric hollow cone. The radiation is radially polarized: at observer location $\vec{x} = (x,y,z)$ (where the z-axis is the straight section axis and the y-axis is vertical) the electric field is nearly parallel to $(x,y,0)$ for $|x|, |y| \ll z$.

The electric field of an observer with $y = 0$ is approximately

$$\vec{E}_a(\vec{x},\lambda) \approx \frac{e\gamma}{\pi\varepsilon_o cR} \frac{-\varphi}{2(1+\varphi^2)} \mathbf{e}_x, \qquad (6)$$

where $\mathbf{e}_x$ is the unit vector in the x-direction, $\varphi \equiv \gamma\theta$, and θ is the angle between $\vec{x}$ and the z-axis. The peak field occurs when $\varphi = 1$, i.e., $\theta_{opening} \sim \theta_{peak} = 1/\gamma$. According to eq. (3), the zero edge length far field result applies for $\lambda \gg d/\gamma^2$, where d is the distance traversed while an electron is deflected through an angle of $1/\gamma$.

For a bunch length exceeding the observed radiation wavelength, radiation from different electrons is essentially temporally incoherent (3). In addition, the electric field in the far-field is transverse to $\vec{x}$. In this case, neglecting the electron beam emittance and energy spread, the photon flux per unit solid angle (in photons/s-relative bandwidth $\Delta\omega/\omega$-steradian) obeys (7)

$$\frac{dF}{d\Omega} = \alpha \frac{\Delta\omega}{\omega} \frac{I}{e} \left(\frac{2R\varepsilon_o c}{e}\right)^2 |\vec{E}(\vec{x},\omega(\lambda))|^2, \qquad (7)$$

where I is the electron current and $\alpha \approx 1/137$ is the fine-structure constant.

For the zero edge length model, the photon flux per unit solid angle in the far field, obtained from eqs. (5) and (7), is

$$\frac{dF_a}{d\Omega}\bigg|_{edge} = \alpha \frac{\Delta\omega}{\omega} \frac{I}{4e\pi^2} \frac{\beta^2 \sin^2\theta}{(1-\beta\cos\theta)^2} \approx \alpha \frac{\Delta\omega}{\omega} \frac{I}{e\pi^2} \frac{\gamma^4 \theta^2}{(1+\gamma^2\theta^2)^2}, \qquad (8)$$

where the approximate expression (3) holds for $\theta \ll 1$ radian. The flux density is zero on the z-axis (where $\theta = 0$) and peaks at the observer angle $\theta = 1/\gamma$.

The flux calculated from eq. (8) was compared with numerical calculations of far-field edge radiation from a bending magnet at the 800 MeV electron storage ring, Aladdin (6), for λ = 1, 10, 100, and 1000 µm. The comparison confirmed that the zero edge length model applies for sufficiently long wavelengths that $\lambda \gg d/\gamma^2$ (i.e., $\lambda \gg$ 40 nm for d ~ 10 cm) (4).

For $\theta_{max} \ll 1$ radian, the flux contained within $\theta < \theta_{max}$ (in photon/s-relative bandwidth $\Delta\omega/\omega$) follows from eq. (8):

$$F(\theta < \theta_{max}) = \alpha \frac{\Delta\omega}{\omega} \frac{I}{e\pi} \left[\ln(1+\gamma^2\theta_{max}^2) - \frac{\gamma^2\theta_{max}^2}{1+\gamma^2\theta_{max}^2}\right]. \qquad (9)$$

The flux within the central bright spot is approximately

$$F(\theta < 3/\gamma) = 1.4\,\alpha(\Delta\omega/\omega)(I/e\pi). \qquad (10)$$

When the diffraction-limited source size, $S_D = \lambda/\theta_{opening}$ exceeds the physical source size, the source is diffraction-limited (8). For $\theta_{opening} \sim 1/\gamma$, the source is diffraction-limited provided that

$$\lambda \gg \max(\sigma_x,\sigma_y)/\gamma, \qquad (11)$$

where σ_x and σ_y are the electron beam transverse dimensions. For the Aladdin 800 MeV electron storage ring with $\max(\sigma_x,\sigma_y) \approx 1$ mm, the edge radiation source is diffraction-limited for $\lambda \gg 0.6$ µm.

For a diffraction-limited source, the brightness of the central spot (in photons/s-relative bandwidth $\Delta\omega/\omega$-sr-m^2) approximately equals the central spot flux divided by λ^2:

$$\text{Brightness} \approx \frac{F(\theta < 3/\gamma)}{\lambda^2} = \frac{1.4\,\alpha(\Delta\omega/\omega)(I/e\pi)}{\lambda^2}. \qquad (12)$$

The brightness depends only on the electron current for a given wavelength. Equation (12) may also be written as:

$$\text{Brightness}\left(\frac{\text{photons}}{\text{s - sr - mm}^2 \text{ - 0.1\%b.w.}}\right) \approx 2 \times 10^{19} \frac{I(A)}{[\lambda(\mu m)]^2}. \qquad (12\text{ a})$$

Equation (12a) approximates numerical calculations for the Aladdin storage ring when the source is diffraction-limited (4).

The angle over which the edge radiation is spatially coherent is $\theta_{coherent} \approx \lambda/\max(\sigma_x,\sigma_y)$ (9). For a diffraction-limited edge radiation source, eq. (11) indicates that $\theta_{coherent} \gg 1/\gamma = \theta_{opening}$. Thus, the central bright spot is spatially-coherent for sufficiently-large wavelengths that the source is diffraction-limited.

III. NEAR-FIELD EDGE RADIATION

In the zero edge length model, the far-field edge radiation consists of a hollow cone with an opening angle of $1/\gamma$. Thus, at an observer distance of R, the far-field radiation cone has radius of R/γ. This is smaller than the diffraction-limited source size, $\lambda/\theta_{opening} \approx \lambda\gamma$, for $R < \lambda/\theta_{opening}^2 = \lambda\gamma^2$. Thus, application of the far-field approximation requires that

$$R > \lambda\gamma^2. \tag{13}$$

Equation (13) may also be obtained from the far-field requirement that the "acceleration" field, E_a, be large compared to the "velocity" field (10) produced by a moving charged particle. For example, with an electron kinetic energy of 800 MeV ($\gamma=1567$) and wavelength of 1 µm, the far field will apply for $R > 2.5$ m.

In order to describe the near-field radiation, the "velocity" field must be computed and summed with the "acceleration" field to obtain the total electric field. The velocity field, $\vec{E}_v$, produced by an electron traversing an infinitely-long straight section until being bent suddenly through an angle $\gg \theta_{opening}$ at $t = 0$ is (10)

$$\vec{E}_v(\vec{x},\omega) = \int_{-\infty}^{\infty} \vec{E}_v(\vec{x},t_o) e^{i\omega t_o} dt_o = \frac{e}{4\pi\varepsilon_o\gamma^2} \int_{-\infty}^{0} \frac{\hat{n}(t)-\vec{\beta}}{(1-\hat{n}(t)\cdot\vec{\beta})^2 R(t)^2} e^{i\omega t_o(t)} dt. \tag{14}$$

Here, $\hat{n}(t)$ is the unit vector directed from the electron to the observer, and $R(t)$ is the distance from the electron to the observer. The contribution from $t > 0$ is insignificant along the straight section axis because of the large value of $1-\hat{n}(t)\cdot\vec{\beta}$ for $t > 0$.

Because eq. (14) displays cylindrical symmetry about the z-axis, we choose the x-axis such that the observer has $y = 0$. Defining $w(t) = \gamma\theta(t)$, where $\theta(t)$ is the angle between $\hat{n}(t)$ and the straight-section axis, eq. (14) may be written, for $t_o(0) \equiv 0$,

$$\vec{E}_v(\vec{x},\lambda) = \left(\frac{e\gamma}{\pi\varepsilon_o cR}\right)\left(\frac{e^{i\pi R_n(1-\varphi^2)}}{\varphi}\right)\int_0^\varphi \frac{\left(w, 0, \frac{1}{2\gamma}(1-w^2)\right)}{(1+w^2)^2} e^{i\pi R_n \varphi(w-1/w)} dw, \tag{15}$$

where $R_n = R/\lambda\gamma^2$ is the normalized observer distance and $\varphi = \gamma\theta$ is the normalized observation angle from the bending magnet edge.

A. Transverse electric field

For observer locations with $\theta \ll 1$, the transverse velocity field in the zero edge length model is approximated by

$$\mathbf{e}_x \cdot \vec{E}_v(\vec{x},\lambda) = \left(\frac{e\gamma}{\pi\varepsilon_o cR}\right)\left(\frac{e^{i\pi R_n(1-\varphi^2)}}{\varphi}\right)\int_0^\varphi \frac{w}{(1+w^2)^2} e^{i\pi R_n \varphi(w-1/w)} dw. \tag{16}$$

The total transverse field is the sum of the transverse velocity and acceleration fields:

$$\mathbf{e}_x \cdot \vec{E}(\vec{x},\lambda) = \left(\frac{e\gamma}{\pi\varepsilon_o cR}\right)\left(\frac{e^{i\pi R_n(1-\varphi^2)}}{\varphi}\int_0^\varphi \frac{w}{(1+w^2)^2} e^{i\pi R_n \varphi(w-1/w)} dw - \frac{\varphi}{2(1+\varphi^2)}\right)$$

$$= \left(\frac{e\gamma}{\pi\varepsilon_o cR}\right)\left(\frac{1}{\varphi}\int_0^\varphi \frac{w}{(1+w^2)^2}\left(e^{i\pi R_n \varphi(1/\varphi-\varphi+w-1/w)} - 1\right)dw\right) \tag{17}$$

Equation (17) applies in the near and far fields. For $R_n = R/\lambda\gamma^2 > 1$, the transverse velocity field is negligible and the far field result applies: the flux is collimated with an opening angle of $\theta_{peak} = 1/\gamma$. For $R_n < 1$ (i.e., $R < \lambda\gamma^2$), the observer is in the near field. The distribution peaks at $\varphi_{peak} \approx R_n^{-1/2}$ (i.e., $\theta_{peak} \approx (\lambda/R)^{1/2}$). The near-field edge radiation is radially polarized, forming a cylindrically-symmetric, hollow distribution with peak intensity at a distance, $r_{peak} = R\theta_{peak} \approx (R\lambda)^{1/2}$, from the straight-section axis.

For $R_n \ll 1$ and $\varphi \sim \varphi_{peak} \approx R_n^{-1/2}$, eq. (17) may be approximated by

$$\mathbf{e}_x \cdot \vec{E}(\vec{x},\lambda) \approx \left(\frac{e\gamma}{\pi\varepsilon_o cR}\right) \left(\frac{e^{-i\pi R_n \varphi^2}}{\varphi} \int_0^\varphi \frac{w}{(1+w^2)^2} e^{-i\pi R_n \varphi/w} dw - \frac{\varphi}{2(1+\varphi^2)} \right), \quad (18)$$

which gives

$$\left| \mathbf{e}_x \cdot \vec{E}(\vec{x},\lambda) \right| \approx \left(\frac{e}{\pi\varepsilon_o cR}\right) \frac{\left|\sin\left(\frac{\pi R\theta^2}{2\lambda}\right)\right|}{\theta}. \quad (19)$$

The photon flux from eq. (19) is

$$\frac{dF}{d\Omega} \approx 4\alpha \frac{\Delta\omega}{\omega} \frac{I}{e\pi^2} \frac{\sin^2\left(\frac{\pi R\theta^2}{2\lambda}\right)}{\theta^2}. \quad (20)$$

This approximate near-field flux distribution is independent of the relativistic factor, γ. Equation (20) approximates the flux distribution given by eq. (17) within 20% for $R_n = R/\lambda\gamma^2 \leq 0.01$, and within 5% for $R_n = R/\lambda\gamma^2 \leq 0.001$.

From eq. (3), the zero edge length model applies for $d \ll \lambda/\theta_{opening}^2$. Thus, in the near field, where $R < \lambda\gamma^2$ and $\theta_{opening} = (\lambda/R)^{1/2}$, the zero edge length model applies for

$$\lambda\gamma^2 > R \gg d, \quad (21)$$

where d is the length required to deflect an electron through an angle $\theta_{opening} = (\lambda/R)^{1/2}$.

In the near field, the radius of the photon distribution, r_{peak}, at observer distance, R, obeys

$$r_{peak}(R) \approx (R\lambda)^{1/2} = \frac{\lambda}{(\lambda/R)^{1/2}} \approx \frac{\lambda}{\theta_{peak}(R)}. \quad (22)$$

Because $\theta_{peak}(R)$ is the opening angle of the radiation, eq. (22) indicates that the near-field radiation forms a diffraction-limited spot, provided that the electron beam emittance does not significantly alter the flux distribution from the single-electron result, i.e., $r_{peak}(R) \gg \max(\sigma_x, \sigma_y)$ and $\theta_{peak}(R) \gg \max(\sigma_{x'}, \sigma_{y'})$, where σ_x, σ_y, $\sigma_{x'}$, and $\sigma_{y'}$ describe the electron beam at the magnet edge.

For an aperture of radius r_{ap} at distance R_{ap}, with $\theta_{ap} \equiv r_{ap}/R_{ap} \gg 1/\gamma$, the central bright spot will intercept the aperture at wavelengths such that $\theta_{ap} < \theta_{peak} \approx (\lambda/R)^{1/2}$. Thus, the flux will be cut off by the aperture for

$$\lambda > \lambda_{cutoff} \approx r_{ap}^2/R_{ap} = \theta_{ap}^2 R_{ap}. \quad (23)$$

The flux for the zero edge length model, obtained from eq. (17), was compared with numerical computations of near field edge radiation from a bending magnet at the 800 MeV, 200 mA electron storage ring, Aladdin. Numerical computations are shown as solid lines in Fig. 1 for a zero-emittance beam, where the Aladdin bending magnet is assumed to have a radius of curvature of 2.083 m, with a linear ramp-up of the field over a fringe field of length 10.8 cm. Similar results were obtained when the fringe field was neglected or modeled as a smooth evanescent field. The results for the zero edge length model are

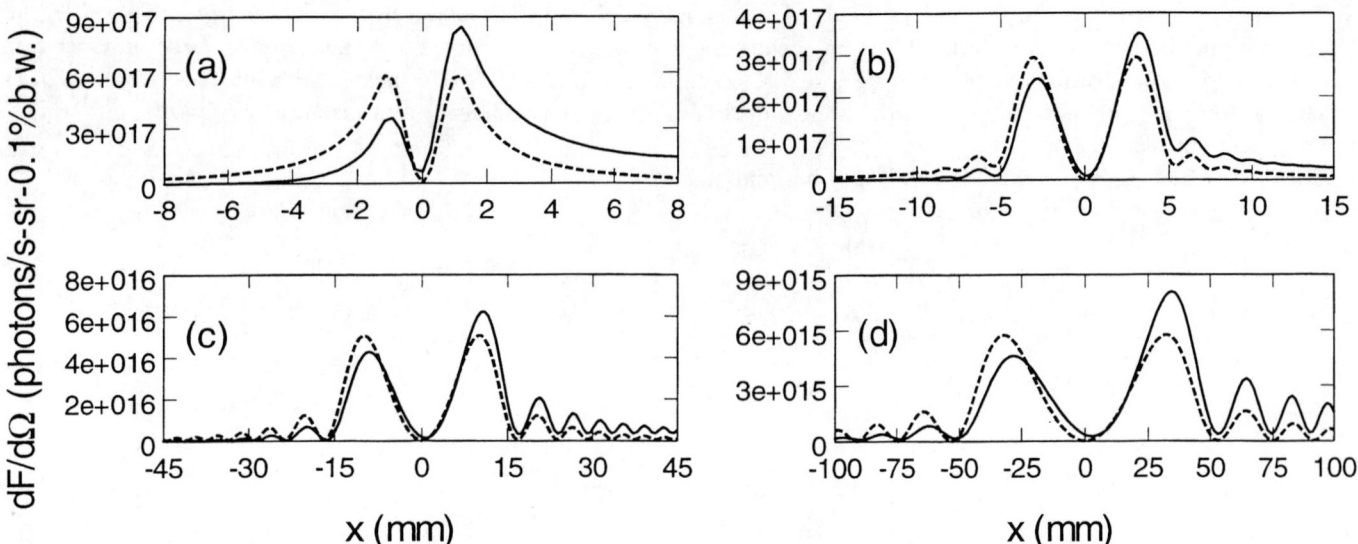

FIGURE 1. Flux density from numerical computations (solid lines) and the zero edge length model (dashed lines), for an observer 1.4 m downstream of an Aladdin bending magnet with an electron current of 200 mA and energy of 800 MeV. The flux is plotted versus horizontal position in the plane of the electron orbit, for electrons deflected by the bending magnet in the positive x-direction. (a) $\lambda = 1$ μm. (b) $\lambda = 10$ μm. (c) $\lambda = 100$ μm. (d) $\lambda = 1000$ μm.

shown as dashed lines. For the wavelengths shown, $\lambda = 1, 10, 100,$ and 1000 μm, the observer is in the near field with $R_n = R/\lambda\gamma^2 < 0.6$.

The numerical computations deviate from cylindrical symmetry because of the finite extent of the region in which an electron is bent (e.g., 10 cm for a 25 mrad deflection). As expected from the zero edge length model, the radiation peaks at $r_{peak} \approx (R\lambda)^{1/2}$. In addition, the average peak height agrees with the zero edge length model. Thus, for this example with $d \ll R < \lambda\gamma^2$, the zero edge length model accurately describes the near-field IR edge radiation except for its deviation from cylindrical symmetry.

B. Longitudinal electric field

In the extreme near field, the longitudinal electric field may be comparable to or larger than the transverse field, so that the Poynting vector may not be approximately parallel to $\vec{x}$, the vector pointing from the bending magnet edge to the observer. In this case, the energy does not flow radially outward from the bending magnet edge, and eq. (7) does not apply. For $\theta \ll 1$, the axial acceleration field given by eq. (5) is

$$\mathbf{e}_z \cdot \vec{E}_a = \frac{e}{\pi\varepsilon_o cR} \frac{\varphi^2}{2(1+\varphi^2)} , \qquad (24)$$

From eq. (15), the axial velocity field is

$$\mathbf{e}_z \cdot \vec{E}_v(\vec{x},\lambda) = \left(\frac{e}{\pi\varepsilon_o cR}\right) \frac{e^{i\pi R_n(1-\varphi^2)}}{2\varphi} \int_0^{\varphi} \frac{1-w^2}{(1+w^2)^2} e^{i\pi R_n\varphi(w-1/w)} dw . \qquad (25)$$

For $R_n \to 0$, eq. (25) reduces to

$$\mathbf{e}_z \cdot \vec{E}_v(\vec{x},\lambda) = \left(\frac{e}{\pi\varepsilon_o cR}\right) \frac{1}{2(1+\varphi^2)} . \qquad (26)$$

This result is accurate for $R_n \ll 0.1$. The axial field is the sum of the axial acceleration and velocity fields, which for $R_n \ll 0.1$ obeys

$$\mathbf{e}_z \cdot \vec{E}(\vec{x}, \lambda) = \frac{e}{2\pi\varepsilon_o cR} . \tag{27}$$

From eq. (19), the transverse electric field peaks at $\theta_{peak} \approx (\lambda/R)^{1/2}$, with approximate peak value

$$\left| \mathbf{e}_x \cdot \vec{E}(\vec{x}, \lambda) \right|_{peak} = \frac{e}{\pi\varepsilon_o cR} \left(\frac{R}{\lambda} \right)^{1/2} . \tag{28}$$

Thus, the transverse field dominates for $R \gg \lambda$. For $R < \lambda$, the longitudinal electric field dominates, the Poynting vector is not parallel to $\vec{x}$, and the flux distribution is not well collimated.

IV. SUMMARY

For sufficiently long wavelengths, the edge radiation along a straight-section axis from the downstream bending magnet may be calculated by the zero edge length model, which neglects the spatial extent of the bending magnet edge field region. The radiation may be characterized in different regions of observer distance, R, which depend on λ and γ. The regions are:

(a) $R > \lambda\gamma^2$: far-field region. At sufficiently-large distances from the bending magnet edge, the radiation is radially polarized and collimated in a cylindrically-symmetric hollow cone with opening angle of $1/\gamma$.

(b) $\lambda \ll R < \lambda\gamma^2$: near-field region. Here, the radiation is radially polarized, forming a hollow cylindrically-symmetric distribution with peak intensity at a distance, $r_{peak} = R\theta_{peak} \approx (R\lambda)^{1/2}$, from the straight-section axis.

(c) $R < \lambda$: extreme near-field region. Within several wavelengths of the bending magnet edge, the radiation has significant longitudinal polarization of the electric field, so that the energy flux is not in the direction of $\vec{x}$, the vector pointing from the bending magnet edge to the observer. The radiation is not well collimated.

The above results, calculated for a zero edge length, apply for sufficiently-large wavelengths such that $d \ll \lambda/\theta_{opening}^2$, where d is the distance required to deflect the electron through an angle of $\theta_{opening}$. In the far field, this criterion may be written as $\lambda \gg d/\gamma^2$, while in the near field this criterion becomes $R \gg d$.

A comparison of the zero edge length model with numerical calculations for the near field and far field of an Aladdin bending magnet verifies that the zero edge length model approximates the IR edge radiation.

ACKNOWLEDGMENTS

The authors appreciate valuable discussions with M. A. Green, T. E. May, R. Reininger and W. S. Trzeciak. This work was supported by NSF grant DMR-95-31009.

REFERENCES

1. Carr, G. L., Hanfland, M., and Williams, G. P., *Rev. Sci. Instrum.* **66**, 1643–1645 (1995).
2. Titov, A., and Yarov, A., *Nucl. Instr. and Meth.* **A308**, 117–119 (1991).
3. Chubar, O. V., and Smolyakov, N. V., *J. Optics (Paris)* **24**, 117–121 (1993); "Generation of intensive long-wavelength edge radiation in high-energy electron storage rings," in *Proc. 1993 IEEE PAC*, 1993, pp. 1626–1628.
4. Bosch, R. A., May, T. E., Reininger, R. and Green, M. A., "Infrared radiation from bending magnet edges in an electron storage ring," in *Proc. SRI '95 (CD ROM), Rev. Sci. Instrum.* **67**, 3346(K) (1996); Bosch, R. A., SRC Technical Note SRC-161, Synchrotron Radiation Center (1995).
5. Bosch, R. A., *Nucl. Instr. And Meth.* **A386**, 525–530 (1997).
6. Green, M. A., Huber, D. L., Rowe, E. M., and Tonner, B., *Rev. Sci. Instrum.* **63**, 1582–1583 (1992).
7. Jackson, J. D., *Classical Electrodynamics*, 2nd. ed., New York: Wiley, 1975, p. 670.
8. Hirschmugl, C. J., Ph.D. thesis, Yale University, 1994.
9. Williams, G. P., *Rev. Sci. Instrum.* **63**, 1535–1538 (1992).
10. Jackson, J. D., *Classical Electrodynamics*, 2nd. ed., New York: Wiley, 1975, p. 657.

Control System for Insertion Devices at the Advanced Photon Source

Oleg A. Makarov, Patric Den Hartog, Elizabeth R. Moog, Martin L. Smith

Advanced Photon Source, Argonne National Laboratory.
9700 South Cass Avenue, Argonne, IL 60439-4800

Abstract. Eighteen insertion devices (IDs) are installed at the Advanced Photon Source (APS), and three more are scheduled for installation by the end of this year. A distributed control system for insertion devices at the APS storage ring was created with the Experimental Physics and Industrial Control System (EPICS). The basic components of this system are operator interfaces (OPIs), input output controllers (IOCs), and a local area network that allows the OPI and IOC to communicate. The IOC operates under the VxWorks OS with an EPICS database and a sequencer. The sequencer runs an ID control program written in State Notation Language. The OPI is built with the EPICS tool MEDM and provides display screens with input and output fields and buttons for gap control of the IDs. Global commands like "open all IDs" are C-shell scripts invoked from the display menu. The algorithms for control and protection of the ID and ID vacuum chamber and the accuracy of gap control are discussed.

INTRODUCTION

The Advanced Photon Source (APS) at Argonne National Laboratory is a national user facility designed to produce insertion-device- and bending-magnet-based synchrotron radiation to be used in forefront research in science and technology (Fig. 1). Of the 18 insertion devices (IDs) currently installed [1-5], 13 are 2.4-meter-long undulators with 3.3 cm periods, three are undulators with periods of 1.8 cm [4], 2.7 cm and 5.5 cm, respectively, one is an 8.5-cm-period wiggler, and one is an elliptical multipole wiggler (EMW) with a 16 cm period. The characteristics of X-rays produced by all of these IDs are adjusted by varying the gap between the upper and lower jaws - arrays of permanent magnets and poles. In addition to permanent magnets producing a vertical magnetic field, the EMW also has electromagnets that produce a horizontal magnetic field of alternating polarity at up to 10 Hz.

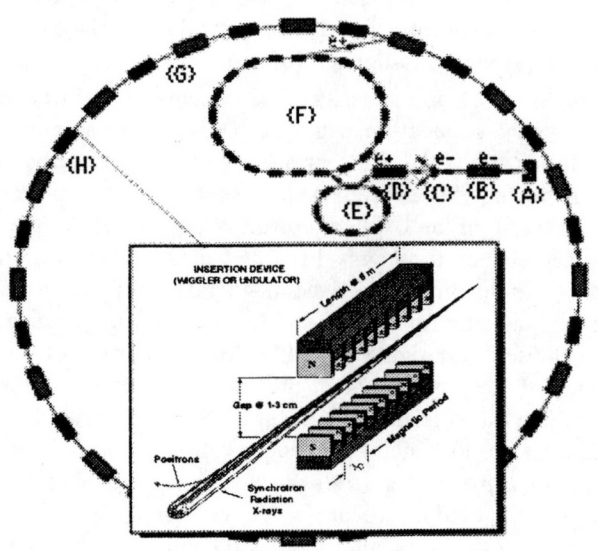

FIGURE 1. Schematic representation of the APS beam acceleration and storage complex: A - electron gun, B - electron linear accelerator (200 MeV), C - positron conversion target, D - positron linear accelerator (450 MeV), E - positron accumulator ring, F - booster/injector synchrotron (7 GeV), G - storage ring (7 GeV nominal energy), H - insertion device.
Insert: an APS insertion device and X-ray beam.

HARDWARE DESCRIPTION

On each ID, two stepper motors are used to change the gap between the upper and lower jaws. On all but the elliptical wiggler, one motor controls the gap at the upstream end, and another - the downstream end. On each end of the ID, one absolute linear and one absolute rotary encoder are installed. The control system includes power supplies, interlock electronics, encoder interfaces, drives for the stepper motors, and IOCs with an MVME-167 single-board computer connected to the local area network (LAN) via ethernet. The IOC is based on VME crate, it includes an Oregon Micro Systems' VME-8 controller, a VAROC interface, and an Acromag AVME9440 interface. The VME-8 board is capable of controlling eight stepper motors. The VAROC board, developed at the European Synchrotron Radiation Facility, interfaces up to 16 absolute encoders complying with the SSI (synchronous serial interface) protocol to the VMEbus. The AVME-9440 board provides 16 digital input and 16 digital output channels.

The EMW is a unique insertion device, which has been installed in sector 11 of the storage ring. It produces circularly polarized X-rays whose CW or CCW polarization is altered by switching the polarity of an electromagnet. A 1000 Amp electromagnet supply is controlled with an Allen-Bradley (AB) crate with binary and analog IO modules. The VME crate communicates with the AB crate through a 6008 SV scanner. The correcting coils of the EMW are controlled through an arbitrary function generator to reduce the magnetic field integrals for the electromagnet during transitions.

STRUCTURE OF THE CONTROL SYSTEM

The control system for insertion devices was created with EPICS [6]. It is a distributed control system with UNIX workstations as OPIs, VME crates as IOCs, and an ethernet-based LAN that allows the workstation and VME crate to communicate.

The IOC operates under the VxWorks OS [7], which is a high performance real-time operating system with a full range of UNIX-compatible networking facilities. The Motorola MVME 167-02B single-board microcomputer with a MC68040 processor is the heart of the IOC. The major part of the control software for an insertion device resides in the IOC. At boot time, after loading the operating system, a start-up script file is executed that contains commands to download the IOC database support programs, a sequencer program, a control program, EPICS databases, and data for magnetic field vs. ID gap. Almost all processing for insertion device control takes place in a *ctrl_ID* program executed by the sequencer. This program is written in State Notation Language [8] and consists of five state sets and an *exit* procedure.

State set *errlog_file* consists of three states: *start*, *init* and *update_file*. The *start* state opens the error logging file, the *init* state checks the connection with the ID database parameter variables, and the *update_file* state updates contents of the *errlog.idXX* file at the Unix file server every 5 minutes if the "errlog" file is open. Also this state terminates the *ctrl_ID* program if the "errlog" file exceeds a size of 500 kilobytes to prevent overbooking of the file storage. State set *speed* consists of two states: *speed_up* and *speed_down*. When the ID moves to a smaller gap (moves in the direction of the magnetic attractive force between the jaws) the *speed_up* state sets a high speed for the stepper motors, and when the ID gap opens (moves opposite to the direction of the magnetic force), the *speed_down* state sets a low speed for the stepper motors at small gaps where the magnetic force between the jaws is strong. State set *encoders* consists of two states: *init_enc* and *offset_enc*. The first one is used to initialize variables for encoder offsets, and the second state is used to write into the "errlog" file the change of each of these offsets or of the encoder selection. State set *buttons* consists of two states: *init_butt* and *butt*. The *init_butt* state is used to initialize the variables and to start the readout of the encoder values, then, after a 1/3 s delay, to initialize the variables in which the encoder values are used. The *butt* state is used for handling encoder and motor errors and events like *button is pressed* or *value is entered*, writing this information into the "errlog" file. State set *main_loop* consists of one state: *main_lp,* which is executed every 0.25 seconds. It checks gap values at the ID, calculates displacements and starts stepper motors, checks errors, and updates ID gap values and energies of X-ray photons. The procedure *exit* is used to close the *errlog.idXX* file and post the message *RESTART "ctrl_ID" PROGRAM* on the *ID Gap Control* display.

Operators and users can monitor status and initiate actions from the control room or from computers located in offices or at beamlines. Security is maintained by the use of a process variable server that allocates control according to predefined access tables. Global commands like *open all IDs* are C-shell scripts invoked from the display menu. During beam injection, users have no access to control the IDs.

OPERATOR INTERFACE

Control displays for IDs are available on UNIX workstations with the Motif Editor and Display Manager (MEDM) [9], which is an EPICS tool.

Basic control functions for the ID are provided by an *ID Gap Control* display, shown in Fig. 2. It allows the user to enter a value either for the desired gap or for the photon energy corresponding to the first harmonic of the undulator radiation.

Pressing the *Move* button starts the ID gap adjustment, and the *Stop* button, when pressed, stops the ID gap adjustment. The current state of the ID, average gap value in mm, and photon energy of the X-rays in keV are displayed above the input fields. Energy calculation uses the lookup table of magnetic field vs. gap for a particular device. Two strings show messages related to the status of the ID control program and to the conversion between the ID gap value and the photon energy of X-rays, for example: *desired position accepted, min gap value substituted, adjustment started, adjustment stopped, adjustment completed, conversion (Gap<-> Energy) OK, energy is too small*, etc. Also, the display contains information about ID type, serial number and its location in the storage ring. The *More* button opens a pull down menu that allows access to additional displays: *ID Taper Control, Call_in List* and *ID Commissioning Limit Control*.

The *ID Info* button invokes execution of a C-shell script that starts an *ID_online_info* program using interactive data language (IDL) and providing graphical and numerical information about IDs. Graphs of *Magnetic Field vs Gap, 1st Harmonic Energy vs Gap, Integrated Moments vs Gap (1st and 2nd integral), Integrated Multipole Moments vs Gap (quadrupole, sextupole, and octupole), RMS Phase Error vs Gap, RMS Peak Field Variation vs Gap, Trajectory Straightness vs Gap, Trajectory Angle vs Gap, Brilliance vs Energy*, as well as numerical data for any ID can be selected interactively. All of the data are collected during magnetic measurements before installation of the ID in the storage ring.

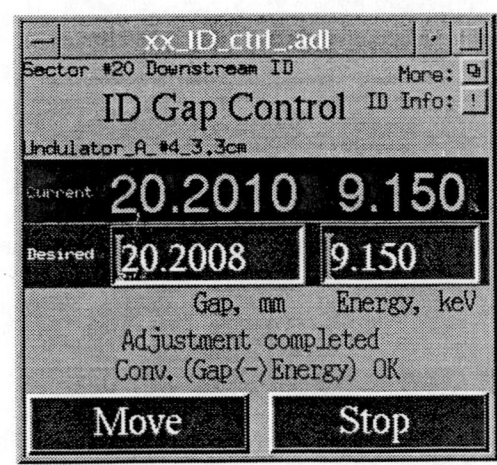

FIGURE 2. ID gap control display.

ALGORITHM FOR GAP ADJUSTMENT

Testing of entered gap values

Each time a new desired gap value (upstream, downstream or parallel) is entered, the *ctrl_ID* program checks that the value is within the range for ID gap and taper. When the entered value cannot be accepted, the program replaces it with the nearest acceptable value. In each case, an appropriate message appears on the *ID Gap Control* display.

Gap adjustment

Because the magnetic force between the ID jaws varies with gap and creates nonlinearity in the gap motion, multiple move commands are used to accomplish a single move request. Each move command is calculated to be 90% of the desired gap change, except when the discrepancy between desired and measured position becomes less than 10 µm. In this case, it moves only 30% of the remaining distance. Convergence to the desired position is usually obtained after several moves. It takes about 20 seconds for adjustment with 1 mm gap change, and about 5 minutes for change from minimal to maximal gap.

Movement of the extreme ends of the ID are coupled to the motion of the opposite end motor because of the placement of the mechanical pivot points. As a consequence, tapering of an ID must be carefully controlled to avoid damage to the storage ring vacuum chanber. The program will reduce the jaws' taper to 10% of the initial value if it exceeds the desired value by more than 50 µm. When the taper request exceeds the current taper by more then 50 µm, and both motors should move in the same direction, the jaws are leveled to the required average gap value before proceeding with the taper. Furthermore, if during adjustment the jaws' taper exceeds 200 µm, its value will appear on the control display. If the current taper exceeds the desired value by more than 400 µm or exceeds a value of 5.1 mm, then the taper will be reduced.

If the current gap value exceeds the maximum allowed gap value by more than 200 µm or becomes 50 µm smaller than the minimum allowed gap value during adjustment, then motion will be paused and new values will be loaded to the stepper motor controller.

Adjustment is considered as completed when both motors have finished motion and a difference between the desired gap value and one measured by linear encoders is less than 0.28 µm for both ends of the device. The resolution of the linear encoders is 0.5 µm.

Encoder and motor errors

The noisy electromagnetic environment at the storage ring occasionally causes encoder errors. Readout of the encoders is executed each 0.25 seconds. Because the gap changes slowly, there are no conditions in which the value would differ from the previous one by more than 1 mm. Therefore, a change in encoder value of more than 1 mm is considered to be an encoder error. If the encoder error occurs during motion, a *Stop* command is issued, and an error message appears on the control display. Electrical noise may force the linear encoder counter to count up/down without motion. The program decides this error has occured when either linear encoder value changes more than 40 µm when there were no gap adjustments. In this case an error message, *reset US(DS)l, err. = <current_US(DS)_linear _encoder_drift>*, appears on the display.

To detect a motor error, the *ctrl_ID* program compares the displacement calculated for that motor from encoder data with the displacement obtained from the change of a step counter. This comparison is made each time the step counter change exceeds 0.5 mm. If the ratio of these two displacements (one based on the change of the step counter, and another based on the change of encoder data) is less than 0.6, this state is considered to be a motor error, a *Stop* command is issued, and an error message appears on the display. The program resets the step counter each time the motor position obtained from it differs from the value calculated with encoder data by more than 0.5 mm, and there is no adjustment. This action is not considered an error.

Monitoring of gap value, taper and limit switches

The upstream and downstream gap values are regularly compared with the software gap limits. When these values exceed the maximal gap limit by more than 0.2 mm, an *US = <value> over limit* or *DS = <value> over limit* message appears on the display. When these values are below the minimal gap limit by more than 0.05 mm, a *US = <value> under limit* or *DS = <value> under limit* message appears on the display.

During adjustment, a taper value (i.e., *US_gap_value - DS_gap_value*) is periodically calculated. If it exceeds 0.2 mm, an *US - DS = <value> mm* message appears on the display. If the taper value exceeds maximal taper + 0.1 mm (i.e., a 5.1 mm value), the *ctrl_ID* program places a *Taper = <value>, out of limit* message on the display.

The stepper motor driver monitors the limit switches. Should these limit switches be tripped, the *ctrl_ID* program would place the *Hard limit, US = <value>* and/or *Hard limit, DS = <value>* messages on the display, switch access to the device from *User* to *Operator* mode, and issue a *Stop* command.

DISCUSSION

The magnetic force between the ID jaws at the minimum gap is comparable with a gravity force for a 3-tonne mass. The storage ring aluminium vacuum chamber is 10 mm high between the jaws of the ID, with a wall thickness as small as 1 mm. The ID jaws cannot be allowed to damage the vacuum chamber. There are three levels of protection: mechanical level - hard stops are adjusted so as to stop motion at a gap of 10.4 mm and at full open gap; hardware level - limit switches that shut off the stepper motors at a gap of 10.7 mm and near maximum gap; software level - software limits (presently set at 11 mm and about 200 mm) that do not allow the jaws to hit the limit switches.

The accuracy of gap adjustments relies on the linear encoders having a 2.5 µm accuracy and a 0.5 µm resolution. At small gap values, the magnetic force between the jaws increases causing deformation

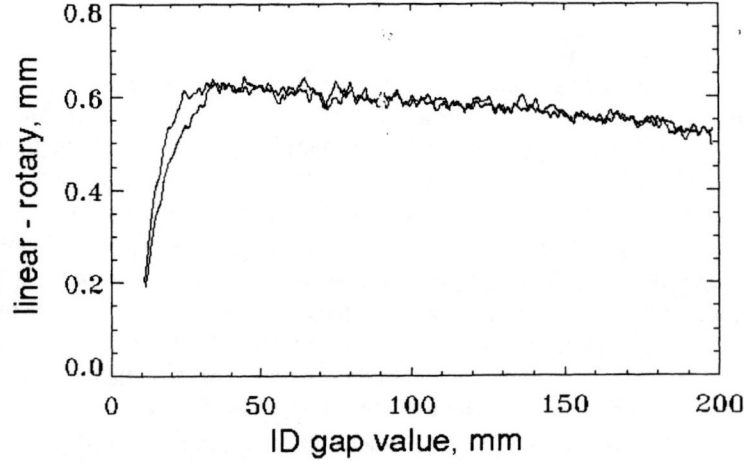

FIGURE 3. Difference between linear and rotary encoder values vs. gap for the Undulator A #4.

of the ID frame and main drive screws. The rotary encoders show values proportional to the drive screw rotation angle and so do not have a linear dependence on gap. The graph in Fig. 3 shows the difference between linear and rotary encoder values vs ID gap for the 3.3-cm-period Undulator A #4. As can be seen in the figure, the deformation exceeds 0.4 mm at minimum gap. Some additional errors in gap adjustment may arise from a jaw cant that appears in the figure as a hysteresis between the two data sets measured during opening and closing of the ID. Cant means a jaw rotation by a small angle around the axis parallel to the beam orbit. Because the linear encoders are not centered over the beam position, cant causes errors in gap measurements. According to our measurements, these errors can be as large as 20 - 40 µm. The rotary encoders have a 2.5 µm resolution, but about 20 µm variation is caused by the belt that connects them to the drive screws. However, they do operate more reliably in the electromagnetically noisy storage ring environment than do the linear encoders.

For further improvement of the reliability and the capabilities of the ID control system, we plan to enhance the control program with new features, such as scanning capabilities, more sophisticated conversion between the ID gap and the photon energy taking into account values of beam emittance and user's aperture, automatic resetting of linear encoders after beam injection into the storage ring, and the use of individual data of elastic deformation for ID frame and lead screws with data from rotary encoders for more precise comparison with gap measurement by linear encoders. To avoid cant-related errors in gap adjustment and to increase accuracy, we are considering installing another set of four linear encoders (with a 0.5 µm accuracy and a 10/256 µm resolution) in the plane of symmetry of the ID.

ACKNOWLEDGMENTS

The authors wish to thank Tim M. Mooney and Ned D. Arnold for their assistance in EPICS programming. Work supported by U.S. Department of Energy, BES - Materials Sciences, under contract no. W-31-109-ENG-38.

REFERENCES

1. Dejus, R. J.; Lai, B.; Moog, E. R.; Gluskin, E., "Undulator A characteristics and specifications : enhanced capabilities." Argonne National Laboratory Report ANL/APS/TB-17, May, 1994.
2. Cai, Z., Dejus, R. J., Den Hartog, P., Feng, Y., Gluskin, E., Haeffner, D., Ilinski, P., Lai, B., Legnini, D., Moog, E. R., Shastri, S., Trakhtenberg, E., Vasserman, I., and Yun, W., "APS Undulator A radiation - first results." *Rev. Sci. Instrum.* **67,** 9, CD ROM, 1996.
3. Gluskin, E., "APS Insertion Devices: Magnetic Performance and Radiation Characteristics," to be published in *Proceedings of the SRI'97 National Conference* (CHESS, the Cornell High Energy Synchrotron Source, June 17-20, 1997)
4. Yang, B.X., Lumpkin, A.H., Goepper, G.A., Sharma, S., Rotela, E., Sheng, I.C., Moog, E., "Status of the APS Diagnostics Undulator Beamline," to be published in *Proceedings of the 1997 Particle Accelerator Conference,* Vancouver, B.C., Canada, May 12-16, 1997.
5. Moog, E.R., Vasserman, I., Borland, M., Dejus, R., Den Hartog, P.K., Gluskin, E., Maines, J., and Trakhtenberg. E., "Magnetic performance of the Insertion Devices at the Advanced Photon Source," to be published in *Proceedings of the 1997 Particle Accelerator Conference,* Vancouver, B.C., Canada, May 12-16, 1997.
6. Dalesio, L., Hill, J., Kraimer, M., Murray, D., Hunt, S., Claussen, M., Watson, C., Dalesio, J., "The Experimental Physics and Industrial Control System architecture: Past, present, and future." presented at the International Conference on Accelerator and Large Experimental Control Systems, 18-22 Oct 1993, Berlin (Germany).
7. *VxWorks Programmer's Guide, 5.1.,* Alameda, CA 94501-1147, USA. December, 1993, Wind River Systems, Inc. Part#: DOC-100000-0002.
8. Kozubal, A. J., Dalesio, L.R., Hill, J.O., and Kerstiens, D.M., "A State Notation Language for Automatic Control," Los Alamos National Laboratory Report LA-UR-89-3564, November, 1989.
9. Kenneth Evans, Jr. "MEDM Reference Manual."
http://www.aps.anl.gov/asd/controls/epics/EpicsDocumentation/ExtensionsManuals/MEDM/MEDM.html, May 1997.

Characterization of the coherent microwave emission from the SURF II synchrotron storage ring

G. T. Fraser, A. R. Hight Walker, U. Arp, and T. Lucatorto

Physics Laboratory
National Institute of Standards and Technology
Gaithersburg, MD 20899
and

K. K. Lehmann

Dept. of Chemistry
Princeton University
Princeton NJ 08544

The temporal profile and frequency spectrum of the microwave emission from the SURF II electron storage ring at the National Institute of Standards and Technology have been studied between 8 GHz and 18 GHz to access the utility of SURF II for long wavelength spectroscopic applications. The microwave emission is dominated by intense ~0.2 ms bursts of radiation, which are random in time at high electron beam currents and nearly periodic in time (t~10 ms) at lower beam currents. The radiation is predominantly polarized in the plane of the ring, consistent with synchrotron emission. The intensity of the radiation is orders of magnitude greater than that calculated for incoherent synchrotron emission. Spectrally resolved measurements using a 2 MHz resolution (FWHM) microwave heterodyne receiver reveal that the radiation bursts consist of intense peaks at frequencies corresponding to even harmonics of the orbital frequency of an electron bunch. The odd harmonics are less intense by more than an order of magnitude. The present measurements are related to previous observations of Rakowsky at SURF II using a beam monitoring electrode.

Characterization of the Elliptical Multipole Wiggler at the Advanced Photon Source

P. Ilinski, C.T. Venkataraman, J.C. Lang and G. Srajer

Advanced Photon Source, Argonne National Laboratory, Argonne, IL 60439

Abstract. An elliptical multipole wiggler (EMW), designed to produce helicity-switching circularly polarized radiation over a wide energy range, has been installed at the Advanced Photon Source (APS). Spectral flux and polarization measurements were performed on the EMW in the photon energy range from 10 to 100 keV for different values of the EMW deflection parameter in the vertical plane. Absolute flux and degree of linear polarization measurements were accomplished by utilizing scattering from a gas in combination with an energy-dispersive detector. In addition, the circular polarization of the beam was confirmed by performing magnetic Compton scattering experiments on a well-characterized iron sample. The obtained results are in reasonable agreement with the calculated values.

INTRODUCTION

Over the past decade, synchrotron radiation has increasingly been used to probe the magnetic properties of materials (1-4). These measurements have generally involved the modulation (or analysis) of the polarization of the incident (or scattered/absorbed) x-ray beams. Techniques, such as circular magnetic x-ray dichroism, core level photoemission, and magnetic Compton scattering, which utilize circularly polarized photons, have attracted particular interest. Because circularly polarized photons couple differently with the magnetic moment of an atom than do neutrons, they are able to provide unique magnetic information not generally accessible by neutron techniques. The development of circularly polarized x-ray diffraction and spectroscopy techniques, however, has been hampered by the lack of efficient sources. Measurements thus far have been primarily taken utilizing off-axis synchrotron radiation from a bending magnet source, which greatly limits incident x-ray flux. The available flux is particularly important for these type of experiments, due to the inherently small nature of the magnetic x-ray cross section. To increase the available circular polarized x-ray flux, an elliptical wiggler was first proposed and built by Yamamoto *et al.* (5). This device consisted of a series of dipole magnets supplemented by a horizontal magnetic field for tilting particle trajectories up and down to obtain circularly polarized radiation along the axis of the wiggler.

Efficient sources of circularly polarized x-rays should provide both the highest possible flux and degree of circular polarization (P_c). Recently, an elliptical multipole wiggler (EMW) has been installed on the 11-ID beamline of the Advanced Photon Source to provide just such characteristics. A comparison of the EMW with other techniques used for the production of circularly polarized x-rays has been given by Lang *et al.* (6). The elliptical multipole wiggler is expected to provide a helicity-switchable (<10 Hz), highly polarized, high flux source of hard x-rays (5 < E < 200 keV). This paper reports the first measurements characterizing the performance of this device between 10 and 100 keV. A detailed description of the elliptical multipole wiggler facility at the APS has been previously reported by Montano *et al.* (7), and only a brief description will be given here. The EMW is based on a design by Gluskin *et al.* (8); it has 16 periods and a period length of 16 cm. The 24 mm wiggler gap yields a horizontal deflection parameter (K_y) of 14.6, corresponding to a critical energy of 32 keV and a total wiggler power of 5.4 kW at 100 mA. An electromagnet controls the vertical deflection parameter (K_x), which can be continuously varied from 0.0 to 1.3. The use of the electromagnet allows for rapid helicity reversal up to 10 Hz. The performance of the EMW was verified by measuring the absolute value of the spectral flux, and degree of linear and circular polarization. The absolute flux and degree of linear polarization measurements were performed by using scattering from a gas in combination with an energy-dispersive detector. This method has been used previously to characterize the spectral flux of undulators (9-11). A direct confirmation that the spectral flux was circularly polarized was made by performing magnetic Compton scattering measurements on an iron sample.

EXPERIMENTAL SETUP

The experimental setup for wiggler diagnostics consisted of three main parts: a water-cooled Cu mask, an assembly for measurement of the absolute flux using a gas scattering method, and an assembly for the magnetic Compton scattering (MCS) measurements (Fig. 1). The first element was a 1-m-long water-cooled Cu mask placed in the 11-ID front-end enclosure at a distance of 22 m from the wiggler. This mask was kept in a He atmosphere and was used to remove a substantial fraction of the total power of the wiggler beam. The entrance aperture of the mask was 75 mm in the horizontal direction, while the exit aperture was 2 mm horizontal by 7 mm vertical.

The mask was followed by the absolute flux measurement setup mounted on an optical table. This consisted of a pinhole-slit assembly and a scattering spectrometer. The pinhole-slit assembly was made of a water-cooled conical pinhole with an entrance diameter of 10 mm, an exit diameter of 0.8 mm, and a length of 210 mm. To define a beam cross section, 1.4-mm-thick tungsten slits (horizontal 150 µm and vertical 75 µm) were placed after the conical pinhole at 28.75 m from the center of the wiggler. The scattering spectrometer employed gas scattering in combination with an energy-dispersive detector. The detector was mounted to measure 90 degree scattering from a well-defined gas volume. The detector could be rotated around the beam to measure horizontal or vertical intensity components. The distance from the scattering volume to the detector was varied between 120 and 300 mm to achieve an appropriate detector count rate. Two solid-state detectors were used: a 3.2-mm-thick, 6-mm-diameter Si(Li) (EG&G ORTEC SLP 06165PS), and a 2-mm-thick, 3x3 mm CdZnTe (Amptek XR-100T-CZT). The amplifiers shaping times were 6 µs and 1.5 µs, respectively.

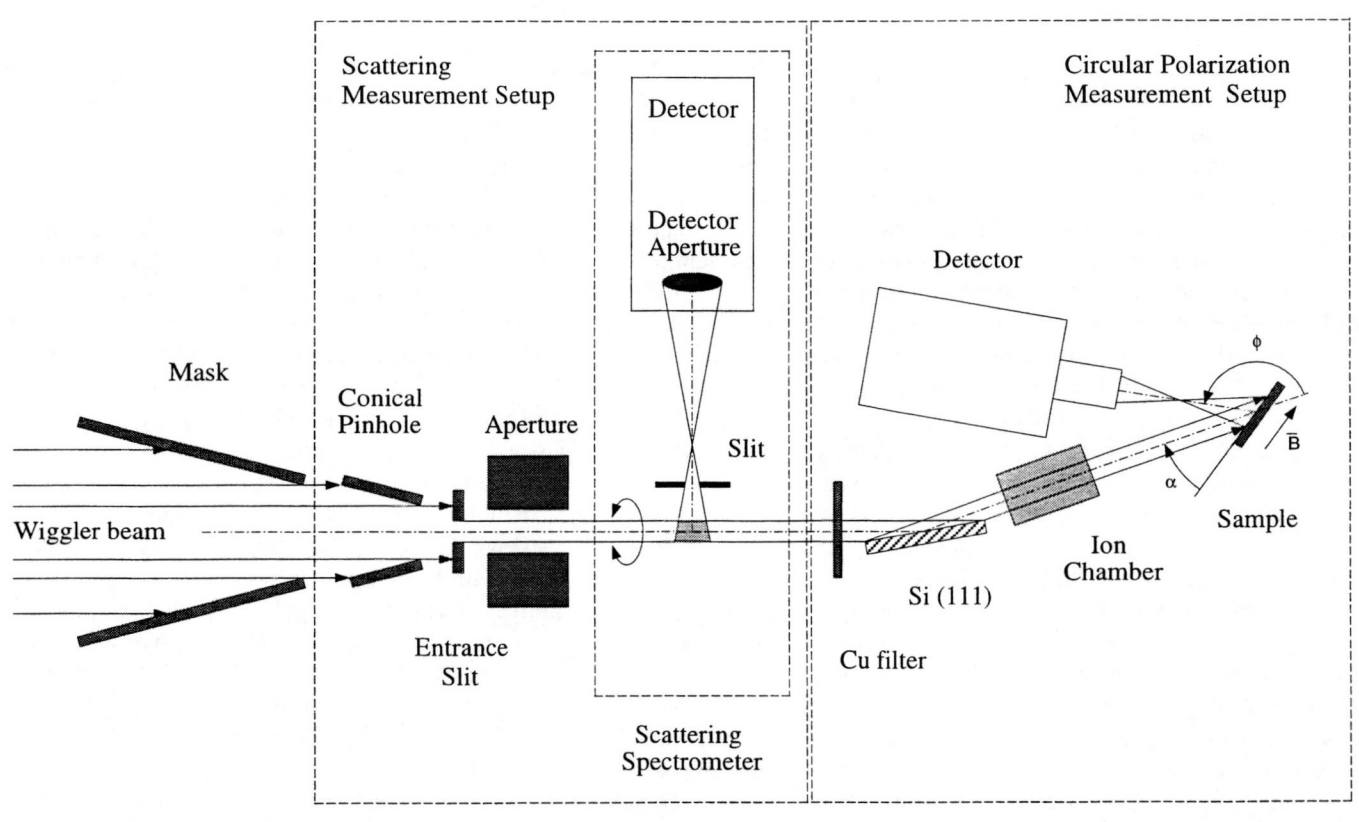

FIGURE 1. Experimental setup.

The detector efficiency is a product of the intrinsic detector efficiency and several correction factors that needed to be minimized or corrected for in the data analysis. First, the entrance aperture placed in front of the detectors was limited to a 2 mm diameter in order to eliminate any radial dependence in the detector efficiency. Escape peak corrections were made assuming that escape through the sides and rear of the detector can be neglected (12). While the escape peak corrections are negligible for the Si(Li) detector above 10 keV, they are quite substantial for the CdZnTe detector. The pressure and temperature of the helium and nitrogen scattering gasses were monitored throughout the measurements in order to obtain the gas concentration. The amount of the scattering atoms is proportional to the gas scattering volume, the cross section of which is defined by the entrance slits size and the length is defined by the detector aperture, detector distance to the scattering volume and an additional 2 mm slit, placed at 31.5 mm from the scattering volume.

A 100-mm-thick 5-mm-diameter tungsten aperture was placed after the entrance slits to reduce radiation scattered from the upstream optics. The pinhole, slits, and tungsten aperture were shielded with lead. The mask was shielded with 20-mm-thick lead, and a 50-mm-thick lead wall was built behind it. The mask and pinhole-slits assembly were connected with a pipe and filled with He gas.

The setup for performing the magnetic Compton scattering measurements is also shown in Fig. 1. For this part of the EMW characterization, the entrance tungsten slits were removed, and the 0.8 mm beam defined by the conical pinhole was used. This beam was monochromated by a Si (111) water-cooled crystal. A 0.75 mm Cu filter was placed prior to the monochromator to reduce the heat load incident on the silicon crystal. The monochromatic beam was then incident on an Fe scattering target placed in contact with the poles of an electromagnet (~3 kG). The angle between the magnet axis and the incident beam, α, was 30 degrees. The x-rays scattered at angle, ϕ, of 150°, were collected using either a single-element (Canberra GL0210P) or a 9-element (Canberra) Ge solid-state detector. The shaping times for the single and 9-element detectors were set to 6 μs and 0.25 μs, respectively.

RESULTS

The results of the absolute flux and degree of polarization measurements are given in Figs. 2 and 3, respectively. Figure 3 shows data obtained from both the spectral flux and magnetic Compton measurements. For the former method, the degree of circular polarization is indirectly obtained from measured degree of linear polarization, assuming an absence of any unpolarized radiation component. The magnetic Compton scattering measurements, on the other hand, directly measure the circularly polarized component. These measurements were able to give the sign of the helicity and also to test the performance of the EMW in the "normal" helicity switching mode of the device.

Gas scattering measurements

The restoration of the spectral flux from the scattering measurement was previously described at ref. (9-11). In Fig. 2, the total spectral flux, which is a sum of the horizontal $I_\parallel$ and vertical $I_\perp$ components, measured in the energy range from 10 to 60 keV, for K_x values of 0.0 and 1.3, are compared to calculations. The wiggler spectra were calculated using the ideal magnetic field approximation for a vertical beam angular divergence of $\sigma_{y'} = 5.3$ μrad (13). The values K_y 14.4 and K_x 1.26 were used for deflection parameters in order to get the best agreement with the calculations using the ideal magnetic field (EWE code) and a measured for K_x values of 0.0 and 1.3 wiggler magnetic field (14). Helium gas was used for the absolute flux measurement because of its well-known scattering cross section (15). Flux measurements for $K_x = 0$ were made using the Si(Li) detector, while those at $K_x = 1.3$ were taken with the CdZnTe detector because of the detector's increased efficiency at higher x-ray energies. The experimental measurements have ~12% rms error, which includes uncertainties in the slits sizes and distances measurements, the number of scattered toward the detector atoms, and detectors efficiency calculation. The measured flux agree with the calculations better than 80% for $K_x = 0$. For the deflection parameter $K_x = 1.3$ the results agree well below 45 keV with an increasing background addition above it.

To evaluate wiggler polarization properties, the horizontal and vertical radiation components were measured for values of the deflection parameter K_x=0.8, 1.0, and 1.3 over the energy range from 10 up to 100 keV. The CdZnTe detector was used in combination with nitrogen gas, which increased the scattering spectrometer efficiency for high energy x-rays, improved the counting statistics and signal-to-background ratio. The distance between the scattering volume and the detector was optimized to obtain a proper count rate at the detector. The degree of linear polarization is given by $P_L = (I_\parallel - I_\perp)/(I_\parallel + I_\perp)$, and, assuming that there is no unpolarized radiation, the degree of circular polarization for wiggler radiation can be calculated using $P_C = \sqrt{1 - P_L^2}$. The degree of circular polarization obtained from the experimental data has less than 5% rms error. In Fig. 3. the resulting smoothed degree of circular polarization is compared with the degree of circular polarization calculated for the ideal magnetic field, the disagreement may be caused by possible presence of the unpolarized component.

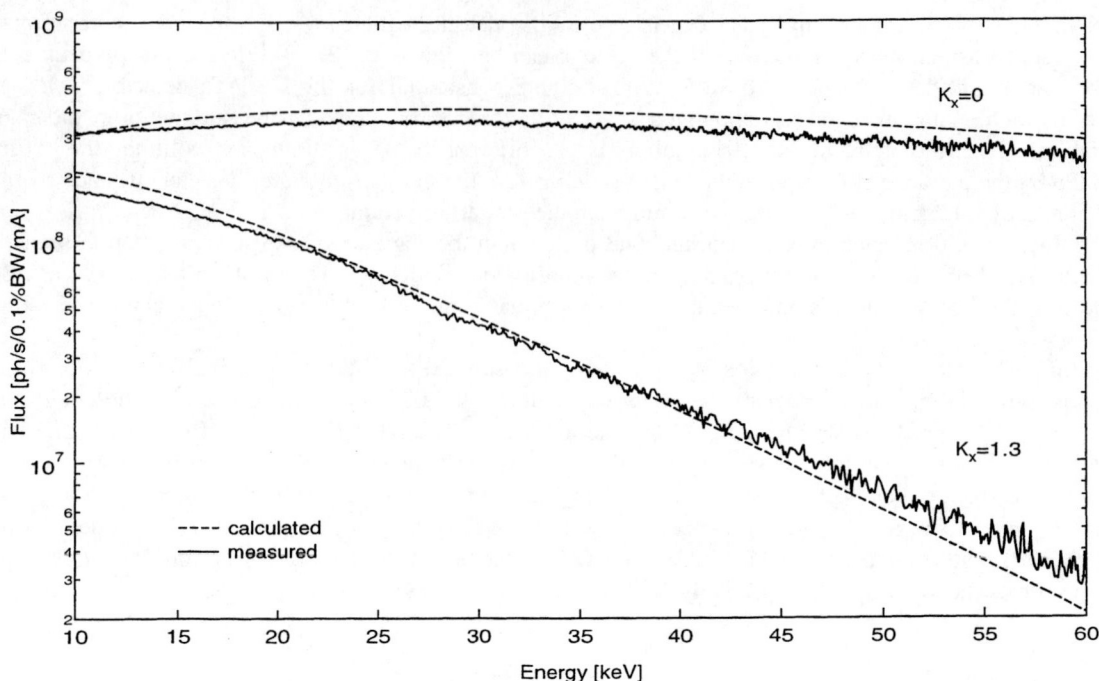

FIGURE. 2. Calculated (dashed) and measured (solid) absolute flux vs. photon energy for $K_x = 0$ and $K_x = 1.3$ through the 75(v)x150(h) μm aperture at 28.75 m.

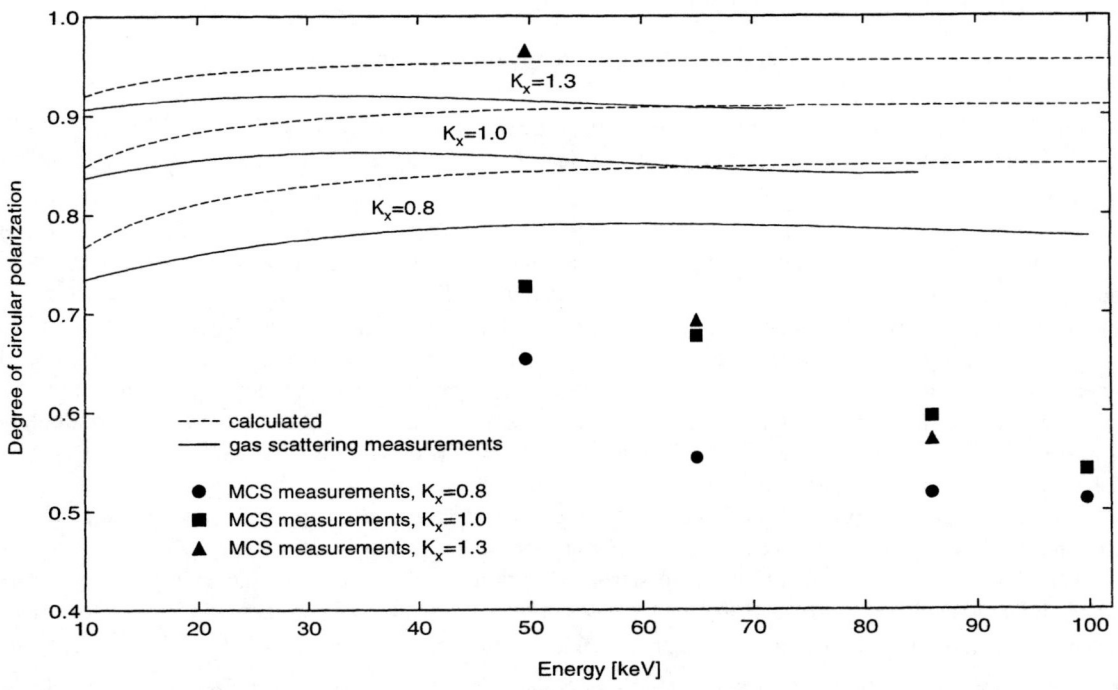

FIGURE. 3. Degree of circular polarization (P_c) vs. photon energy for K_x values of 0.8, 1.0, and 1.3.

Magnetic Compton scattering measurements

The degree of circular polarization was obtained from the magnetic Compton scattering measurements using the following expression for the total Compton scattering cross section (2),

$$\frac{d\sigma}{d\Omega d\omega} = \frac{1}{2} r_o^2 \frac{E_2}{E_1} \left[f_1 J(p_z) + f_2 g P_c S(\alpha) J_{mag}(p_z) \right],$$

$$f_1 = 1 + cos^2 \theta + \frac{E_1 - E_2}{mc^2}(1 - cos\theta) + P_L sin^2 \theta,$$

$$f_2 = cos\theta - 1, \text{ and}$$

$$S(\alpha) = \pm \left[cos\theta cos\alpha + \frac{E_2}{E_1} cos(\theta - \alpha) \right],$$

where r_0 is the classical electron radius, E_1 and E_2 are the incident and scattered x-ray energies, θ is the scattering angle, α is the angle between the magnetic field direction and the incident beam, P_L is the degree of linear polarization of the incident beam, and $g = E_1/mc^2$. The Compton profiles, $J(p_z)$ and $J_{mag}(p_z)$, are the projections of the total and magnetic ground-state electron momentum densities along the scattering vector; when integrated over the momentum, p_z, they yield the number of total and magnetic electrons, respectively. The integrals of $J(p_z)$ and $J_{mag}(p_z)$ were taken to be 22.34 and 2.07 over a range of p_z ±10 a.u. The sign of the $S(\alpha)$ term depends on the direction of the magnetic field. Therefore, by reversing the magnetization of the sample (or the incident beam helicity) the magnetic term can be isolated. The degree of circular polarization, P_c, then can be obtained by comparing the magnetic term with the total cross section.

The values of P_c measured over an energy range from 45 to 100 keV for K_x values of 0.8, 1.0, and 1.3 are also shown in Fig. 3. The pinhole position for the measurements was set to the axis of the wiggler by scanning the vertical beam profile. The data were obtained by switching either the magnetic field direction applied to the sample or the helicity of the incident photons. In the magnetization reversal mode (~ 0.2 Hz), equal P_c magnitudes were obtained for $\pm K_x$ values confirming the linearity of the magnetic field produced by the wiggler electromagnet. Furthermore, the P_c value obtained in helicity switching mode (~ 0.2 Hz) was identical to that obtained from magnetization reversal verifying the full magnetic saturation of the sample.

The values for P_c obtained from MCS measurements are in reasonable agreement with the gas scattering measurements for the lower energies but are substantially lower at the higher energy portion of the spectrum. Differences in beamline components used for the two measurements can partially explain the lower measured degree of circular polarization. First, the size of the beam apertures were different, which isolated different portions of the angular distribution of the EMW radiation. For the MCS measurements the angular acceptance of the conical pinhole was $0.4/\gamma$. This would cause a reduction in the degree of circular polarization due to the resulting increase of the linear polarization contribution. Theoretical calculations with this size of aperture reduce the expected degree of circular polarization by ~5%, therefore, the P_c values obtained from the gas scattering would scale accordingly. At higher energies, a strong trend indicating decreasing P_c as the energy is increased is observed in the measurements, which is not reproduced in the theoretical curves. We can partially explain this discrepancy as arising from the increasing effective aperture size with increasing energy because of the partial transparency of the conical pinhole at the higher energies. Including these effects in our calculations, we find that the overall decrease in the measured energy range should be about 5-10%, leaving an additional unexplained 15% discrepancy between the MCS and the gas scattering measurements. Systematic errors such as corrections for the energy-dependent absorption in the sample and multiple scattering have not been included in analysis and could explain some of the shortfall.

CONCLUSION

For measurements that probe difference signals on the order of less than a percent of the total signal, statistical accuracy usually requires long data acquisition times. We found that use of a 9-element solid-state detector at 86 keV on this high flux wiggler line, with K_x=0.8 and ring current of 93 mA, rendered a statistical accuracy of 3% in the difference signal for an Fe sample in 30 minutes. The very reduced data collection times coupled with the switchable photon helicity makes this beamline extremely well suited to the study of materials with weak magnetic moments, of hard ferromagnets, and of magnetic properties at cryogenic temperatures.

ACKNOWLEDGMENTS

This work supported by U.S. Department of Energy, BES-Material Sciences, under contract No. W-31-109-Eng-38. We acknowledge G. Jennings and P. Montano for their support in the setup of the multi-element detector for the MCS experiment.

REFERENCES

1. Blume, M. and Gibbs, D., *Physical Review B* **37**, 1779-1789 (1988).
2. Cooper, M. J., Laundy, D., Cardwell, D. A., Timms, D. N., Holt, R. S., and Clark, G., *Physical Review B* **34**, 5984-5987 (1986).
3. Hannon, J. P., Trammell, G. T., Blume, M., and Gibbs, D., *Physical Review Letters* **61**, 1245-1248 (1988).
4. Schütz, G., Knulle, M., Wienke, R., Wilhelm, W., Wagner, W., Kienle, P., and Frahm, R., *Zeitschrift für Physik B* **73**, 67-75 (1988).
5. Yamamoto, S., Kawata, H., Kitamura, H., Ando, M., Saki, N., and Shiotani, N., *Phys. Rev. Lett.* **62**, 2672 (1989).
6. Lang, J. C., Srajer, G., and Dejus, R. J., *Rev. Sci. Instrum.* **67**, 62-67 (1996).
7. Montano, P. A., Knapp, G. S., Jennings, G., Gluskin, E., Trakhtenberg, E., Vasserman, I. B., Ivanov, P. M., Frachon, D., Moog, E. R., Turner, L. R., Shenoy, G. K., Bedzyk, M. J., Ramanathan, M., Beno, M. A., and Cowan, P. L., *Rev. Sci. Instrum.* **66**, 1839-1841 (1995).
8. Gluskin, E., Frashon, D., Ivanov, P. M., Maines, J., Medvedko, E. A., Trakhtenberg, E., Turner, L. R., Vasserman, I., Erg, G. I., Evtushenko, Y. A., Gavrilov, N. G., Kulipanov, G. N., Medvedko, A. S., Petrov, S. P., Popik, V. M., Vinokurov, N. A., Friedman, A., Krinsky, S., Rakowsky, G., and Singh, O., "The Elliptical Multipole Wiggler Project," in *Proceedings of the 1995 Particle Accelerator Conference*, Dallas, Texas, 1426-1428 (1995).
9. Ilinski, P., Yun, W., Lai, B., Gluskin, E., and Cai, Z., *Rev. Sci. Instrum.* **66**, 1907-1909 (1995).
10. Hahn, U., Schulte-Schrepping, H., Balewski, K., Schneider, J. R., Ilinski, P., Lai, B., Yun, W., Legnini, D., and Gluskin, E., *J. Synchrotron Rad.* **4**, 1-5 (1997).
11. Cai, Z., Dejus, R. J., Hartog, P. D., Feng, Y., Gluskin, E., Haeffner, D., Ilinski, P., Lai, B., Legnini, D., E.R. Moog, Shastri, S., Trakhtenberg, E., Vasserman, I., and Yun, W., *Rev. Sci. Instrum.* **67**, CD ROM (1996).
12. Reed, S. J. B. and Ware, N. G., *J. Phys. E* **5**, 582-584 (1972).
13. Ilinski, P., Dejus, R. J., Gluskin, E., and Morrison, T. I., "Some Practical Aspects of Undulator Radiation Properties," in *Optics for High-Brightness Synchrotron Radiation Beamlines II*, Denver, Colorado, Proc. SPIE 2856, 16-25 (1996).
14. Dejus, R. J., Argonne National Laboratory, unpublished, 1997.
15. Veigele, W. J., *Atomic Data Tables* **5**, 51-111 (1973).

Computation of Undulator Tuning Curves

Roger J. Dejus

Advanced Photon Source, Argonne National Laboratory
9700 S. Cass Avenue, Argonne, IL 60439, USA

Computer codes for fast computation of on-axis brilliance tuning curves and flux tuning curves have been developed. They are valid for an ideal device (regular planar device or a helical device) using the Bessel function formalism. The effects of the particle beam emittance and the beam energy spread on the spectrum are taken into account. The applicability of the codes and the importance of magnetic field errors of real insertion devices are addressed. The validity of the codes has been experimentally verfied at the APS and observed discrepancies are in agreement with predicted reduction of intensities due to magnetic field errors. The codes are distributed as part of the graphical user interface XOP (X-ray OPtics utilities), which simplifies execution and viewing of the results.

This work was supported by the U.S. Department of Energy, BES-Materials Sciences under contract No. W-31-109-ENG-38.

Magnetic Field Characterization of the NIST Undulator

Lewis E. Johnson, Greg Denbeaux, John M. J. Madey,
and Karl D. Straub

Duke University
LaSalle St. Extension
Durham NC 27708

and

National Institute of Standards and Technology
Gaithersburg, MD 20899

A 3.64 m undulator was constructed by the Brobeck Division of Maxwell Laboratories for FEL experiments at NIST in Washington, DC. The Duke University FEL Lab has since acquired the undulator for use as a soft x-ray source. We report on our effort to transform the undulator into a high performance soft x-ray insertion device through careful characterization of the existing magnet blocks, sorting and trimming.

Soft X-ray Sources on the Duke Storage Ring

Lewis E. Johnson, Greg Denbeaux, Nelson Hower,
John M. J. Madey, Karl D. Straub

Duke University
LaSalle St. Extension
Durham NC 27708

and

National Institute of Standards and Technology
Gaithersburg, MD 20899

The Duke University 1GeV Storage Ring was designed for use as a third generation driver for sources of intense radiation. The NIST undulator will provide a high brightness source of quasi-coherent first harmonic radiation in the wavelength range between 37 - 45Å. In addition, the Bend magnets of the storage ring produce radiation centered around 11.6Å. We report on the current status of the soft x-ray undulator and beamline.

Resorting the NIST Undulator using Simulated Annealing for Field Error Reduction

Greg Denbeaux, Lewis E. Johnson, John M. J. Madey,
Karl D. Straub

Duke University Free Electron Laser Laboratory
LaSalle St. Extension
Durham NC 27708

We have used a simulated annealing algorithm to sort the magnets in the NIST undulator in order to optimize the spectrum of light emitted. This algorithm sorted the order of the magnets in order to minimize the magnetic field error at each pole, and to minimize the net angular deflection of the electrons passing through the undulator. Without sorting, the RMS magnetic field error was more than 2.5 %, and there was a maximum angular deflection of more than 400 microradians. After the sorting, the RMS magnetic field error was less than 0.8%, and there was a maximum angular deflection of less than 40 microradians. We show both the expected spectrum emitted from the undulator in its original order, and also the improved spectrum from the undulator after optimization.

MAKING BEAMS: NOVEL OPTICS DESIGN, FABRICATION, TESTING, AND USE

Cryogenic high-heat-load optics at the Advanced Photon Source (Invited)

C. S. Rogers

*Advanced Photon Source,
Argonne National Laboratory, Argonne, IL 60439*

Abstract. Cryogenically cooled silicon monochromators have found wide application at the Advanced Photon Source (APS) and other third-generation synchrotron radiation facilities. Currently, 17 insertion device beamlines at the APS are implementing cryogenic, silicon double-crystal monochromators (DCM) as the first optical element. Recently, several silicon crystal monochromators internally cooled with liquid nitrogen have been tested on the sector 1-ID undulator beamline at the APS. Rocking curves at various energies were measured simultaneously in first and third order from a Si(111) DCM in the Bragg reflection geometry at a fixed undulator gap of 11.1 mm. The crystal exhibited sub-arc second thermal broadening of the rocking curve over a first order energy range from 6.0 to 17.0 keV up to a maximum incident power of 561 W in a 2.5 V x 2.0 H mm^2 beam. It has been demonstrated that cryogenic silicon monochromators can handle the highest power beams from hard x-ray undulators at the APS without significant thermo-mechanical distortion.

I. INTRODUCTION

Great strides have been made over the last several years in developing high-heat-load optics for insertion device beamlines at the Advanced Photon Source (APS). A review of the high-heat-load monochromator development program at the APS is given in Ref. 1. Since commissioning began in early 1996, all of the operational elements necessary for cryogenically cooled optics have been installed and tested on beamline 1-ID. These elements include an internally cooled first crystal, cryogenically compatible monochromator mechanism, and a continuous liquid nitrogen pumping system. Also, a nitrogen gas liquefier has been installed on sector 1. The liquefier collects the boiled-off gas from the pumping system heat exchanger, liquefies it, and returns it to the heat exchanger bath, thus making a completely closed-loop cooling system. Of the 20 sectors currently under development at the APS, 17 have so far opted for using cryogenically cooled silicon monochromators on their insertion device beamlines. The remaining undulator beamlines are employing a water-cooled mirror as the first optical element and/or diamond monochromators (2). The incidence angle for total external reflection mirrors is very small, typically less than 0.2°. This greatly reduces the heat flux projected onto the mirror allowing it to be water cooled. The heat load on downstream optical components is reduced due to the high energy cutoff of the mirror. Single-crystal diamond possesses a very large thermal conductivity even at room temperature, the expansion coefficient is smaller, and it absorbs less of the incident x-ray power than silicon making it attractive for high-heat-flux monochromators (3). However, diffraction-quality diamond of the required size and orientation is difficult to obtain, which has limited its use to date.

It was recognized early in the planning stages for third-generation, hard x-ray synchrotrons that the power emitted by insertion devices, undulators in particular, would probably be the limiting factor affecting the quality (spectral brilliance and flux) of the delivered monochromatic beam. Overwhelmingly, the material of choice for hard x-ray monochromators is single-crystal silicon. Large, highly perfect boules of silicon are readily available due to the extensive use of silicon in the semiconductor industry. Also, fabrication techniques are well understood, silicon is UHV compatible and impervious to radiation damage. Additionally, the acceptance angle for Bragg diffraction (Darwin width) is well matched to the opening angle of radiation at the APS. The Darwin width for the Si(111) reflection at 8 keV is about 36 microradians, which compares favorably with the APS undulator vertical opening angle of about 20 microradians. For these reasons, great emphasis has been placed on developing silicon monochromators that can perform under the enormous power density of an APS undulator. Because of the very narrow acceptance angle, the optical performance of single-crystal monochromators is greatly degraded by thermo-mechanical strain. The most successful monochromator design has proven to be cryogenically cooled silicon.

Cryogenic cooling of high power density synchrotron x-ray optics was first suggested in 1985 for mirrors and in the following year for crystal monochromators (4,5). As was pointed out in those papers, the advantage of operating single-crystal silicon optical components at cryogenic temperatures is twofold: (1) the thermal conductivity, k, increases by nearly an order of magnitude in going from room temperature to liquid nitrogen temperatures, while (2) the coefficient of thermal expansion, α, decreases from its room temperature value of 2.6 x 10^{-6} K^{-1}, going through zero at about 125 K, and remaining slightly negative. It can be shown that the thermal strain gradient in the scattering plane is directly proportional to the ratio, α/k. Consequently, the thermal gradients and resulting strain are much lower for silicon monochromators operated at cryogenic temperature compared to operation at room temperature. Cryogenic silicon monochromators were first tested on high power wiggler and focused wiggler beamlines at HASYLAB and the National Synchrotron Light Source (NSLS) (6,7). Cryogenic monochromators have subsequently found wide use at the European Synchrotron Radiation Facility (ESRF) (8). The goal for the cryogenic optics program at the APS has been to develop cryogenically cooled silicon monochromators that will deliver near theoretical performance over the widest possible functional range of insertion devices.

II. CRYSTAL MONOCHROMATORS

All but two of the insertion device beamlines at the APS utilize an undulator as the primary source. Consequently, most of the development effort has been directed at designing optics for the undulator beamlines. The APS undulator emits a very narrow, well-collimated beam with a very high power density. Accepting just the central-cone of radiation from undulator A yields about 780 W and a peak power density of about 179 W/mm^2 at the monochromator (29 m from the source) at a magnetic gap of 11.1 mm for 100 mA stored current. This corresponds to a first harmonic energy of 3.28 keV.

A double-crystal monochromator is typically the first optical element of most beamlines at the APS. Consequently, it is often exposed to the full central-cone power emitted by the source. The DCM utilizes diffraction to select and pass the desired energy band from the broad incident spectrum. Most of the remaining energy is absorbed by the first monochromator crystal. The quality of the diffracted beam is very sensitive to the perfection of the crystalline lattice. Any thermo-mechanical strain introduced into the crystal degrades the quality of the monochromatic beam. Some of the characteristics that a monochromator crystal assembly must possess are: high radiation resistance, UHV compatibility, and the ability to undergo many thermal cycles. Also, the cooling manifold and piping cannot transmit significant stress or vibration to the optic.

The thermal design considerations for cryogenic monochromator crystals are significantly different than for room-temperature crystals. The two main differences are that the thermal conductivity of silicon is about an order of magnitude larger at liquid nitrogen temperatures and the transport properties of liquid nitrogen are not as good as water, the most common heat transfer fluid. The primary resistances to heat flow from the crystal to the fluid are conduction resistance through the silicon and convection resistance through the solid-fluid interface. For room-temperature crystals, the conduction resistance often dominates. Therefore, it is desirable to minimize the thermal path length by placing the coolant channels very close, about 1 mm or less, to the diffraction surface in order to decrease the temperature rise. However, the situation is reversed in the cryogenic case. The cooling channels should be placed much further from the diffraction surface due to the higher thermal conductivity and smaller heat transfer coefficient. Locating the cooling fluid further from the diffraction surface allows the heat to diffuse laterally throughout the crystal, thereby lowering the peak surface temperature and reducing the thermal flux at the cooling channel interface and the likelihood of boiling.

Two general classes of cryogenic crystals have been tested at the APS: thin and thick. Thin crystals are designed so that only a portion of the x-ray power is absorbed. Much of the hard x-ray energy passes unattenuated through the silicon. Roughly speaking, thin crystals are usually < 1 mm thick.

The advantages of thin crystals are:

- Less absorbed power
- Less liquid nitrogen consumption
- Smaller thermal gradients normal to surface

The disadvantages are:

- Heat flow restricted by thin membrane
- Fabrication is more difficult
- More susceptible to mechanical strain

The advantages of thick crystals are:

- Easier to fabricate
- Less susceptible to mechanical strain
- Cooling geometry is not limited
- Beam expansion geometries can be readily used

The disadvantages are:

- Most of the x-ray power is absorbed
- More liquid nitrogen consumption
- Crystal heat exchanger must be larger
- Larger thermal gradients normal to surface

III. EXPERIMENTS

Prior to beamline operations at the APS, cryogenically cooled monochromator crystals manufactured at the APS were tested on a focused wiggler beamline at the ESRF in a tripartite collaboration between the APS, ESRF, and the Japanese synchrotron facility SPring-8 (9,10). In these experiments, a thin, cryogenic crystal was tested up to a maximum absorbed power of 186 W with a peak power density at normal incidence of 521 W/mm^2. The crystals had slightly more mounting strain than the crystals recently tested at the APS, however the combined thermo-mechanical rocking curve broadening was less than 2 arc seconds at the highest power.

The crystal shown in Fig. 1 was installed and tested in May 1996 on APS sector 1-ID (11). The crystal consists of a monolithic block of silicon incorporating a thin diffraction element and integral cooling channels. A relatively thin crystal was desired so that a large fraction of the incident beam power would be transmitted, hence reducing the absorbed power in the component. The thin element of the crystal is fabricated by milling slots in the top and bottom faces leaving a region approximately one-half mm thick. A third slot was milled in the downstream face to allow the transmitted beam to pass through. A maximum horizontal beam size of 2.5 mm can be accommodated. The downstream face of the crystal is visible showing the slot that allows the transmitted x-ray beam to pass through. The seal between the Invar manifold and the silicon is made via In-coated metal C-rings. Sealing pressure is maintained by using spring washers on the clamping screws. The mounted crystal assembly is supported on a kinematic plate that allows for unconstrained thermal expansion while preserving the absolute position of the thin diffraction element relative to the x-ray beam. One of the technical difficulties was the development of an ultrahigh-vacuum seal between the coolant manifold and the optical component that is radiation hard, can be thermally cycled, and introduces minimal strain into the crystal. This problem is exacerbated by the fact that the desired thickness of the diffracting crystal is less than one millimeter. The In-coated metal C-rings coupled with highly polished seal faces has proven to be a reliable, strain-free cryogenic seal.

A similar cryogenic crystal was tested in October 1996. The only difference being that there was no slot fabricated into the top surface of the crystal. The thin element was fabricated by milling the downstream slot about 0.5 mm below the top face. This design is simpler to fabricate and it is possible to polish the diffraction surface. Some thermal performance is sacrificed because the heat exiting the lateral edge of the thin element has less material to diffuse into, but this has not proven to be a problem. One difficulty with the "slotted design" where the diffraction surface is below the crystal top surface as shown in Fig. 1, is that the acid etching process can create a radius on the crystal surface in the diffraction slot perpendicular to the beam direction. This can lead to a deformed beam shape due to the variation of optical path length across the beam footprint.

For the experiments on APS sector 1, the beam emitted by the undulator passed through a temporary commissioning window positioned at 23.5 m and consisting of 0.50 mm of graphite, 0.17 mm of CVD diamond, and 0.50 mm of Be. The

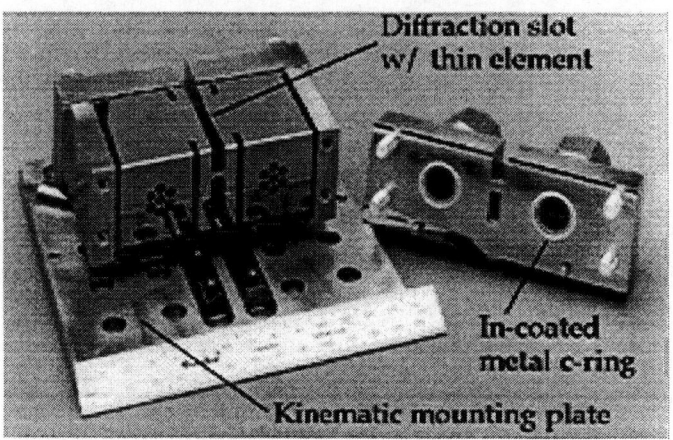

FIGURE 1. Photograph of the cryogenic crystal and coolant manifold.

fraction of the power absorbed in the commissioning window was about 12 % for a 2.5 V x 2.0 H mm² beam at an undulator gap of 11.1 mm. Horizontal and vertical white beam slits were located at 26.75 m, and the monochromator was at 28.5 m. A pair of ionization chambers were placed at 34 m to monitor the diffracted beam intensity. An Al filter was placed between the ionization chambers so that the first- and third-order reflections could be recorded simultaneously.

The performance of a thick part of the crystal was also investigated. The thick crystal data were taken from the top surface of the monochromator crystal just laterally adjacent to the thin element. Obviously, the cooling geometry for the thick crystal data is not optimum because the heat flows predominantly to only one set of coolant channels; the other set is thermally isolated by the thin element. Rocking curve widths (FWHM) as a function of photon energy are shown in Fig. 2 for a fixed undulator gap of 11.1 mm corresponding to a deflection parameter, K, of 2.57. This situation simulates far worse heat loads than would normally be encountered at higher energies because the undulator gap was kept at 11.1 mm, corresponding to a first harmonic energy of 3.28 keV, for all of the rocking curves and was not opened to track the harmonic as the diffracted photon energy was increased, which would normally be the case. As the gap is opened, the emitted power rapidly decreases. For example, at a first harmonic energy of 8 keV, corresponding to a gap of about 18.3 mm, the incident power and peak power density are only about 40 percent of that at a gap of 11.1 mm. Consequently, for typical operation in which the gap (i.e., harmonic) is matched to the diffracted photon energy, the monochromator should perform equally as well at much higher currents.

The data for the thin crystal were collected with a 2.5 V x 2.0 H mm² and 2.0 V x 2.0 H mm² (normal incidence) beam with an incident measured maximum power of 561 W and 313 W, respectively. The beam size for the thick crystal data was 1.9 V x 3.0 H mm² with a maximum measured power of 495 W. The storage ring current ranged from 61 to 96 mA for the thin crystal data and from 89 to 95 mA for the thick crystal data. The liquid nitrogen volume flow rate ranged from 6.5 to 10.6 l/min. at a head pressure of 40 psia.

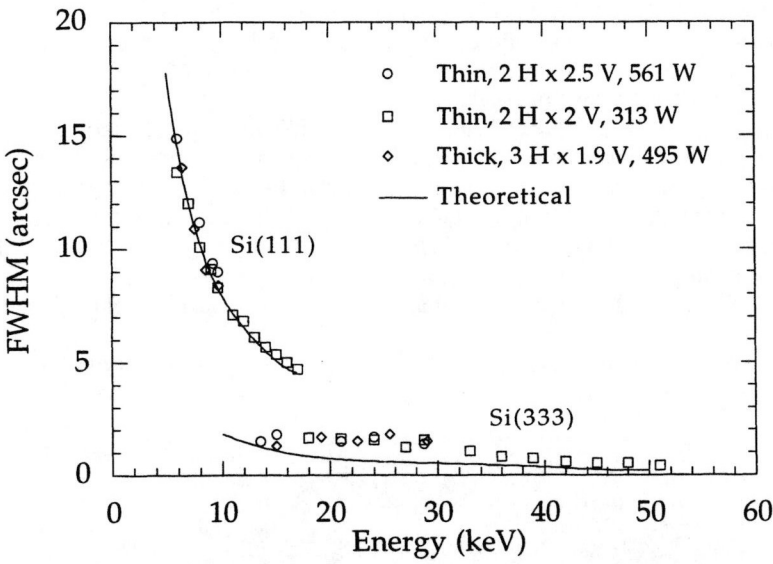

FIGURE 2. First- and third-order rocking curve widths (FWHM) as a function of diffracted energy at a fixed undulator gap of 11.1 mm for several beam sizes. The beam incident on the thin part of the crystal measured 2.5 V x 2.0 H mm² with a power of 561 W in one case and 2.0 V x 2.0 H mm² with a power of 313 W in a second case. The beam on the thick part of the crystal measured 1.9 V x 3.0 H mm² with a power of 495 W.

IV. SUMMARY

The cryogenically cooled monochromator and liquid nitrogen pumping system have been successfully tested and commissioned using the undulator on APS sector 1 ID and have provided the highest quality monochromatic beam. The thin, cryogenically cooled monochromator crystal has been tested under worst-case conditions with the APS undulator and displayed a thermal strain of no more than 1.0 to 1.5 arc seconds at the minimum undulator gap and a current up to 96 mA. The rocking curve broadening attributable to the manifold mounting stress was no more than about 1 arc second. An important benefit of the thin crystal is that it absorbs only a portion of the incident beam power. About 50 percent of the power was absorbed from a 2.5-mm-square beam at an undulator gap of 11.5 mm and a Bragg angle of 19.24°. The thick crystal performed much better than our expectations, and, due primarily to its lower mechanical strain in the diffraction volume, it exhibited sub-arcsec broadening.

The future direction for cryogenic optics research at the APS is to explore cryogenic cooling of other optical components, such as mirrors and multilayers. Additionally, methods necessary to extend monochromator designs to higher storage ring currents (up to a maximum of 300 mA) will be explored. These techniques may include, enhanced heat exchangers using porous matrices, and beam expansion geometries, such as inclined and variable asymmetric crystals (12-14).

ACKNOWLEDGMENTS

The author wishes to express his deepest gratitude to the following people who have been involved in the cryogenic monochromator program at the APS: Dr. D. M. Mills, Dr. W. K. Lee, Dr. P. B. Fernandez, Dr. T. Graber, Dr. L. Assoufid and Dr. S. Krasnicky.

This work was supported by the U.S. Department of Energy, BES-Materials Sciences, under contract number W-31-109-ENG-38.

REFERENCES

1. Lee, W. K., D. M. Mills, L. Assoufid, R. C. Blasdell, P. B. Fernandez, C. S. Rogers, and R. K. Smither, *Opt. Engr.*, **34** (2), 418, 1995.
2. Yun, W., A. Khounsary, B. Lai, and E. Gluskin, *Argonne National Laboratory Technical Bulletin*, ANL/APS/TB-2, 1992.
3. Blasdell, R. C., L. A. Assoufid, and D. M. Mills, *Argonne National Laboratory Technical Bulletin*, ANL/APS/TB-24, 1995.
4. Rehn, V., *International Conference on Insertion Devices for Synchrotron Sources*, R. Tatchyn, and I. Lindau, eds., **582**, 238, SPIE - Soc. of Photo-Optical Instrumentation Engineers, Stanford, CA, 1985.
5. Bilderback, D. H., *Nucl. Meth. in Phy.* **A246**, 434, 1986.
6. Joksch, S., G. Marot, A. Freund, and M. Krisch, *Nucl. Instrum. & Meth. Phy. Res.*, **A306**, 386, 1991.
7. Marot, G., M. Rossat, A. Freund, S. Joksch, H. Kawata, L. Zhang, E. Ziegler, L. Berman, D. Chapman, J. B. Hastings, and M. Iarocci, *Rev. Sci. Instrum.*, **63**, 477, 1992.
8. Marot, G., and M. Rossat, *High Heat Flux Engineering, Proc. SPIE*, **1739**, 464, 1992.
9. Rogers, C. S., D. M. Mills, W.-K. Lee, G. S. Knapp, J. Holmberg, A. Freund, M. Wulff, M. Rossat, M. Hanfland, and H. Yamaoka, *Rev. Sci. Instrum.*, **66**(6), 3494, 1995.
10. Rogers, C. S., D. M. Mills, P. B. Fernandez, G. S. Knapp, M. Wulff, M. Hanfland, M. Rossat, A. Freund, G. Marot, J. Holmberg, and H. Yamaoka, *Rev. Sci. Instrum.*, **67**(9), CD-ROM, 1996.
11. Rogers, C. S., D. M. Mills, W.-K. Lee, P. B. Fernandez, and T. Graber, *High Heat Flux Engineering III*, **2855**, A. M. Khounsary, ed., SPIE - Soc. of Photo-Optical Instrumentation Engineers, 1996.
12. Rogers, C. S., D. M. Mills, L. Assoufid, and T. Graber, *Rev. Sci. Instrum.*, **67** (9), CD-ROM, 1996.
13. Rogers, C. S., and L. Assoufid, *Rev. Sci. Instrum.*, **66**(2) part II, 1996.
14. Smither, R. K., and P. B. Fernandez, *Nucl. Meth. in Phy.*, **A246**, 434, 1986.

Silver Bonded, Internally Water-Cooled Monochromators for CHESS Wiggler Beamlines

Karl W. Smolenski, Qun Shen and Park Doing

Cornell High Energy Synchrotron Source (CHESS), Cornell University, Ithaca, New York 14853

Intense synchrotron radiation from high power wiggler sources has long been a difficult high-heat-load problem to the design of properly cooled x-ray optics. Large, high power and very intense beams thermally distort crystal optics, reducing throughput and broadening rocking curves. An internally cooled silicon monochromator has been fabricated which demonstrated the capability of diffracting wiggler radiation of unprecedented power without significant degradation of the beam. Cooling water flows through rectangular cooling channels 1 mm wide, 1 mm below the diffracting surface, fed by a manifold bonded to the underside of the diffracting crystal. A novel silver diffusion bond was used to ensure leak-tight UHV performance. Recent test results at wiggler station F2 show a linear behavior of the x-ray flux with increasing storage ring current up to a total power of 3 kW and a peak surface power density of 5 W/mm^2. The improved monochromator has led to an increase of x-ray flux by a factor of six over previous contact-cooled designs and shows that internal water-cooling can be an effective solution to high-heat-load problems at high power wiggler stations.

INTRODUCTION

There has been tremendous interest and effort in the past decade in producing high quality x-ray monochromators that are free of thermal distortions caused by highly intense and powerful radiation at various synchrotron facilities. For undulator beam lines with high power densities, it has been shown by several groups (1, 2, 3) that liquid nitrogen cooled silicon or diamond crystals are to date the best solutions to the high heat load problem because of the relatively small beam size (3 - 5 mm^2) and moderately low total incident power (< 600 W). These designs are rapidly become the standard for undulator beamlines worldwide.

For high power wiggler beam lines, however, there has not been a clear consensus on how to handle the high total power (> 2 kW) and moderate specific power densities (< 20 W/mm^2). Among the possible remedies, liquid metal cooling (4) has been tested but the cost and complexity of a liquid metal pumping system have so far prevented the wide-spread use of this novel technology. Liquid nitrogen cryogenic cooling (5) has been successful at modest powers (1500 W) but ultimately the low heat capacity and limited single phase working range of liquid nitrogen limits its use in high power wiggler applications. Simple internally water-cooled monochromators with channels of roughly 1 mm scale (mini-channels) have proven surprisingly effective as tested at ESRF (6) and the Photon Factory (7). The design presented here follows closely from these two earlier works. Water-cooled monochromators with complex cooling channel designs such as micro-channels (8) and pin-post cells (9) perhaps hold the most promise due to the potential for improved convective heat transfer if the difficulties in fabrication of these crystals, which have led to residual strain, can be solved.

One of the key issues in making internally cooled monochromator crystals is the radiation-resistant, strain-free bonding of the coolant manifold to the top diffracting crystal. Various methods, such as die attach paste (10), glass frit bonding, and direct silicon bonding (6) have been tried and utilized with a varied degree of success. In this article, we present a simple and effective monochromator design constructed at the Cornell High Energy Synchrotron Source (CHESS) with straight rectangular water-cooling channels. The adhesion between the water manifold and the top diffracting crystal is achieved through a novel high-temperature metal diffusion bond, which is highly radiation resistant. The monochromator has been tested successfully at a high power wiggler station F2 at CHESS and the results indicate that our fabrication method is very promising in producing high quality, internally-cooled silicon monochromators.

WIGGLER HEAT LOAD AT CHESS

The wiggler at the F2 station of CHESS is a 24-pole hybrid permanent magnet structure with a peak field of 1.2 Tesla. At the storage ring energy of 5.3 GeV, it produces a white synchrotron radiation spectrum with a critical energy of 23 keV and a total power of 10 kW in 4 horizontal milliradians at 150 mA of stored positron current. The radiation is split evenly between two stations on F-line, F1 and F2. At present running conditions of 150 mA, the total power incident on the first monochromator crystal at the F2 station is 3.0 kW and the peak power density is 30 W/mm^2 normal to the beam at a distance of 21 meters from the source. Upgrades to the Cornell Electron Storage Ring (CESR) over the next 1-2 years will double the beam current from 150 mA to 300 mA, presenting an enormous heat load problem for x-ray optics at these wiggler stations.

Until recently the first crystal design at F2 consisted of a 1 mm thick wafer placed on a water-cooled copper block with a liquid film of Ga-In eutectic in between to promote efficient thermal contact. This scheme, while sufficient for bend magnet beamlines, cannot withstand the full intense wiggler beams, leading to severe thermal bump distortion. This distortion of the first crystal reduces the overall throughput of the beamline optics due to broadened rocking curve widths and reduced peak photon fluxes. An improved internally cooled design is clearly needed to fully take advantage of future CESR increases in stored-beam currents.

Numerous groups have tested advanced monochromator designs at the F2 station including water-cooled pin-post crystals (9), water-cooled microchannel designs (8), and cryogenically cooled designs (5). Based on these past test results, it has become clear that water-cooling is preferred to liquid nitrogen cooling in high total power applications because of the larger heat capacity of water and the reduced cost to implement.

DESIGN AND FABRICATION

We have tried to draw on this wealth of test experience in the design of a simple, robust, and low-cost monochromator. The two part monochromator (fig. 1) is cooled by a series of 26 water-cooling channels 1 mm wide, 1 mm below the top illuminated surface. The 5 mm thick channeled top piece is bonded to a silicon coolant manifold using a novel silicon bonding technique using a thin silver interlayer. The geometric parameters of the channels were based both on heat transfer analysis and on currently available saw widths which would ease the fabrication of the crystals.

FIGURE 1. View of completed monochromator (top) with unfinished water manifold (right) and mini-channel diffracting top piece (left) before bonding and cutting to size.

The two pieces are sliced from a silicon (111) oriented boule and the mating faces are polished. The mini-channels are plunge cut with a diamond saw and the coolant manifold passages are machined with diamond coated core drills and grinding pins. After machining, the polished surfaces are protected with a thin layer of wax and etched in a 95% HNO_3 / 5% HF solution for 30 minutes to remove the surface damage. The wax is then removed and the polished surface is solvent cleaned in an ultrasonic bath. Since the monochromator operates in an UHV chamber the utmost attention is paid to obtaining helium leak-tight bonds between the manifold and the channeled. Previous work on bonding techniques (11) has led us to select a robust metallurgical bond between the two pieces with the assurance of a leak-tight bond unaffected by radiation damage. Immediately before bonding, the silicon pieces are rinsed in dilute HF solution. A 25μm foil of silver (99.9% pure) is placed between the two pieces and the assembly is heated in an UHV furnace to 850°C for 15 minutes while a pressure of 7000 Pa (1 PSI) is applied to form the bond. Coolant tubes are attached to the water manifold using indium wire seals and the assembly is leak checked to 1×10^{-9} mbar-l/sec before installation in the monochromator vacuum chamber.

EXPERIMENTS

The monochromator was tested on the F2 wiggler station with 12 keV x-rays. The crystal is exposed to a peak surface power density of 5 W/mm² over a 42 mm x 25 mm beam footprint. In order to expose the crystal to the highest possible heat loading all slits upstream of the monochromator were fully opened. The center of the crystal (with minimum distance from cooling channel to surface) was accurately centered on the wiggler beam. The coolant was filtered, demineralized water, chilled to 25°C, flowing initially at 9.6 l/min with an inlet pressure of 114 KPa. The pressure drop across the crystal is less than 7 KPa. A second perfect silicon (111) crystal was used in a nondispersive arrangement to measure the throughput and the rocking curve of the double crystal system.

The rocking curve width and peak photon flux from the crystal are used as a judge of the effectiveness of the improved internal cooling scheme. Over the course of a machine fill (about 90 minutes) the (111) rocking curve was repeatedly measured at 12 keV, using an ion chamber, as the machine current decayed from 150 mA to 90 mA. Plotting the peak photon flux and rocking curve width versus stored beam current, which is proportional to the incident power, gives some insight into the nature of the strains in the crystal. The offset of the rocking curve width at zero current from the intrinsic value gives a measure of the residual fabrication and bonding strain, and any strain due to the deformation caused by the pressurized internal coolant, which is normally negligible. The slope of the rocking curve width is an indication of the severity of the thermal distortion of the crystal caused by the wiggler heat load. Deviations from a linear response of flux versus current indicates that heat load strain is becoming significant compared to fabrication strain and at some beam current the flux throughput will saturate completely.

Figure 2 compares the peak photon flux of the internally cooled monochromator to the previous contact cooled design, showing an improvement at all currents of as much as a factor of six, and the nearly linear response of the peak photon

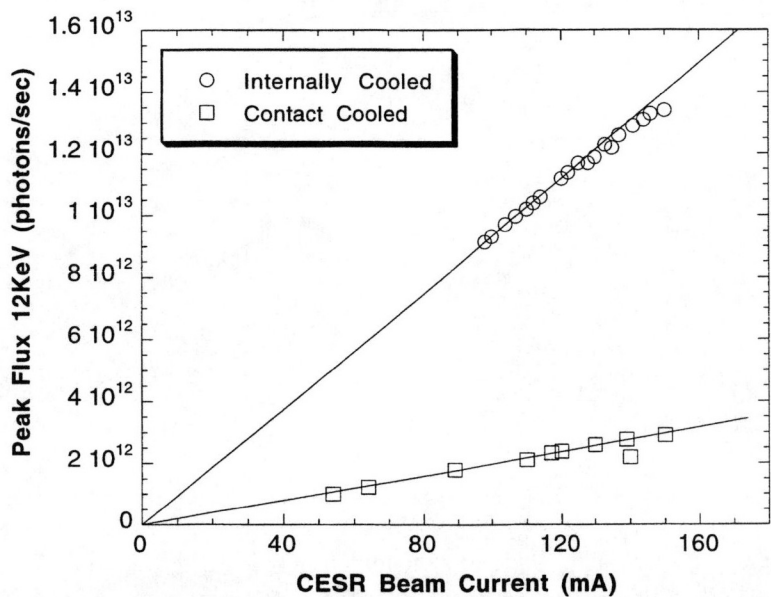

FIGURE 2. Comparison of peak 12 KeV photon flux from internally water-cooled monochromator with previous, contact cooled design.

flux with increasing beam current up to 135 mA. Above 135 mA a slight degradation in the performance of the internally cooled monochromator is noted in both the peak flux and the rocking curve width (fig. 3), indicating the heat load strain is becoming significant at these currents. To check this hypothesis, the coolant flow rate was reduced to 3.8 l/min (with a 50 KPa inlet pressure) to decrease the effectiveness of the cooling. As expected the performance degradation now begins at a lower beam current (about 105 mA) with reduced peak fluxes and a corresponding increase in the rocking curve widths.

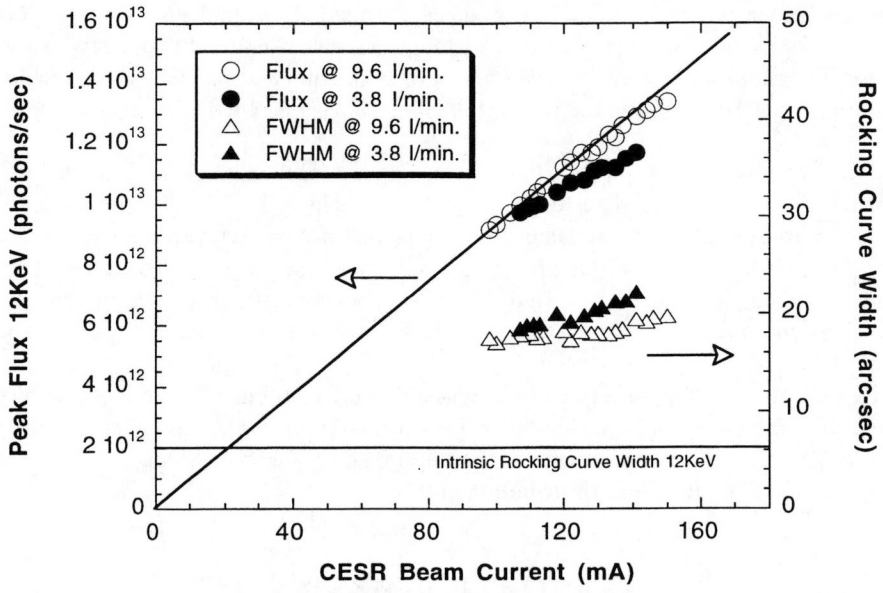

FIGURE 3. Effect of reduced water flow rates on the monochromator performance at high currents.

The energy band width of the x-ray beam was measured by an energy scan at the selenium K absorption edge. An example of such scans is shown in figure 4. It shows that the energy width with a 5 mm vertical slit in front of the monochromator is about 4 eV, which is significantly smaller than the calculated value for the full wiggler beam. We believe that this discrepancy is due to the overall bending distortion in the first monochromator crystal caused by the bonding process.

A further test of the new monochromator performance was conducted with a sagittal focusing second crystal and a vertically focusing mirror. The doubly focused beam was then passed through a circular collimator-aperture with a diameter of 0.3 mm. The flux through the collimator was measured to be 0.7×10^{11} photons/sec at 98 mA and 12.5 keV, which is about an order of magnitude higher than using the standard contact cooled first monochromator crystal.

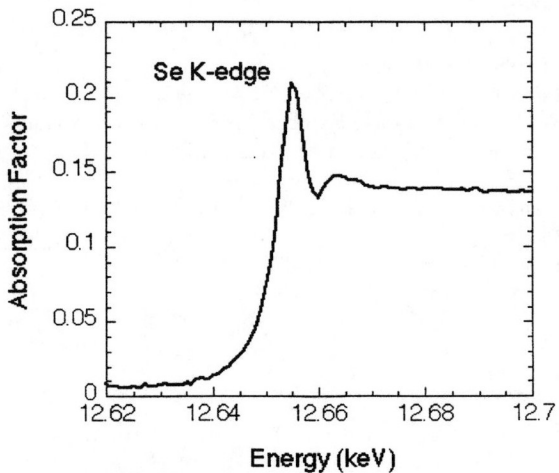

FIGURE 4. Selenium K absorption edge scan.

DISCUSSION AND CONCLUSIONS

Overall performance of the new monochromator, though significantly improved over past designs, is still limited by fabrication strain. Throughput is roughly a factor of two less than theoretical predictions of the synchrotron radiation spectrum and improvements in the bonding step are sought to reduce residual strains. It is expected that lower temperature diffusion bonds (i.e.: gold vs. silver interlayers) and use of thinner 1μm evaporated metal layers, instead of foils, could lead to reduced strain and improvements in monochromator performance. Structurally sound gold diffusion bonds have been performed in-house at temperatures as low as 450°C, but have so far not proven leak tight. It may be possible to reduce the fabrication strain by bonding the manifold to a much thicker top piece containing the channels which would later be cut down to final dimension. This technique has been used in the past with die-attach paste bonded monochromators and has been proven effective.

At currents greater than 135 mA improvements to the cooling channel geometry (channel width, fin design, etc.) could lead to further improvements in the overall performance of the monochromator since the heat load strain is compounded with the bonding strain. Preliminary finite element calculations predict improved performance with decreasing channel width due to increases in the coolant's heat transfer coefficient. Present techniques used to cut and etch the water channels may limit us to relatively large channel dimensions. Currently we have fabricated, but not tested, monochromators with channels as small as 0.4 mm and potentially wet chemical anisotropic etching could be used to form even narrower channels for use in future monochromators.

These results indicate that simple internally water-cooled monochromators are a cost effective and technically promising solution for wiggler optics cooling at present power levels and the foreseeable future. With improvements in the fabrication strain, water cooling will be competitive with liquid gallium and cryogenic cooling in high power load applications but with significantly reduced capital equipment and operating costs.

ACKNOWLEDGMENTS

We would like to acknowledge Randy Headrick for his initial suggestion of the diffusion bonding technique. We also thank B. W. Batterman and Don Bilderback for useful discussions and the CHESS operators for their technical support. This work is supported by the National Science Foundation through CHESS under Award No. DMR-9311772. The facilities used to perform the diffusion bonding are supported by the Material Science Center at Cornell under NSF Grant No. DMR-9121654.

REFERENCES

1. Marot, G., *Optical Engineering* **34**, 426-431 (1995).
2. Rogers, C., Mills, D., Lee, W.-K., Knapp, G., Holmberg, J., Freund, A., Wulff, M., Rossat, M., Hanfland, M., and Yamaoka, H., *Review of Scientific Instruments* **66**, 3494-3499 (1995).
3. Freund, A., *Optical Engineering* **34**, 432-440 (1995).
4. Smither, R., Forster, G., Bilderback, D., Bedzyk, M., Finkelstein, K., Henderson, C., White, J., Berman, L., Stefan, P., and Oversluizen, T., *Review of Scientific Instruments* **60**, 1486-1492 (1989).
5. Rogers, C., Mills, D., Assoufid, L., and Graber, T., *Review of Scientific Instruments* **67**, 1-8 (1996).
6. Yamaoka, H., Häusermann, D., Freund, A., Krumrey, M., and Kvick, Å, *Nuclear Instruments and Methods* **A351**, 559-564 (1994).
7. Oversluizen, T., Matsushita, T., Ishikawa, T., Stefan, P. M., Sharma, S., and Mikuni, A., *Review of Scientific Instruments* **60**, 1493-1500 (1989).
8. Arthur, J., Tompkins, W., Troxel Jr., C., Contolini, R., Schmitt, E., Bilderback, D., Henderson, C., White, J., and Settersten, T., *Review of Scientific Instruments* **63**, 433-436 (1992).
9. Tonnessen, T., and Arthur, J., *SPIE Proceedings* **1739**, 622-627 (1992).
10. Bilderback, D., *Review of Scientific Instruments* **60**, 1977-1978 (1989).
11. Smolenski, K., Conolly, C., Doing, P., Kiang, B., and Shen, Q., *SPIE Proceedings* **2856**, 246-257 (1996).

Performance of the Double Multilayer Monochromator On the NSLS Wiggler Beam Line X25

Lonny E. Berman and Zhijian Yin

Brookhaven National Laboratory, National Synchrotron Light Source, Upton, New York 11973

Steven B. Dierker and Eric Dufresne

University of Michigan, Department of Physics, Ann Arbor, Michigan 48109

Simon G.J. Mochrie and Ophelia K.C. Tsui

Massachusetts Institute of Technology, Department of Physics, Cambridge, Massachusetts 02139

Stephen K. Burley, Fong Shu and Xiaoling Xie[1]

Rockefeller University, Laboratories of Molecular Biophysics, Howard Hughes Medical Institute, New York, New York 10021

Malcolm S. Capel and Robert M. Sweet

Brookhaven National Laboratory, Biology Department, Upton, New York 11973

Abstract.
A tunable, double multilayer x-ray monochromator has recently been implemented on the National Synchrotron Light Source (NSLS) X25 wiggler beam line. It is based on a parallel pair of tungsten-boron-carbide multilayer films grown on silicon substrates and purchased from Osmic, Inc. of Troy, Michigan, USA. It acts as an optional alternative to the conventional double silicon crystal monochromator, and uses the same alignment mechanism. Two other NSLS beam lines also have had this kind of monochromator installed recently, following the lead of the NSLS X20C IBM/MIT beam line which has used a double multilayer monochromator for several years. Owing to the 100 times broader bandwidth of a multilayer x-ray monochromator, compared with a silicon monochromator, the multilayer monochromator has the obvious advantage of delivering 100 times the flux of a silicon monochromator, and thereby makes more efficient use of the continuous synchrotron radiation spectrum, yet preserves the narrow collimation of the incident synchrotron beam. In particular, multilayer x-ray bandwidths, on the order of 1%, are well-matched to x-ray undulator linewidths. Performance results for the X25 multilayer monochromator are presented, comparing it with the silicon monochromator. Of note is its short- and long-term performance as an x-ray monochromator delivering the brightness of the wiggler source in the presence of the high-power white beam. Detailed measurements of its spatial beam profile and wavelength dispersion have been made, and it is shown how its resolution could be improved when desired. Finally, its peculiar, anisotropic resolution function in reciprocal space, and its bearing upon x-ray crystallography and scattering experiments, will be discussed, and highlighted by the results of a protein crystallography experiment.

I INTRODUCTION

Most experimental applications of synchrotron radiation require monochromatic beams, which are usually produced by perfect silicon or germanium crystal monochromators. Such monochromators diffract a photon energy bandwidth of typically 10^{-4}, producing concomitant relative energy or wavevector resolutions. Often, such fine resolution is unnecessary for monochromatic beam applications such as small-angle scattering, absorption, and emission experiments. Broader bandwidth diffractive optics, such as layered synthetic microstructures (henceforth referred to as multilayers), consisting of a periodic film of alternating heavy and light layers grown atop a substrate for which

---

[1] Current address: Vertex Pharmaceuticals, Inc., Cambridge, Ma.

the resultant bilayers form "Bragg planes", can be designed to diffract a photon energy bandwidth of 10^{-2} or greater [1,2]. This results in a "monochromatic" beam that is at least 100 times as intense as a beam monochromated by a perfect silicon or germanium crystal, making more efficient use of the continuous synchrotron radiation spectrum. In particular, multilayer bandwidths are very well matched to the radiated linewidths of modern x-ray undulators, and hence multilayers offer the prospect of perfect line filtering for such insertion devices [3]. While mosaic crystal monochromators can be conceived to diffract broad bandwidths [4,5], they do not preserve the synchrotron source brightness. Since multilayers in principle function as perfect, specularly-reflecting diffractors, they do preserve the brightness.

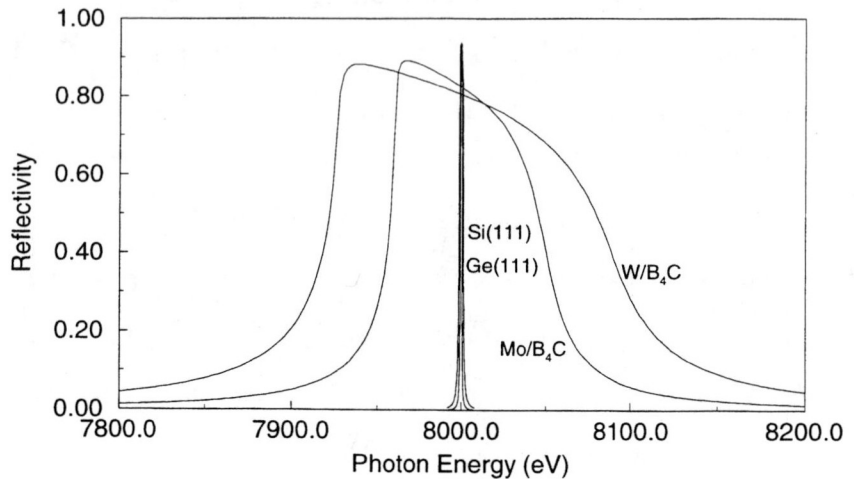

FIGURE 1. The theoretical x-ray reflectivities, at 8 keV, of semi-infinite perfect crystal Bragg peaks of Si(111) and Ge(111), and first-order multilayer Bragg peaks of Mo/B_4C and W/B_4C (with fabrication parameters described in the text), are compared.

A comparison of the diffractive properties of perfect multilayers vs. perfect single crystals is shown in Fig. 1. First-order reflectivity curves are presented, for a nominal photon energy of 8 keV, for a Si(111) crystal, a Ge(111) crystal, a semi-infinite Mo/B_4C multilayer of d-spacing 25Å (for which the thickness of the heavier Mo component of each bilayer is 30% of the 25Å bilayer thickness), and a semi-infinite W/B_4C multilayer with identical d-spacing and thickness attributes to the Mo/B_4C multilayer [6]. Whereas the relative bandwidth of the Si(111) curve is 1.3×10^{-4} and that of the Ge(111) curve is 3.3×10^{-4}, the relative width of the Mo/B_4C curve is 1.2×10^{-2} and that of the W/B_4C curve is 2.0×10^{-2}, about 100 times the widths of the Si(111) and Ge(111) curves. The bandwidth B of a perfect Bragg diffractive optic is given by

$$B = Cd^2 \frac{|F|}{v} \tag{1}$$

where C is a proportionality constant, d is the d-spacing of the Bragg planes, F is the structure factor for the Bragg reflection, and v is the unit cell volume. The ratio of $|F|$ to v is simply an effective x-ray scattering density for the material and Bragg reflection at hand. It is easy to see that the Ge(111) bandwidth is larger than the Si(111) width by virtue of its higher scattering density (the d-spacings of these crystal Bragg planes are about the same), and similarly for the W/B_4C width in comparison with the Mo/B_4C width. The bandwidths of the multilayers are broader than those for the crystals by virtue of their larger d-spacings. The reflectivity of the perfect semi-infinite multilayers can be calculated using the same Darwin-Prins formula that is applicable to perfect semi-infinite crystals, with a suitable definition of a structure factor for the multilayer's Bragg planes which takes into account the bilayer thickness and composition profile [1]. Reference 7 presents a comprehensive review of the state-of-the-art of the fabrication and use of x-ray multilayers.

II X25 DOUBLE MULTILAYER MONOCHROMATOR

A double multilayer monochromator, consisting of a pair of W/Si multilayers, was implemented on the IBM/MIT X20C bending magnet beam line at NSLS over 10 years ago, primarily for time-resolved x-ray scattering experiments for which high monochromatic beam intensity was desired but high monochromaticity was unnecessary [8]. This beam line has been highly sought and oversubscribed by users since.

Single multilayer elements were tested for their diffractive properties under very high power, and power density, loads on the NSLS X25 wiggler beam line [9], using its focussed white beam, a few years later [10]. Tests showed that commonly-used multilayers such as W/Si and W/B_4C, grown on Si substrates, could well survive incident power densities of as much as 150 W/mm^2 and have their diffractive properties preserved over the course of beam exposure, provided that they could be suitably cooled and contained in a proper inert gas or vacuum environment in order to prevent chemical attack of the multilayer coating.

Subsequently, x-ray scattering experiments were carried out on X25 using a single W/Si multilayer element (borrowed from the X20C beam line) as a monochromator, placed in the experimental hutch just upstream of the sample and cooled in the same manner (at room temperature) described in Ref. 10. Success with this simple arrangement was attained in high angle scattering experiments [11], but not in small angle scattering experiments [12], owing to the substantial low angle diffuse scattering from the single multilayer element (which was impacted by white x-ray beam) and close distance of the sample and detector to it. It rapidly became clear that an experimental arrangement similar to that employed on beam line X20C, i.e. placing the multilayer monochromator in the shielded beam line enclosure far upstream of the experimental hutch, and using a double multilayer monochromator setup instead of a single one (to suppress diffuse scattering) which would additionally afford convenient energy tunability without needing to tilt or elevate the experimental diffractometer downstream, would be highly desirable on X25.

Several W/B_4C multilayer elements were purchased from Osmic, Inc. of Troy, Michigan, USA [13] for use on X25 initially and other NSLS beam lines later. Each element consists of 120 bilayers, each of thickness (d-spacing) about 25Å with the heavy W component's thickness being about 30% of the d-spacing. The multilayer films were grown on optically-flat Si crystal wafers of roughness < 1Å and slope error < 1 arcsec, and with dimensions 50mm×25mm×5mm. The elements were designed to fit readily into the standard X25 double crystal monochromator tank using the same exact fixtures and cooling block that silicon crystals are normally mounted on, thereby facilitating interchange. At a typical photon energy of use of 9 keV, the multilayer Bragg angle is 1.6 deg. The standard offset height of the X25 monochromatic beam from the white beam is 25mm. To preserve this height with a pair of diffractive elements set to a Bragg angle of 1.6 deg requires that they be separated, along the beam path, by 45 cm. The largest separation permitted within the monochromator tank is less than half of this, 20 cm. Accomodation of the multilayer pair within the tank required us to reduce the offset height of the doubly-diffracted monochromatic beam to be nominally 6mm which, for the 1.6 deg Bragg angle, cut down the separation along the beam path to a manageable 11 cm value, and at the same time permitted a reasonable range of tunability. The 50mm multilayer element length is just large enough to intercept the full incident beam height that is reflected from the upstream focussing mirror, at the 1.6 deg Bragg angle. The X25 monochromator design [9] allows for scanning of the perpendicular spacing between the two diffractive elements. This feature is particularly useful when multilayers are employed, for the spacing can be scanned, as the photon energy is tuned, in order to maximize the beam footprint intercept and thus the photon throughput.

With this upgrade of the monochromator tank to accomodate multilayers, small angle scattering x-ray photon correlation spectroscopy [12], and other experimental applications requiring a multilayer monochromator, became feasible on X25.

III BEAM PROFILE AND WAVELENGTH DISPERSION

Upon the first installation of a pair of W/B_4C multilayer elements in the X25 monochromator tank, measurements of the doubly-diffracted flux, beam profile, and wavelength dispersion functions were made, with the beam line set up in its focussed mode of operation (upstream toroidal mirror placed in the white beam path, focussing nominally at 1:1 magnification). With the NSLS x-ray ring operating at 2.584 GeV, 210 mA current, and with the X25 27-pole wiggler operating at a gap of 24mm (giving a nominal field of 1.1 T), a focussed "monochromatic" beam flux of 1.4×10^{14} ph/sec was measured at 8 keV. An ionization chamber was used to determine the flux, and was carefully calibrated with filters to ensure linearity of response. In fact, it was not possible to measure the unattenuated flux with the chamber because of non-linearity of response in that flux range, so the final flux determination was made via extrapolation of the attenuated fluxes. Under these operation conditions of the NSLS x-ray ring and X25, the standard Si(111) double crystal monochromator typically delivers 2.5×10^{12} ph/sec at 8 keV. Thus the flux derived with use of the multilayer monochromator represented a gain of 56 compared with that from the Si(111) monochromator. It was noticed that most, but not all, of the incident beam was intercepted by the multilayers; with a more careful alignment, the potential flux gain could have been even higher. The observed flux gain compared with the Si(111) double crystal monochromator is roughly consistent with expectation, based upon comparison of the diffracted bandwidths and reflectivities (Fig. 1), and by taking account of the fact that two successive reflections are involved, not one.

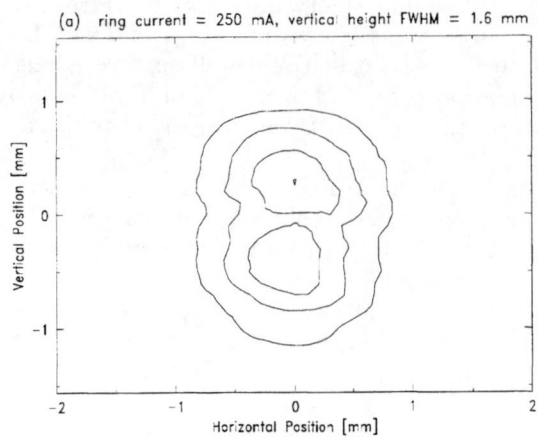
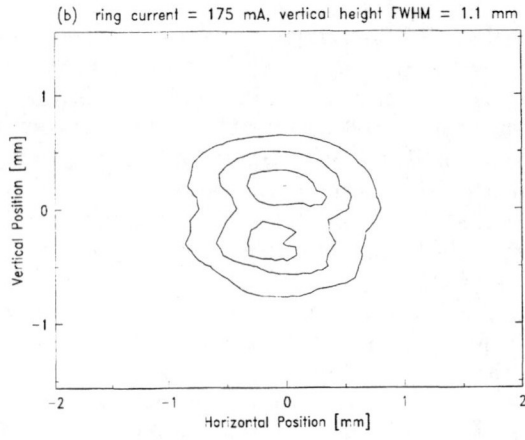

FIGURE 2. Spatial double multilayer diffracted beam profiles, at the nominal beam focus position in the X25 experimental hutch, are shown at 8 keV with ring currents of (a) 250mA and (b) 175mA. The contours are at 25% increments.

An important aspect of performance of the multilayer monochromator relates to its withstanding of the incident heat load. At a machine current of 300 mA, the x-ray beam power reflected from the upstream focussing mirror and incident upon the monochromator is about 120 W, with a density (normal to the beam direction) of 5 W/mm^2. It is straightforward to show that, for a freely-supported wafer, an integrated thermal bowing slope error of about 25 arcsec in the silicon substrate of the multilayer would be induced to span the beam footprint, with the silicon nominally at room temperature, and irrespective of the substrate's thickness [14,15]. A silicon monochromator has an intrinsic rocking curve angular width that is somewhat lower than this, even for low-index Bragg reflections, and this implies that the thermal slope error is rather severe in this case. Fortunately, the intrinsic rocking curve angular width of the W/B$_4$C multilayer predicted by dynamical diffraction theory is much broader, about 125 arcsec at 8 keV, and one might therefore expect that the thermal slope error is not problematic. This however is not truly the case, as revealed in Fig. 2, which shows spatial profiles of the multilayer-monochromated beam at the focal position in the experimental hutch at different machine currents, obtained by raster-scanning, across the beam, a pinhole placed before an ionization chamber detector. The bowing slope error in the first multilayer element spreads out the ray paths of the diffracted beam, smearing the vertical beam profile, and this effect decreases as the machine current (and thence the power load) decreases [16]. The extent of the vertical profile smearing shown in Fig. 2 (1.6mm FWHM at 250mA current) is approximately consistent with the calculated thermal bowing, given the 10m distance from the monochromator to the focal position. Extrapolation of the vertical profile widths to zero current confirms that they are principally due to thermal bowing and not to intrinsic slope error.

Profiling the beam spatially is a suitable means of determining the thermal slope error of the first multilayer in this case, because deconvolution from a measurement of the multilayer's rocking curve would be complicated by the fact that the intrinsic rocking curve width is relatively broad to begin with. Advantage is taken of the long distance from the multilayer to the beam focus, in gaining sensitivity with the profiling technique. It could be used to provide feedback to an adaptive correction system that might be designed to compensate the thermal slope error [16]. Such a system has not been implemented at X25 for multilayers, but has for silicon crystals [17], and has been shown to produce a vertical beam focus size of as small as 0.3mm FWHM when the collection solid angle of the upstream focussing mirror is filled [16]. Consequently, compared with silicon crystals, the gain in flux afforded via the use of multilayers at X25 is offset somewhat by the broader spot profile which is left uncompensated.

The reflectivity of all W/B$_4$C multilayer elements purchased from Osmic had been measured by Osmic prior to delivery, using a Cu K$_\alpha$ x-ray source. Each of the multilayers in the most recent pair we obtained has an absolute peak reflectivity of 0.65 and a diffracted bandwidth (for the first-order Bragg peak) of 1.5×10^{-2}. The diffracted bandwidth of the multilayer pair was measured at X25 using a Si(111) analyzer crystal in the experimental hutch, and the observed dispersion function (measured at 9 keV) is shown in Fig. 3. Its main peak has a relative ($\Delta E/E$, where E is the nominal photon energy) FWHM of 1.1×10^{-2}, which is consistent with the product of two identical functions each with a width of 1.5×10^{-2}. The interference function side fringes about the main peak arise from the finite number of bilayers (120 for each element in this case). The calculated reflectivity curve for this double multilayer monochromator, based upon its fabrication parameters and assuming ideal interfaces, is shown in Fig. 3, and agrees poorly with experimental measurement in terms of both peak reflectivity and width. The calculation

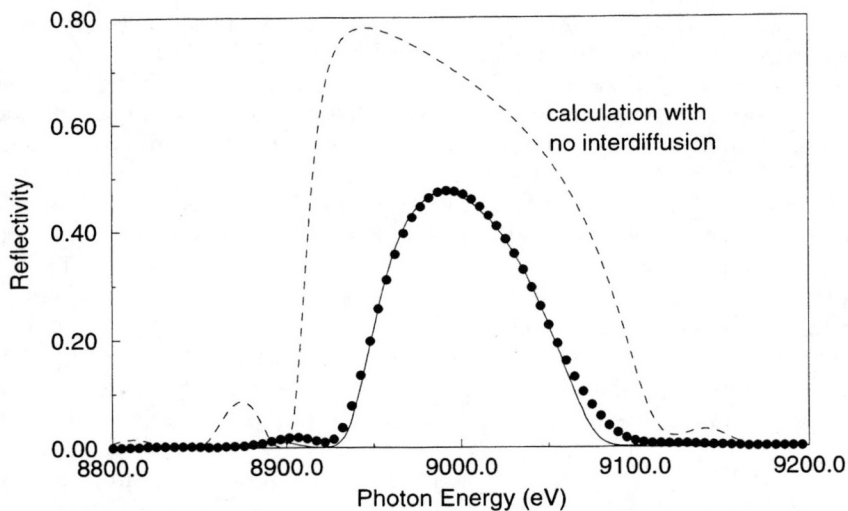

FIGURE 3. The measured double multilayer dispersion function at 9 keV (dotted curve) is compared with theory, for the ideal case of no interdiffusion at the interfaces (dashed curve) and for the case of an interdiffusion thickness with a σ of 4 Å (solid curve).

was repeated by assuming that an interdiffusion thickness, between layers, was present with a σ value of 4Å, and this gave much better agreement with experiment as also shown in Fig. 3. This level of interdiffusion is consistent with the expectation of the vendor for this kind of multilayer. As it bears upon the reflectivity, it can be understood as a Debye-Waller effect, due to static disorder, that modulates the calculated structure factor for the case of ideal interfaces. Under dynamical diffraction, this leads mainly to a narrowing of the width of the reflectivity curve, and secondarily in some instances to a lowering of the peak reflectivity.

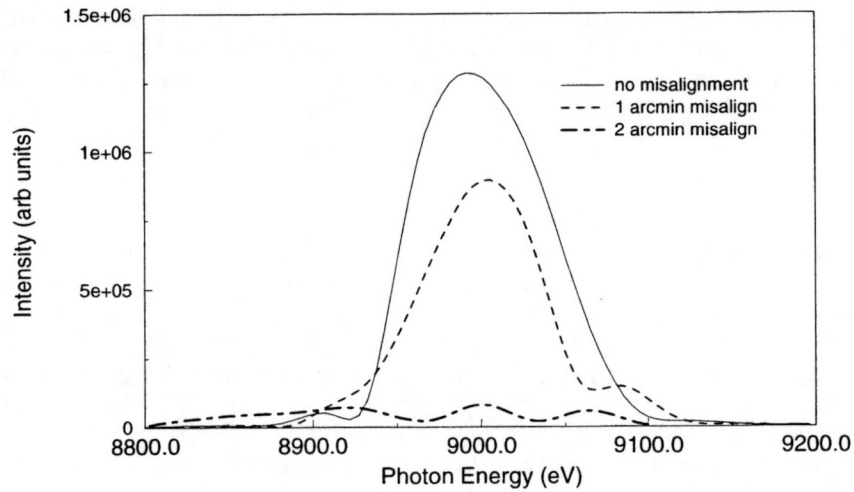

FIGURE 4. The measured double multilayer dispersion functions at 9 keV are shown for the cases of no misalignment of the multilayers (solid curve), a misalignment of 1 arcmin (dashed curve), and of 2 arcmin (chain-dashed curve).

An important issue concerns whether a relative resolution function width narrower than 1.1×10^{-2} FWHM, produced by the present pair of multilayers, is attainable, without compromise of peak reflectivity. From Eqn. 1, it seems that two pathways can be pursued to obtain a narrower bandwidth multilayer element. One involves reducing the ratio of structure factor to unit cell volume. This could be attained through use of a lower density "heavy layer", e.g. via the substitution of Mo for W (and which is available commercially from Osmic) or via a reduction of the thickness fraction of the "heavy layer". The other alternative is to reduce the d-spacing of the bilayer, whose effect on the bandwidth is rather sensitive because of the proportionality to d^2. In practice, the performance of these multilayers, mainly in terms of their peak reflectivity, might be compromised upon pursuit of most of these bandwidth reduction methods, due to a greater relative impact of the interdiffusion and roughness at the layer interfaces which are rather difficult to improve beyond present capabilities (a few Å) [7,13].

An alternative method of effective bandwidth narrowing takes advantage of the fact that multiple reflections are involved in most monochromator systems, including X25's. Just as resolution narrowing has been demonstrated with multiple silicon crystal reflections, whereby the individual silicon reflections can be slightly misaligned from parallel leaving a narrower effective transmission in terms of relative angle or wavelength spread (as well as reduced intensity) [18,19], so can the principle be applied to multiple multilayer reflections. This is illustrated in Fig. 4, which shows the measured dispersion function at X25, at 9 keV, using different misalignment angles between the two multilayer elements. Marginal improvement is demonstrated, from a FWHM of 1.1×10^{-2} to 0.9×10^{-2}, upon introduction of a misalignment angle of 1 arcmin. A misalignment angle of 2 arcmin leaves a much narrower main peak but comparable neighboring interference function side fringes, for in this case the main peak of one multilayer element overlaps the first side fringe of the other, and vice versa. The multiple fringes which result, of similar intensities, compromise the objective of bandwidth narrowing via misalignment of the multilayers. It is envisioned that this problem could be minimized via fabrication of multilayers with more bilayers to suppress the interference function fringes.

With intermittent use of the W/B_4C multilayer elements in the X25 monochromator tank over the past few years, their reflectivity properties have not been noted to degrade with continued exposure to the X25 x-ray beam. Only slight discoloration has been observed on the multilayer films at the beam footprints, mainly on the first multilayer element (which sees the white beam). The high vacuum of the X25 monochromator tank, which typically operates at a pressure of as low as 10^{-8} Torr, is largely responsible for minimizing the contamination of the films.

IV APPLICATION TO X-RAY CRYSTALLOGRAPHY AND SCATTERING

The usefulness of a multilayer monochromator in an x-ray diffraction experiment requires consideration of the resulting resolution function in reciprocal space. For simplicity, let us consider incident and diffracted (from a sample) beams to have δ function angular profiles within and perpendicular to the scattering plane; at synchrotron sources, the angular widths of the incident and scattered beams (with the latter usually defined by slits or an analyzer crystal) result in relative wavevector resolution width contributions of 10^{-4} to 10^{-3}. The wavelength spread of a monochromator can be represented in reciprocal space by a pair of Ewald spheres with a common incident beam direction, and tangent at the origin of reciprocal space. The spheres have different radii corresponding to the low and high energy limits (long and short wavelength limits) of the wavelength dispersion function. For any given scattered beam direction, the finite wavelength spread gives rise to a spread of the reciprocal space resolution function in the longitudinal direction (the direction of momentum transfer), according to

$$\frac{\Delta q}{q} = \frac{\Delta \lambda}{\lambda} \qquad (2)$$

where q denotes the momentum transfer magnitude and λ denotes the wavelength. It is clearly evident that, as q increases, so does Δq; at higher momentum transfers, the longitudinal reciprocal space resolution becomes poorer in absolute terms. This is shown schematically in Fig. 5.

Since the relative wavelength resolution from a multilayer monochromator is about 10^{-2}, this contribution to the resolution function in reciprocal space, at synchrotron sources, is generally much larger than the contributions arising from the angular spreads of the incident and scattered beams, by a factor of 10 to 100. The resolution function can be thought of as a longitudinal streak, rather than a sharp point as is normally the case with use of a silicon or germanium monochromator. This has important implications on the application of a multilayer monochromator in x-ray diffraction experiments, with data interpretation requiring careful consideration of the resolution function. Several diffraction experiments executed at NSLS beam lines X20C and X25 have needed to consider carefully the details of this highly anisotropic resolution function in order for the data to be properly understood; a case example, a small angle scattering "speckle" experiment carried out at X25, is described in Ref. 20.

We will now focus attention on the application of a multilayer monochromator for rotation-diffraction experiments of the sort that often are performed in macromolecular crystallography. These involve rotating a single crystal in a monochromatic beam and recording the Bragg diffraction spots on a film, imaging plate, or charge-coupled device (CCD) area detector. As in any crystallography experiment, the success of a rotation-diffraction measurement depends on the capability of separating and indexing Bragg reflections, and reducing their integrated intensities. The poorer the resolution in reciprocal space, the more challenging these tasks are. The situation is shown in Fig. 5, which overlays the reciprocal lattice points corresponding to a single crystal with the Ewald spheres corresponding to the low and high energy limits of the multilayer monochromator resolution function. A rotation measurement entails rotation of the crystal's reciprocal lattice, about the origin, and a Bragg peak will appear on the detector

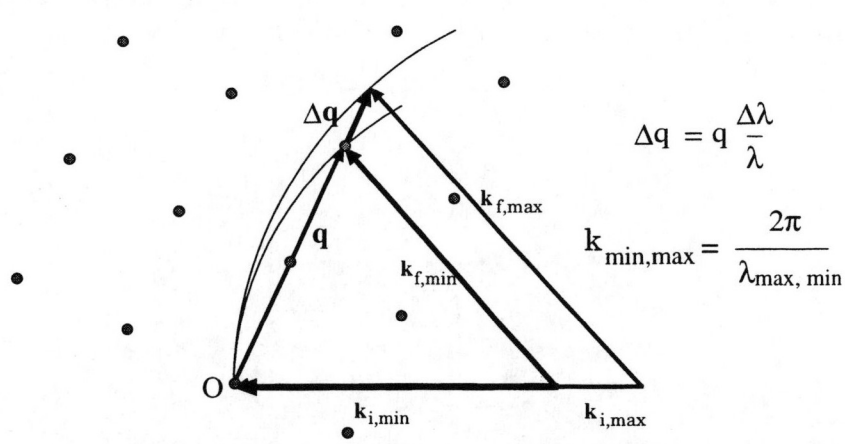

FIGURE 5. Reciprocal space representation of the scattering kinematics and resolution with use of a multilayer monochromator. $k_{i,min}$ and $k_{f,min}$ are the incident and scattered wavevectors, respectively, corresponding to the long wavelength limit of the monochromator resolution, and $k_{i,max}$ and $k_{f,max}$ are the incident and scattered wavevectors, respectively, corresponding to the short wavelength limit. Superimposed is the reciprocal lattice corresponding to a single crystal whose diffraction pattern is to be measured.

plane in a scattered beam direction that is determined by the intersection of a lattice point with any of the allowed Ewald spheres that fall within the resolution function (with the two spheres shown in Fig. 5 being the limiting spheres). As the crystal (and thereby its reciprocal lattice) is rotated, a lattice point will first cross one limiting sphere, giving rise to a diffracted beam aimed from the center of the sphere toward that point where it contacts the sphere, and ultimately the lattice point will cross the other limiting sphere, creating a diffracted beam aimed in a different direction from the center of the second sphere toward the point of contact with the sphere. The locus of points on the detector plane which traces satisfaction of the Bragg condition for all intermediate Ewald spheres (intermediate photon energies within the resolution function width) will be a radial streak with increasing length at higher q values.

Separation of streaks along the radial direction on the detector plane requires that the separation between the limiting Ewald spheres in the longitudinal direction in reciprocal space must be smaller than the distance between adjacent reciprocal lattice points in that direction. This distance is just $2\pi/a$, where a is the unit cell dimension in that direction. Thus we require that

$$\Delta q < \frac{2\pi}{a} \tag{3}$$

Given that the momentum transfer q at the Bragg condition is $2\pi/d$ with d being the crystal d-spacing corresponding to the appropriate reciprocal lattice point, we find that

$$\frac{\Delta q}{q} < \frac{d}{a} \tag{4}$$

This essentially says that the limit in relative d-spacing resolution, d/a, is determined by the relative momentum transfer resolution $\Delta q/q$ or (from Eqn. 2) the relative wavelength spread $\Delta\lambda/\lambda$.

This was put to the test in a protein crystallography experiment on X25, with use of the original double multilayer pair which gave a relative wavelength resolution of 1.5×10^{-2} FWHM. A crystallized heterodimer of two eukaryotic transcription factors, TAF42/TAF62, whose structure had already been solved on an NSLS bending magnet beam line [21], was studied at X25 during single-bunch operation of the NSLS x-ray ring, for which the X25 intensity is close to that available on an NSLS bending magnet beam line with a similar optical configuration under normal multi-bunch operation of the x-ray ring. The longest unit cell dimension of this crystal is 111Å, and it therefore should have been feasible to observe resolvable Bragg peaks to a d-spacing resolution as small as about 2Å. Figure 6 shows a diffraction pattern for which the crystal rotation range was 1deg with an exposure time of 5 sec, recorded with a MAR Research 300mm diameter imaging plate camera using a wavelength of 1.4Å and a detector distance of 175mm. Under these conditions, the nominal d-spacing resolution at the perimeter of the plate was about 1.7Å. Achievement of comparable Bragg spot intensities using the bending magnet beam line, equipped with a silicon crystal monochromator, required exposure times of more than 1 min (during normal multi-bunch ring operation). This by itself proves that dramatic increases in intensity can be afforded with the use of a broad bandwidth multilayer monochromator compared with a narrow bandwidth silicon monochromator.

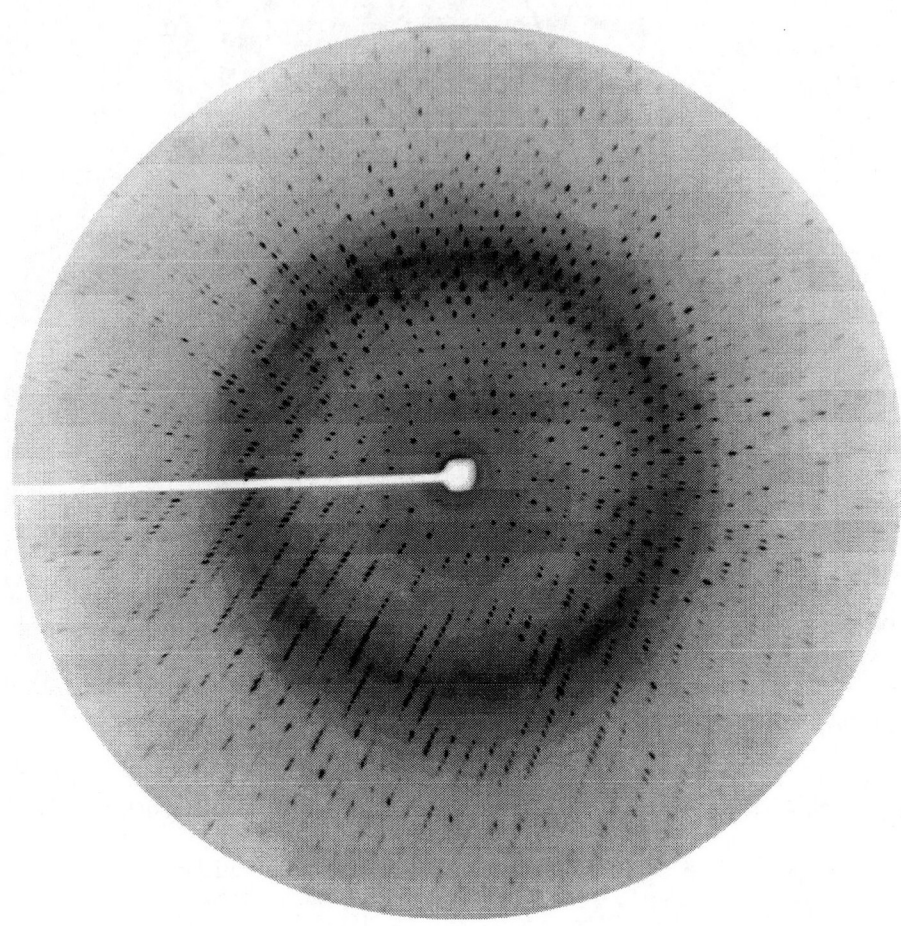

FIGURE 6. Rotation-diffraction pattern from a crystal of TAF42/TAF62, using the multilayer monochromator at a wavelength of 1.4 Å. Note the radial streaks of the Bragg spots, which get longer away from the center of the pattern.

One can see, in Fig. 6, that the Bragg spots are indeed radially streaked, more so at the edges of the pattern than closer to the center. In order to make use of this pattern in a structure solution, the Bragg spots must be indexable and their integrated intensities reduceable. The data reduction must take account of the radial elongation of the spots brought about by the anisotropic reciprocal space resolution function, and be able to separate closely spaced or slightly overlapping spots. The crystallography data reduction computer program MOSFLM [22] was used for this purpose. Two data sets were collected, one for which the minimum d-spacing resolution was 3Å and the other for which it was 2Å. The latter set, being measured to a smaller d-spacing resolution (i.e. to higher q values), was expected to pose more of a data reduction challenge, because of a larger fraction of closely spaced Bragg spots at higher q values. This was indeed confirmed, via comparison of the determined values of R_{sym} for each data set. R_{sym} is a figure of merit that bears upon the uniformity of intensities of symmetry-equivalent Bragg spots; the lower the value of R_{sym}, the better the quality of the data set. The lower resolution (to 3Å) data set gave an R_{sym} of 4.3%, and the higher resolution (to 2Å) set gave an R_{sym} of 9.4%. The lower resolution set is regarded as rather good, whereas the latter might be of questionable value to a structure solution; the 2.4Å resolution data set collected on the bending magnet beam line where the crystal structure was solved gave an R_{sym} of 7.6%. It is clear that use of a multilayer monochromator with a slightly narrower wavelength resolution than for the one employed would have resulted in a superior high-resolution data set. Plans are afoot to repeat the experiment with use of the new multilayer pair presently in hand, which has a relative wavelength resolution of 1.1×10^{-2} FWHM.

V OUTLOOK

Broad bandwidth multilayer monochromators can be employed on synchrotron x-ray beam lines to replace narrow bandwidth silicon or germanium crystal monochromators, resulting in substantial intensity gains (as much as 100) at the expense of photon energy and reciprocal space resolution. Experimental applications that do not require high resolution, such as small angle x-ray scattering and some crystallography, might profit from the use of such monochromators, for they may make studies of weakly diffracting samples feasible. A bending magnet beam line equipped with a multilayer monochromator may deliver a comparable or higher intensity than an insertion device beam line equipped with a silicon monochromator. This has been demonstrated on the NSLS X20C bending magnet beam line for several years, and multilayer monochromators have recently been installed on bending magnet beam lines X27A and X27C. On insertion device beam lines such as X25, a multilayer monochromator could allow the capability to pursue experiments that could not be pursued with use of a silicon monochromator, even with the enhanced brightness of the insertion device.

VI ACKNOWLEDGMENTS

We gratefully acknowledge discussions with and advice from Brian Stephenson, Jim Wood, Jerry Hastings, and Peter Siddons. Technical assistance was provided by Anthony Lenhard, Rick Greene, and Tom Langdon. This work was supported in part by the US Department of Energy under Contract No. DE-AC02-76CH00016. E.D. acknowledges the support of the Natural Sciences and Engineering Research Council of Canada. S.K.B. is an investigator in the Howard Hughes Medical Institute.

REFERENCES

1. Underwood, J.H. and Barbee, T.W., *Appl. Opt.* **20**, 3027-3034 (1981), and references therein.
2. Bilderback, D.H., Lairson, B.M., Barbee, T.W., Ice, G.E., and Sparks, C.J., *Nucl. Instrum. Methods* **208**, 251-261 (1983).
3. This has been demonstrated with a double multilayer monochromator installed on the microfocus undulator beam line at ESRF, described in Deschamps, P., Engström, P., Fiedler, S., Riekel, C., Wakatsuki, S., Høghøj, P., and Ziegler, E., *J. Synchrotron Rad.* **2**, 124-131 (1995).
4. Freund, A.K., *Nucl. Instrum. Methods Phys. Res.* **A266**, 461-466 (1988).
5. Hohlwein, D., Siddons, D.P., and Hastings, J.B., *J. Appl. Cryst.* **21**, 911-915 (1988).
6. X-ray reflectivity calculations for single crystals were made with use of computer software written by the first author. All x-ray reflectivity calculations for multilayers, that are shown in figures in this article, were made with use of computer software operating under the auspices of the Lawrence Berkeley Laboratory Center for X-Ray Optics World Wide Web site, http://www-cxro.lbl.gov.
7. Ziegler, E., *Opt. Eng.* **34**, 445-452 (1995).
8. Stephenson, G.B., *Nucl. Instrum. Methods Phys. Res.* **A266**, 447-451 (1988).
9. Berman, L.E., Hastings, J.B., Oversluizen, T., and Woodle, M., *Rev. Sci. Instrum.* **63**, 428-432 (1992).
10. Ziegler, E., Marot, G., Freund, A.K., Joksch, St., Kawata, H., Berman, L.E., and Iarocci, M., *Rev. Sci. Instrum.* **63**, 496-500 (1992).
11. Song, S., Mochrie, S.G.J., and Stephenson, G.B., *Phys. Rev. Lett.* **74**, 5240-5243 (1995).
12. Dierker, S.B., Pindak, R., Fleming, R.M., Robinson, I.K., and Berman, L.E., *Phys. Rev. Lett.* **75**, 449-452 (1995).
13. J.L. Wood, Product Manager, Osmic, Inc., 1788 Northwood Drive, Troy, Michigan 48084, USA.
14. Smither, R.K., Forster, G.A., Bilderback, D.H., Bedzyk, M., Finkelstein, K., Henderson, C., White, J., Berman, L.E., Stefan, P., and Oversluizen, T., *Rev. Sci. Instrum.* **60**, 1486-1492 (1989).
15. Hart, M., *Nucl. Instrum. Methods Phys. Res.* **A297**, 306-311 (1990).
16. Berman, L.E., *Rev. Sci. Instrum.* **66**, 2041-2047 (1995).
17. Berman, L.E. and Hastings, J.B., *Proc. SPIE* **1739**, 489-501 (1993).
18. Hart, M., Rodrigues, A.R.D., and Siddons, D.P., *Acta Cryst.* **A40**, 502-507 (1984).
19. Berman, L.E., Durbin, S.M., and Batterman, B.W., *Nucl. Instrum. Methods Phys. Res.* **A241**, 295-301 (1985).
20. Tsui, O.K.C., Mochrie, S.G.J., and Berman, L.E., *J. Synchrotron Rad.*, to be published (1997).
21. Xie, X., Kokubo, T., Cohen, S.L., Mirza, U.A., Hoffman, A., Chait, B.T., Roeder, R.G., Nakatani, Y., and Burley, S.K., *Nature* **380**, 316-322 (1996).
22. Written by A. Leslie, and for which treatment of the wavelength dispersion is based on formalism described in Greenhough, T.J. and Helliwell, J.R., *J. Appl. Cryst.* **15**, 338-351 (1982); *ibid* 493-508 (1982); A. Leslie, "Data Collection and Processing," *Proceedings of the CCP4 Study Weekend at Daresbury Laboratory*, 29-30 January 1993, compiled by L. Sawyer, N. Isaac, S. Bailey, pp. 44-51.

Inelastic X-ray Scattering at Modest Energy Resolution

K.D. Finkelstein[1], J.Z. Tischler[2] and B.C. Larson[2]

[1]*Cornell High Energy Synchrotron Source, Cornell Univ. Ithaca, N.Y. 14853*
[2]*Solid State Division, Oak Ridge National Laboratory, P.O. Box 2008,
Oak Ridge, Tenn. 37831-6030*

Abstract

We report results from the development of an inelastic scattering spectrometer designed to take advantage of high energy synchrotron radiation available at CHESS. The device allows a large increase of the effective scattering volume in the sample by permitting measurements to be made in an energy range up to 25KeV. The highest useable energy appears limited by the efficiency of the analyzers under consideration.

At 20KeV a novel 4-bounce, sagittal focusing monochromator passes 10e11 photons/second with Darwin width limited energy resolution. In the scattering plane, the monochromator images the electron beam producing a small scattering source for the analyzing optics. Analyzer systems under study include a cooled mosaic crystal in para-focusing geometry, and an adjustable spherically bent silicon crystal respectively for parallel and point-by-point collection of the energy loss spectrum.

This paper discusses the optical configurations, presents results from our early measurements, and suggests directions for improvements.

INTRODUCTION

Recent experiments demonstrate that inelastic x-ray scattering is a useful tool to study dynamic correlations between the electrons in metals. In a good survey of the field Schulke[1] points out that because they interact weakly, x-rays are well suited for measuring the dynamic structure factor $S(\mathbf{Q},\omega)$. The discussion which follows is concerned with instrumentation for studying valence electron excitations where the momentum transfer, Q is comparable to the inverse distance between electrons, and the energy transfer ω is generally below the excitation energy for inner-shell transitions.

By modest energy resolution we mean that the spectrometer total (incident beam plus analyzer) resolution is on order 2 eV. Significant development efforts, like those at the NSLS beamline X-21[2], can yield energy resolution below 1eV, but the maximum energy available is well below the goal for this work. At high incident energy, the monochromator energy width limits resolution unless one is willing to suffer a severe loss in intensity. However, experience with aluminum shows that spectral features associated with many-body effects are broadly distributed in energy loss[3]. In addition, as the atomic number rises, more of the scattering is contributed by localized electrons which generally results in a broadening[4,5] of plasmon and higher energy components.

Most previous x-ray work has focused on simple metals with atomic number less than or equal to 13 (aluminum). A considerably wider range of materials have been examined by electron energy loss spectroscopy (EELS)[6]. Because of multiple scattering effects these data are difficult to interpret at large Q; the regime essential for understanding short-ranged exchange correlations.

The following sections describe the spectrometer illustrated in Figure (1). First we discuss the monochromator, then experimental considerations related to the sample and incident beam, and then the analyzer. We conclude by discussing future directions for this project.

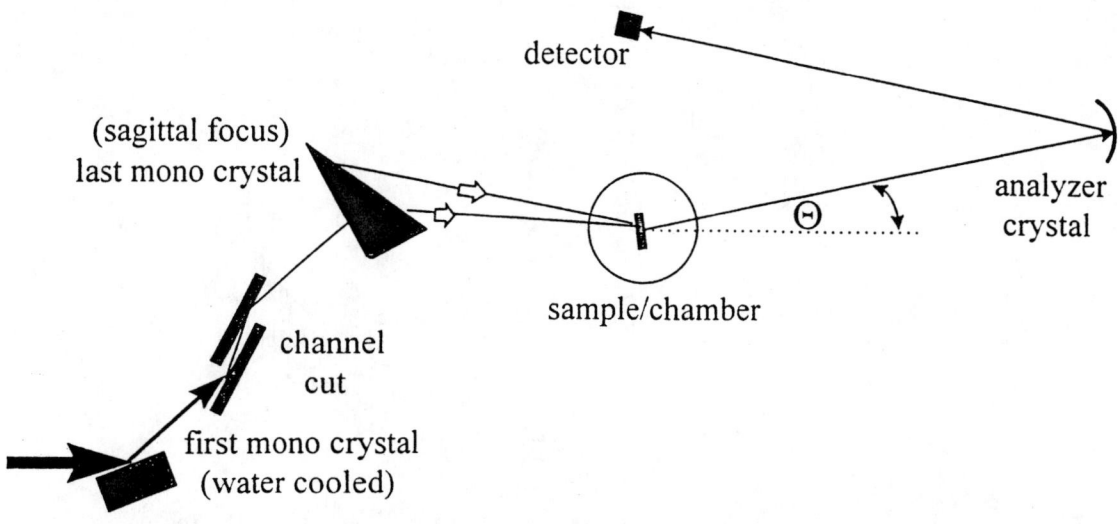

FIGURE 1. Schematic illustration of the inelastic scattering x-ray spectrometer discussed in the text. The 4-bounce monochromator is composed of silicon (111) crystals. Rays leaving the sagittal focusing crystal are horizontally focused at the sample. The scattered signal is collected by a spherically bent silicon analyzer near-backscattering angles.

THE MONOCHROMATOR

Given the characteristics of the CHESS source and the required incident beam energy resolution, we have chosen a modified sagittal focusing silicon (111) monochromator to maximize flux at the specimen. A standard 2-bounce monochromator has an energy resolution which depends on the vertical angular spread of the beam. At 20KeV, the 0.5mm (FWHM) CHESS C line source supplies approximately 90µradians of divergence producing an energy width of about 18eV. This width is reduced by insertion of a monolithic silicon (111) channel cut crystal, in anti-parallel orientation between the water-cooled first and the sagittally bent (last) crystal. The channel cut passes only the Darwin width in angle, producing a beam with energy (full) width equal to about 2.5eV. The 4-bounce throughput is about 11% of the 2-bounce monochromator, in agreement with the product of the reduction in bandwidth and the measured channel-cut reflectivity (approximately 80%). The energy band-pass can be further reduced (as much as a factor of two at the expense of intensity) by detuning the last crystal relative to the first. Because the channel-cut preserves the direction of rays from the first crystal, one could further reduce the energy width by choosing a smaller d-spacing reflection for the channel cut. We have chosen instead to maximize the intensity by using (111) reflections.

Figure (2) shows the channel cut designed to operate between 17.5 and 24 KeV. It passes the full vertical beam and up to 65mm (approximately 6 milli-radians) of horizontal bend magnet radiation. It was fabricated using a 16mm thick disk cut from a 100mm diameter boule of dislocation free, float zone silicon with resistivity > 1000 Ohm-cm.

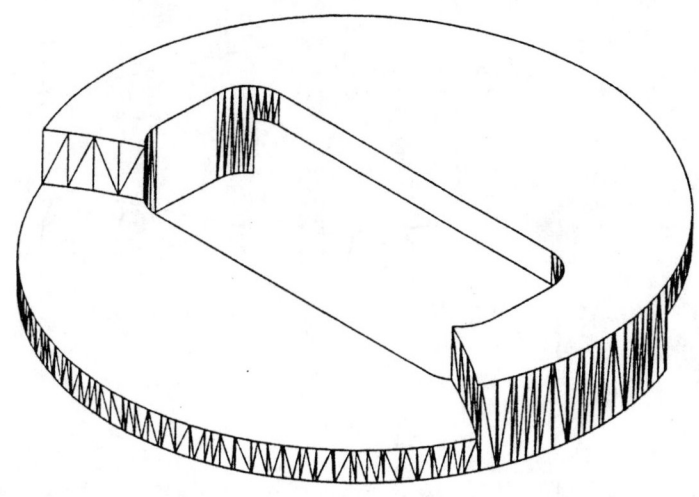

FIGURE 2. Design of the monolithic channel-cut crystal is shown. The outer diameter is 100 mm and the vertical offset (gap) between diffracting surfaces is 4 mm. The beam is incident on the left surface. A 65mm wide beam can diffract at between 4.7 and 6.5 degree Bragg angles.

Without the channel cut, at 20KeV and 150mA particle beam current, the double crystal monochromator focuses slightly over 10e12 photons/sec in a spot size approximately 1.2mm vertical by 2mm horizontal. Insertion of the channel cut reduces both the intensity <u>and</u> the vertical spot size to less than 1mm. This second effect is a reasonable consequence of the high angular selectivity of the 4-bounce system. By passing only the Darwin width in vertical angle, the channel cut acts to create an image of the source at the sample. Another useful characteristic of the anti-parallel arrangement is that harmonics in energy are rejected because they are shifted in angle relative to the fundamental due to a difference in index of refraction.

The channel cut is positioned upstream of (rather then after) the sagittally bent crystal to avoid a loss in throughput due to an increase in vertical angular width that results from the focusing. This effect becomes significant when a wide horizontal beam is focused at high energy (small Bragg angle) because the tight sagittal bend radius.

This scheme, as illustrated in Figure (1), does not allow a simple means to perform energy scanning. This problem will be addressed by modifications discussed in the last section.

THE SAMPLE AND INCIDENT BEAM CONSIDERATIONS

We are interested in scattering measurements from materials with atomic numbers up to about 50, and K absorption edge energies to above 25KeV. To limit absorption and fluorescence it will be an advantage for the incident energy to be just below, or well above the absorption edge. Optimizing the sample and incident beam for inelastic scattering involves considerations quite similar to those for diffraction from mosaic crystals[7]. In this case, absorption limits the scattering and the signal is inversely proportional to the absorption coefficient (μ). For symmetric reflection from (through) a surface, the signal depends on $1/2\mu$ (e^{-1}/μ) and Table (1) indicates the benefit (in some cases the necessity) of working at higher energy.

TABLE 1. The absorption length which increases with x-ray energy is given. The scattered signal is limited by absorption and therefore inelastic measurements benefit from high incident energy. The underlined value at 24KeV is just below the k-absorption edge in palladium.

Energy	scandium	palladium
8KeV	17.6μm	4.2μm
12	54.9	12.8
16	126	28.5
20	242.5	52.8
24	416.1	<u>86.9</u>
28	659	20.3

Generally measurements are made using single crystals to probe well defined directions in the sample and also because diffuse scattering surrounding Bragg peaks is easier to avoid. An appropriate choice to scatter in transmission or reflection geometry may lead to significant practical benefits. In reflection, thickness is not important so sample preparation is simplified. However, at shallow angles of incidence, the sample must be accurately lined up with the beam. A beam spot that walks across the sample with changes in orientation can result in shifts of the analyzer energy scale. For transmission geometry, the optimum sample thickness is $\cosine(\Theta/2)/\mu$ (where Θ is the scattering angle)[7] making sample thickness critical. But in transmission alignment is easier, and one can access multiple directions in reciprocal space simply by rotating about the surface normal.

Other important considerations related to the sample include: minimizing the beam spot size on the sample and accounting for changes in scattering volume and beam footprint as seen by the analyzer. Slits placed downstream of the sample can limit the detected scatter from air and the sample chamber walls. Finally, it may be useful to include the capability to control, and in particular to lower the sample temperature.

THE ANALYZER SYSTEM

Inelastic scattering measures the double differential cross section ($d^2\sigma/(d\Omega d\omega)$) for x-rays scattered by electrons. This is proportional to the product of the usual Thomson cross section and the dynamical structure factor. In contrast to Bragg diffraction, this structure factor has no sharp angular dependence so the scattering per unit solid angle is small and the analyzer/detector angular definition should be large enough to gather as much signal as possible within the Q-resolution required. On the other hand, the analyzer must have an energy resolution comparable to the incident beam so that the total resolution is sufficient to resolve features of the valence electron excitations. For our purposes, a total energy resolution of the system should of order 2eV.

Mosaic crystals, used in parafocusing geometry[3] can disperse the energy loss spectrum in position so that a low noise position sensitive detector (such as a position sensitive proportional counter or PSPC) can collect the data in parallel. However, we have found the efficiency of the mosaic crystal analyzer to be compromised at higher energy because of low reflectivity. Following the analysis by Ice and Sparks[8] one finds that high order reflections are required to attain the appropriate energy resolution. A rule of thumb for mosaic graphite crystals is that with a 1 meter focal length a dispersion of several eV/mm requires a reflection (0,0,L), where L is even and approximately equal to the energy in KeV.

In Figure (3) we use the analysis by Freund[9] to calculate the peak reflectivity for a series of reflections in graphite. The energy dependence is weak, and by using our rule of thumb (stated above) we find that the analyzer efficiency degrades significantly with increasing energy.

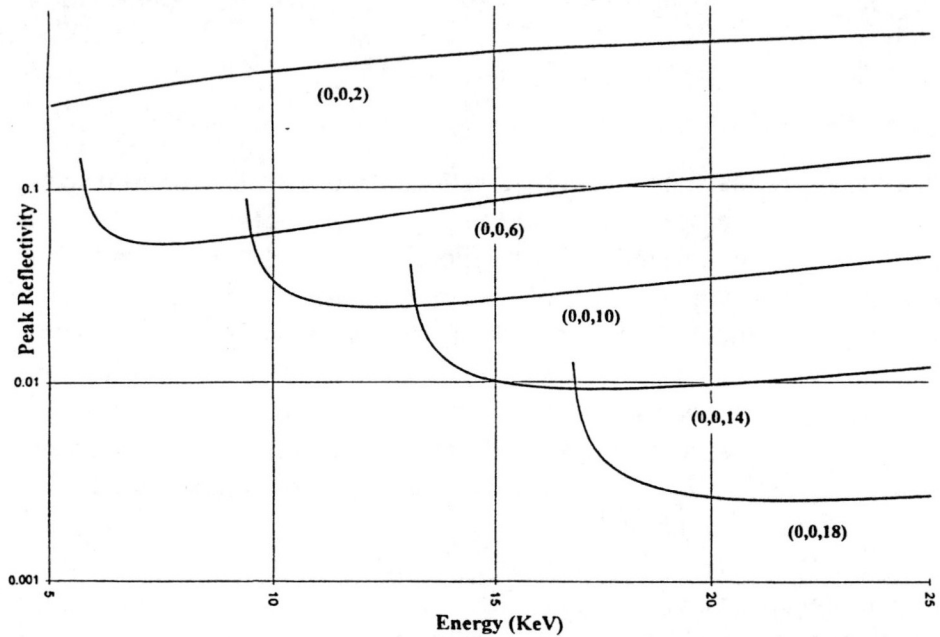

FIGURE 3. The peak reflectivity verses energy calculated for several pyrolytic graphite reflections at room temperature. Crystal thickness is 3mm. The upward tails at low energy correspond to Bragg angles approaching 90 degrees. The calculation is based on theory[9] which does not sufficiently account for diffraction near back-scattering angles.

For high order reflections, this situation can be improved somewhat by cooling the analyzer crystal. Using the thermal factor in [9], the (0,0,18) peak reflectivity will increase by a factor of 5 at liquid nitrogen temperature, however it will remain less than about 0.02.

One may instead choose a more perfect analyzer crystal, but sacrifice the benefits of parallel data collection. We are experimenting with spherically bent silicon crystals which have long been used for high resolution spectroscopy[10]. These crystals can have unit reflectivity, but only for highly monochromatic radiation because the bandwidth accepted (near back reflection) is very small. A general expression for the bandwidth, given in Equation (1), includes contributions from the Darwin width, strain (or variation in lattice constant), and angular size of the incident radiation.

$$\Delta\lambda/\lambda = 4r_e d^2 F/(\pi V) + \Delta d/(2d) + \Delta\theta \cot(\theta), \tag{1}$$

where λ is the x-ray wavelength, d, F, and θ are the d-spacing, structure factor and Bragg angle for the reflection under consideration. V is the unit cell volume and r_e is the classical electron radius while $\Delta d/d$ is the strain in the crystal and $\Delta\theta$ is the incident beam angular size. For example, without strain the silicon (999) reflection will pass only about 30meV at 18KeV.

We are currently examining the possibility of tailoring the crystal strain to better match this bandwidth to the required resolution. In this effort, we have designed and tested an apparatus using the uniform force applied by a vacuum created behind a thin wafer supported on its edges by an o-ring. The bender/crystal holder illustrated in Figure (4), reproducibly bends a 100mm diameter, 250micron thick silicon wafer to a 0.75 meter radius of curvature. An air piston and mercury manometer are used to set and measure the vacuum force. This system will be reported on in detail in a future publication.

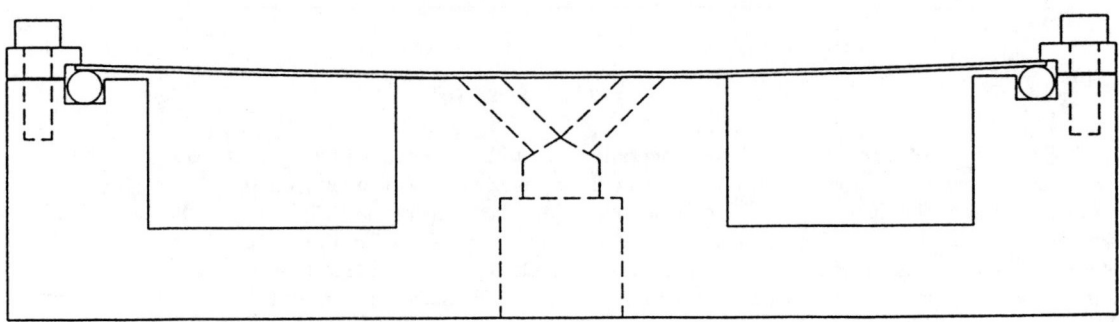

FIGURE 4. A cut-away view of the spherical bender discussed in the text. The force of vacuum is used to bend a 100mm diameter crystal of thickness ¼ micron to a 0.75 meter radius. The crystal disk sits on an o-ring forming a volume that is evacuated and held at controlled pressure.

An example of data collected on scandium metal may be used to illustrate the first measurements made with this system. The energy loss spectrum is collected after the following setup. The incident energy is calibrated, and the beam focused on a thin polystyrene film placed at the sample position in an evacuated chamber. The sample, the silicon (0,0,16) analyzer and the detector are symmetrically placed on a Rowland circle. The analyzer curvature is set to a minimum radius, and the vacuum force (and curvature) reduced until a minimum in rocking curve width is recorded as the analyzer and detector are scanned in $\Theta-2\Theta$ fashion. The metal sample then replaces the plastic and a series of $\Theta-2\Theta$ scans are summed to produce the raw inelastic scattering spectrum in Figure (5). A series of measurements, over a range of Q-values are performed by adjusting the orientation of the sample and analyzer arm.

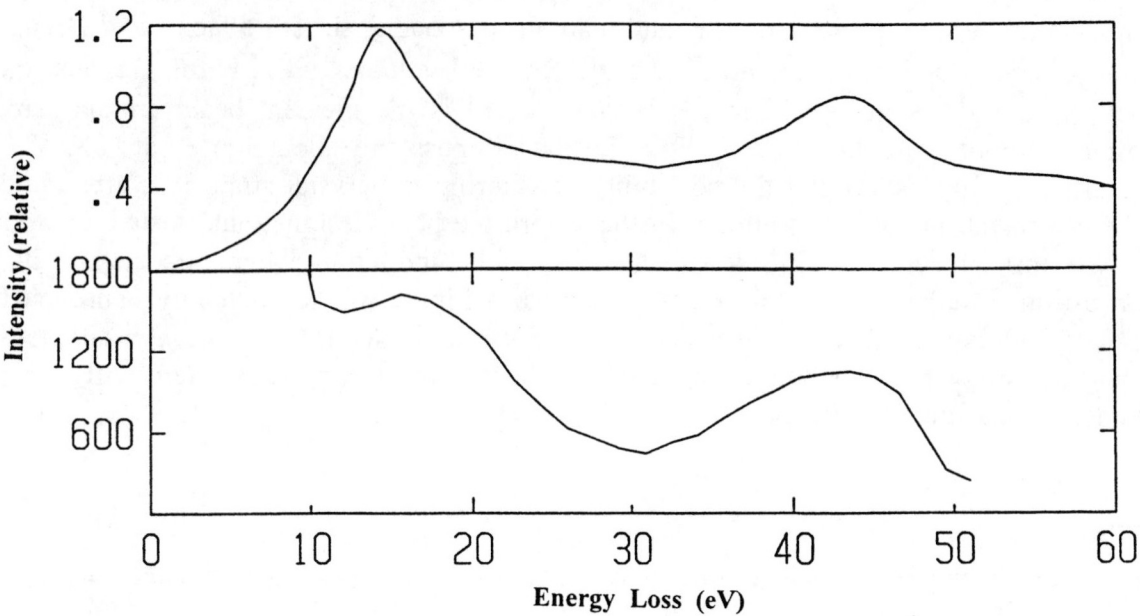

FIGURE 5. The lower panel shows an X-ray energy loss spectrum accumulated by repeated scans of the analyzer and detector as described in the text. 18 scans, accumulated over about 12 hours have been summed. The sample is scandium metal, the incident beam energy is 18.39KeV, and the scattering angle is 7 degrees. For comparison, an EELS spectrum, taken from reference 11, is plotted in the upper panel. The position, in energy loss, and relative size of three features discussed in that reference are in reasonable agreement with the x-ray data. We have not removed the quasi-elastic peak (centered at zero energy loss) in the x-ray data.

The data were collected at a constant incident x-ray energy of 18.39KeV, at a scattering angle of 7 degrees corresponding to Q = 1.14 Angstrom^{-1}. The upper and lower panels in Figure (5) compare the dynamic structure factor extracted from EELS measurements near Q = 0[11] with the raw data collected with this new x-ray spectrometer. No attempt has been made to remove the quasi-elastic scattering that is centered at zero x-ray energy loss.

FUTURE DIRECTIONS AND CONCLUSIONS

Immediate improvements to the spectrometer should come from work in several areas. First, we are building a Θ–2Θ arrangement to control the motion of the monochromator first crystal and channel cut. Automating the alignment of the first two optical elements will make energy scanning straight-forward. The 4-bounce monochromator will provide a fixed-offset beam height by a simple adjustment of the position and angle of the last crystal.

Further improvement will come by control of the change in lattice constant (strain) as the scattered beam penetrates into the analyzer. At the energies and back-scattering angles under consideration, the extinction depth for high order reflections in silicon is greater than one hundred microns and the absorption length is many times larger. In this regime of diffraction, x-rays behave much like neutrons. Klar and Rustichelli[12] have calculated the behavior of neutrons in ideally curved and strained crystals. They show that reflection widths may be significantly increased, without sacrificing reflectivity, in thick and uniformly strained crystals.

In our case, this would mean larger bandwidth accepted by the analyzer, and an increase in spectrometer efficiency. We are presently studying this effect using crystals of varying thickness, by measuring reflectivity as the bend radius is adjusted.

In conclusion, we are developing, testing and improving an inelastic x-ray scattering spectrometer designed to use high energy synchrotron radiation available at CHESS. This work has attempted to optimize the system in three critical areas. A novel 4-bounce monochromator is used to produce high intensity, focused radiation with Darwin width equivalent energy resolution. By using high energy x-rays, scattering from the sample increases as (absorption length)$^{-1}$. Finally we have built a continuously adjustable, spherically bent focusing analyzer and are presently examining the possibility of increasing its efficiency for collecting the scattered signal.

Acknowledgments

We gratefully acknowledge the CHESS staff without whose assistance this work could not have been accomplished. We further thank Arthur Moore and Advanced Ceramics Corp., Lois Pollack of the Cornell MicroKelvin Laboratory, Gene Ice of Oak Ridge National Laboratory, and Ron Kemp of Cornell Physics Technical Operations Lab for their material contributions to this project. Finally we acknowledge B.W. Batterman for his support and useful discussions. This work is supported by the National Science Foundation through CHESS, under Grant No. DMR-93-11772.

REFERENCES

1. W. Schulke, Chapter 15, pg. 565 in *Handbook on Synchrotron Radiation, Vol. 3*, editors G.S. Brown and D.E. Moncton, North-Holland, Elsevier Science Publishers B.V. (1991).
2. W.A. Caliebe, C.-C. Kao, M. Krisch, T. Oversluizen, P. Montanez, and J.B. Hastings, see article in this volume.
3. B.C. Larsen, J.Z. Tischler, E.D. Isaacs, P. Zschack, A. Fleszar, and A.G. Eguiluz, Phys. Rev. Lett. 77, 1346 (1996).
4. D.L. Misell and A.J. Atkins, Philos. Mag. 27, 95 (1973).
5. C. Wehenkel and B. Gauthe, phys. Stat. Sol. (b) 64, 515 (1974).
6. H. Raether, *Excitations of Plasmons and Interband Transitions by Electrons*, Springer, Berlin (1980).
7. R.W. James, Chapter 2, *The Optical Principles of the Diffraction of X-Rays*, Ox Bow Press, Woodbridge, Connecticut (1982).
8. G.E. Ice and C.J. Sparks Jr., Nuc. Instr. and Meth. A291, 110 (1990).
9. A.K. Freund, Nuc. Instr. and Meth. A266, 461 (1988).
10. V. Stojanoff, K. Hamalainen, D.P. Siddons, J.B. Hastings, L.E. Berman, S. Cramer, and G. Smith, Rev. Sci. Instrum. 63 (1), 1125 (1992).
11. B. Brousseau-Lahaye, C. Colliex, J. Frandon, M. Gasgnier, and P. Trebbia, Phys. Stat. Sol. (b) 69, 257 (1975).
12. B. Klar and F. Rustichelli, Il Nuovo Cimento, 13b, 249 (1973).

X-ray Monochromators With Sub-MeV Resolution*

T.S. Toellner, M.Y. Hu, W. Sturhahn,
P. Hession, E. Alp, and J. Sutter

Argonne National Laboratory
Advanced Photon Source
9700 S. Cass Ave., Argonne, IL 60439

Recent developments in tunable x-ray monochromators allow one to achieve sub-meV energy resolutions in the 10-30 keV energy range. These monochromators are designed to take advantage of the high-brightness of undulators to efficiently extract the available spectral flux. The design and testing of such high-resolution monochromators will be presented along with an application that uses nuclear resonant scattering to measure lattice excitations with sub-meV resolution.

This work is supported by U.S. Department of Energy, BES Materials Sciences, under contract W-31-109-ENG-38.

Test Results of a Diamond Double-crystal Monochromator at the Advanced Photon Source

P.B. Fernandez, T. Graber, S. Krasnicki, W.-K. Lee, D.M. Mills, C.S. Rogers, and L. Assoufid

Advanced Photon Source, Argonne National Laboratory, Argonne, IL 60439

Abstract. We have tested the first diamond double-crystal monochromator at the Advanced Photon Source (APS). The monochromator consisted of two synthetic type 1b (111) diamond plates in symmetric Bragg geometry. We tested two pairs of single-crystal plates: the first pair was 6 mm by 5 mm by 0.25 mm and 6 mm by 5 mm by 0.37 mm; the second set was 7 mm by 5.5 mm by 0.44 mm. The monochromator first crystal was indirectly cooled by edge contact with a water-cooled copper holder. We studied the performance of the monochromator under the high-power x-ray beam delivered by the APS undulator A. We found no indication of thermal distortions or strains even at the highest incident power (280 watts) and power density (123 W/mm^2 at normal incidence). The calculated maximum power and power density absorbed by the first crystal were 37 watts and 4.3 W/mm^2, respectively. We also compared the maximum intensity delivered by the diamond monochromator and by a silicon (111) cryogenically cooled monochromator. For energies in the range of 6 to 10 keV, the flux through the diamond monochromator was about a factor of two less than through the silicon monochromator, in good agreement with calculations. We conclude that water-cooled diamond monochromators can handle the high-power beams from the undulator beamlines at the APS. As single-crystal diamond plates of larger size and better quality become available, the use of diamond monochromators will become a very attractive option.

I. INTRODUCTION

The x-ray beams delivered by the undulators at the Advanced Photon Source can have a total power of several kilowatts, with a peak power density of 160 W/mm^2 at 30 m from the source. To preserve the brilliance of these beams, the first optical components of the x-ray beamline have to be designed to work under the extreme power loads without showing significant thermal distortions. Thermal management approaches that have been implemented in the design of the monochromator first crystal include: the use of crystals with internal cooling geometries that are optimized for maximum heat transfer (1); the use of more efficient cooling fluids, such as liquid gallium (2); modifications of the symmetric Bragg reflection geometry to reduce the power density on the crystal (3); and improvement of the thermal properties of the diffracting material, for example by using cryogenically cooled silicon crystals (4) or room-temperature single-crystal diamonds (5).

At room temperature, the thermal conductivity of diamond is about ten times larger than that for silicon, while the linear expansion coefficient of diamond is two times smaller. For the same absorbed power, cooling geometry and coolant, we expect that the temperature gradients and the thermal distortions in diamond will be considerably less than in silicon. Another advantage is that the absorption of x rays in diamond is less than in silicon: a 0.25-mm-thick diamond crystal will absorb 30% of 8 keV x rays, while a silicon crystal of the same thickness will absorb 97%. Thus, a thin diamond crystal will absorb a smaller fraction of the incident synchrotron beam, with considerably smaller thermal gradients and strain compared to a similarly cooled silicon crystal (6). Also linked to the low absorption is the possibility of beam multiplexing, i.e., allowing the incident white beam to be used at several experimental stations (5, 7). The use of single-crystal diamonds as high-heat-load monochromators has been implemented at the European Synchrotron Radiation Facility (7) and is planned for beamlines at SPring-8.

While the thermal characteristics of diamond are superior to those of silicon, the opposite is true for the x-ray diffraction performance. For energies above 6 keV, the photon flux delivered by a double-crystal diamond (111) monochromator will be about two times lower than the flux from a silicon (111) monochromator. The reduction in flux is due to the smaller Darwin width and lattice constant, which result in a narrower energy bandpass. This reduction may be an acceptable trade-off when the ease of cooling a diamond crystal is considered and/or when beams with a narrow energy width are required. Another drawback of using diamond crystals is the current unavailability of perfect single crystals of appropriate size. The largest commercially available plates at this point are 7 mm by 5 mm (8). These plates are cleaved from synthetic type 1b stones and then are ground and polished to the desired thickness. The plates typically exhibit several arcseconds of mosaic spread and/or strain.

In this paper we describe the test of a high-heat-load double-crystal diamond (111) monochromator in Bragg reflection geometry. The tests were carried out on the Sector 1 insertion device beamline at the Advanced Photon Source, using x-ray beams produced by the 2.4-m-long undulator A. This beamline is operated by the Synchrotron Radiation Instrumentation Collaborative Access Team.

II. EXPERIMENTAL SETUP

Our double-crystal monochromator consisted of two type 1b, synthetic (111) diamond plates. The plates were manufactured by Drukker International, from synthetic stones grown by De Beers, and they were supplied by Harris Diamond Corporation (8). We tested two sets of plates. Plates in the first set were 6 mm by 5 mm in size; the first crystal was 0.25 mm and the second 0.37 mm thick. Plates in the second set were 7 mm by 5.5 mm in size, and both crystals were 0.44 mm thick. All the plates had a slight asymmetry of the (111) planes, ranging from 0.99 to 3.16 degrees. We assessed the quality of the diamonds by taking x-ray topographs at 8 keV. The data indicate that the crystals are not perfect and that the mosaic spread/strain is of the order of 5 or 6 arcseconds over the full face of the plates. Residual strain in the crystal is due to impurities that can distort the perfect lattice, generate variations in the lattice spacing, and cause dislocations. The mosaic spread arises from the variation of the lattice orientation between the different growth sectors (9, 10). While the added mosaic spread/strain can result in a loss of brilliance in the diffracted undulator beam, it can also increase the bandpass for applications in which the flux is important, thus making up for some of the loss in throughput compared to a Si(111) monochromator.

For our first run (May 1996), the diamond first crystal straddled a 2-mm-wide trough on a water-cooled copper block, which resulted in a 2-mm-wide by 6-mm-long area available for diffraction. For the second set of tests (January 1997), we increased the trough in the copper block to 3 mm; a sketch of the first crystal mount is shown in Fig. 1. The thermal contact between the diamond and the copper was achieved by using a thin layer of Ga/In eutectic (80% gallium, 20% indium); the crystals were held in place by the surface tension of the eutectic layer.

The distance from the undulator x-ray source to the first crystal was 32.6 m for the May 1996 run and 29.4 m for the January 1997 run. White beam slits were located at 26.8 m and defined the beam size at the first crystal position: 1.4 mm horizontal by 1.8 mm vertical for the first run, and 2 mm horizontal by 1.2 mm vertical for the second run. Through these apertures, we accepted 62% and 83% of the undulator central cone of radiation in the horizontal plane, for the first and second runs, respectively, and over 97% in the vertical plane in both cases. An x-ray window was located between the front end and the beamline, and transmitted over 86% of the power produced by the undulator at a gap of 11 mm. The window consisted of a 0.5 mm (May 1996) or 0.3 mm (January 1997) graphite filter, followed by a 0.17 mm diamond window and two 0.25 mm beryllium foils. This window assembly was the only filter in the beam upstream of the monochromator.

Two ion chambers downstream from the monochromator vacuum tank recorded the intensity of the diffracted beam. The first ion chamber was filled with helium at atmospheric pressure; space charge effects from the undulator beam prevented operation in air for this counter. The second detector operated in air, behind an aluminum filter that absorbed x rays from the first order reflection of the (111) diamond crystals. Thus, the second ion chamber recorded the intensity of the higher order reflections, mainly diamond (333). Figure 2 shows sample rocking curves for the (333) reflection, taken during our first run. The discrepancy between the theoretical and measured width of the rocking curves is due to the mosaic spread/strain of the diamond crystals. The measured width of the (333) reflection is a direct indication of the quality of the region of the crystals being sampled by the beam. For example, from the experimental (333) widths (see Fig. 3 below), we deduce that the mosaic

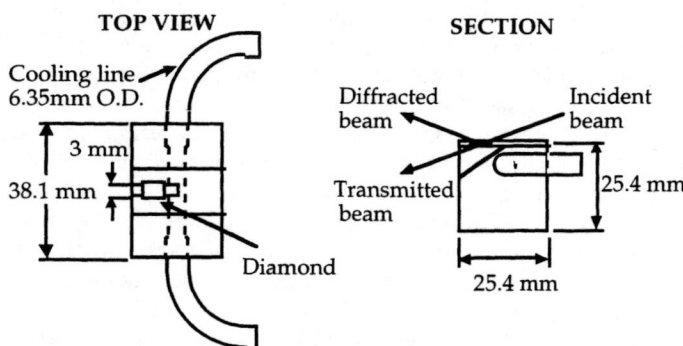

FIGURE 1. Top view and section of the water-cooled first crystal mount of our diamond double-crystal monochromator. The copper block was nickel plated to prevent the diffusion of the Ga/In eutectic into the copper. The trough in the block allows the transmission of the incident x-ray beam.

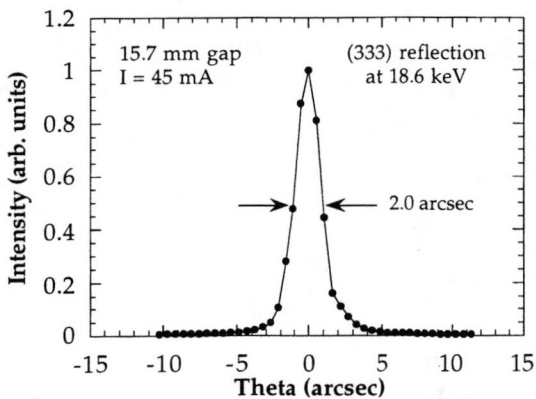 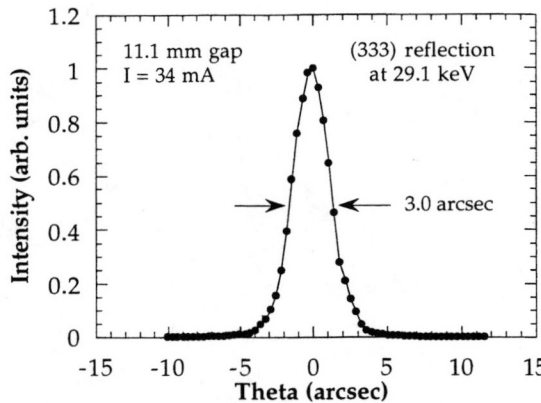

FIGURE 2. Sample double-crystal rocking curves for the diamond (333) reflection (May 1996 data). The theoretical widths for the (333) reflection are 0.55 arcsec at 18.6 keV and 0.33 arcsec at 29.1 keV. The undulator gap and the storage ring current are indicated.

spread/strain is 2 to 4 arcseconds for the central 1.4-mm-wide strip of the crystals used in our first run. This measured widening of the (333) rocking curve is consistent with the 5 or 6 arcseconds of mosaic spread/strain observed over the full face of the crystals, as measured with the topography setup.

III. DATA

The purpose of our experiments was to test the diamond double-crystal monochromator under the high-power undulator beam. We typically gauge the performance of a cooled first crystal by tracking changes in the FWHM of the rocking curve of the diffracted beam as a function of the power incident on the crystal. The incident power can be increased by closing the undulator gap or increasing the storage ring current. Thermal stress and deformation in the crystal will result in an increase of the FWHM of the rocking curve.

We calculated the expected thermal deformation of the monochromator first crystal. We modeled our experimental configuration and used finite element analysis (FEA) to predict the thermal performance of the cooled diamond crystal at E = 17 keV, with the undulator gap at 11 mm and storage ring current I = 100 mA, for the parameters of the January 1997 run. The FEA calculation predicted a maximum temperature difference of 0.4 °C along the tangential direction (along the beam direction) and of 0.2 °C through the thickness of the crystal. The bowing and strains induced in the crystal by these temperature gradients resulted in a maximum tangential slope error (local bending of the diffraction planes) of 0.8 arcseconds. This slope error is much smaller than the theoretical (111) rocking curve width of 4.3 arcseconds, but larger than the calculated 0.23 arcseconds width for the (333) reflection at 51 keV. Because the thermal broadening would add in quadrature to the measured 2 to 4 arcseconds of mosaic spread/strain in the diamond crystals (5), the FEA results indicated that we should not see a resolvable thermal broadening of the double-crystal rocking curve.

In our first run, we took data at two (111) diffraction energies, E = 6.2 and 9.7 keV, and several undulator gaps, 11.1, 15.7, and 21 mm. The corresponding first harmonic energies at these gaps were 3.3, 6.2, and 9.7 keV. For E = 6.2 keV, we took data on the second harmonic at closed gap (11.1 mm) and on the first harmonic at 15.7 mm, while for E = 9.7 keV, we used the third harmonic at 11.1 mm and the first harmonic at open gap (21 mm). The measured maximum power incident on the first crystal was 200 watts, with a calculated power density of 108 W/mm^2 (normal incidence), at 11.1 mm gap and 90 mA. The corresponding calculated power and power density absorbed by the first crystal were 11.6 watts and 1.8 W/mm^2, respectively, at 9.7 keV. The power incident on the first crystal is largest for the 11.1 mm gap configuration, but the power absorbed in the diamond is largest when the undulator gap is 21 mm. This counter-intuitive effect is due to the spectral distribution of the x-ray beam and to the fact that the diamond transmits most of the photons that have energies above 20 keV. The calculated maximum power and power density absorbed by the first crystal were 14.5 watts and 2.4 W/mm^2, respectively, at 21 mm gap and 96.5 mA, for E= 9.7 keV.

Figure 3 shows the FWHM of the rocking curve of the diffracted beam as a function of energy, for the (111) and (333) reflections. The theoretical double-crystal rocking curve widths, calculated for perfect single crystals, are also shown. The scatter in the data is due to an unexplained relative motion between the x-ray beam and the crystals; the extreme sensitivity to the position of the beam is due to the less-than-perfect crystalline quality of the diamonds. An increase in the width as a function of increasing incident power (closing undulator gap or increasing ring current) would indicate a worsening

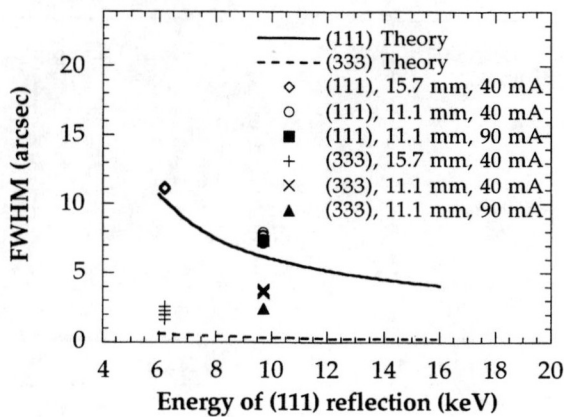

FIGURE 3. Measured and calculated values for the FWHM of the diamond double-crystal rocking curve as a function of energy of the x-rays diffracted from the (111) planes (May 1996 data). Data were taken simultaneously for the (111) and (333) reflections; the energy of the x-rays diffracted by the (333) planes is three times the abscissa value. The undulator gap and the storage ring current are indicated.

performance of the monochromator. The data do not show such an increase. Thermal broadening of the rocking curve, if present, is probably less than 1 or 2 arcseconds.

In our second run, we measured the widths of the diffracted beam as a function of energy for a fixed gap (11 mm), see Fig. 4, and we also changed the undulator gap so that the diffraction energy corresponded to either the first or third harmonic of the undulator spectrum, see Fig. 5. At some energies, data were taken at two different undulator gaps; the gap is smaller when the energy coincides with the third harmonic. In Fig. 5, the first harmonic gaps range from 17 to 26.5 mm, and from 11 to 15 mm for the third harmonic. The maximum power and power density (normal incidence) incident on the first crystal were 280 watts (measured) and 123 W/mm^2 (calculated), respectively, at 11 mm gap and 86 mA. The calculated maximum power and power density absorbed by the first crystal were 37 watts at 17 keV (11 mm gap and 82 mA), and 4.3 W/mm^2 at 8 keV (18.5 mm gap and 77 mA), respectively. The scatter in the data in Fig. 4 and 5 is again due to a small motion in the incident x-ray beam relative to the imperfect diamond crystals. The data are encouraging from a thermal perspective, because there is not an observable increase in the width of the (333) rocking curve with increasing energy (increasing absorbed power) or decreasing undulator gap (increasing incident power).

We were also interested in comparing the flux through our diamond (111) monochromator and a silicon (111) double-

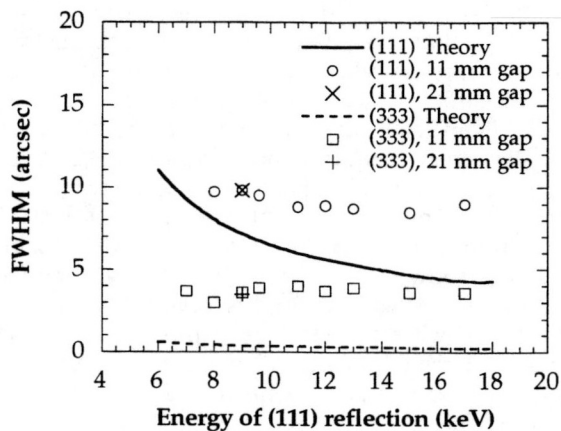

FIGURE 4. Measured and calculated values for the FWHM of the diamond double-crystal rocking curve as a function of energy of the x-rays diffracted from the (111) planes (January 1997 data). Data were taken simultaneously for the (111) and (333) reflections; the energy of the x-rays diffracted by the (333) planes is three times the abscissa value. The storage ring current ranged from 87 to 82 mA.

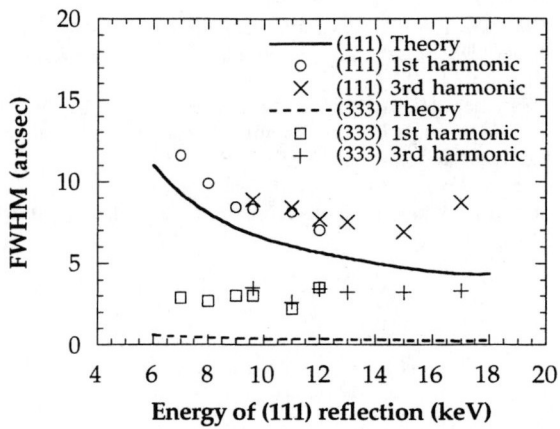

FIGURE 5. Measured and calculated values for the FWHM of the diamond double-crystal rocking curve as a function of energy of the x-rays diffracted from the (111) planes (January 1997 data). Data were taken simultaneously for the (111) and (333) reflections; the energy of the x-rays diffracted by the (333) planes is three times the abscissa value. For a given energy, the gap setting of the third harmonic datum is smaller than that of the first harmonic point. The storage ring current ranged from 81 to 77 mA.

crystal monochromator. For this comparison, we used a liquid-nitrogen-cooled silicon monochromator during the May 1996 run. This silicon monochromator had been successfully tested at the same beamline and had shown excellent performance under high thermal loads (11). Using the same beam size and undulator gap, we measured the maximum intensity of the (111) reflection at 6.2 and 9.7 keV for both monochromators. We found the maximum intensity of the silicon monochromator to be 1.9 and 1.4 times that of the diamond monochromator at 6.2 and 9.7 keV, respectively. The theoretical calculations for perfect crystals (6) predict a ratio of 1.6 at 6.2 keV and 1.9 at 9.7 keV. The reversed trend between the measured and calculated ratio may be explained as follows. At 9.7 keV, the imperfections in the diamond crystals account for a larger fraction of the double-crystal rocking curve width than at 6.2 keV. The reduction in the measured ratio at the higher energy is then due to a larger percent increase in flux for the diamond monochromator at 9.7 keV compared to 6.2 keV. The flux measurements are in reasonable agreement with the calculations and confirm the expected loss in monochromatic beam flux.

IV. CONCLUSIONS

We have installed and tested the first double-crystal diamond monochromator at the APS. Our data indicate that this water-cooled monochromator will perform well even under the highest heat loads on an undulator beamline. The drawbacks are the inherent reduction in flux compared to a silicon monochromator and the lack of absolutely perfect diamond single crystals of appropriate size. Even so, a diamond monochromator might prove a good choice for certain applications, for example, if a narrow energy bandpass or beam multiplexing is desired.

ACKNOWLEDGMENTS

We wish to thank Mr. Joseph Maj for his assistance in topographing the diamonds, Mr. Al Paugys for his help in assembling the first crystal mount, Mr. Mark Keeffe and Mr. Bill McHargue for their help in configuring the beamline for these tests, and Dr. Dean Haeffner for his assistance on beamline matters. We also want to acknowledge the assistance of Mr. Mark Fraser (Harris Diamond Corporation) in obtaining the diamond plates used in these tests. This work was supported by the U.S. Department of Energy, BES-Materials Sciences, under contract number W-31-109-ENG-38.

REFERENCES

1. John Arthur, *Optical Engineering* **34**, 441 (1995), and references therein.
2. R.K. Smither, W.K. Lee, A.T. Macrander, D.M. Mills, and C.S. Rogers, *Rev. Sci. Instrum.* **63**, 1746 (1992).
3. W.K. Lee, A.T. Macrander, D.M. Mills, C.S. Rogers, R.K. Smither, and L.E. Berman, *Nucl. Instrum. Methods* **A320**, 381 (1992).

4. Gerard Marot, *Optical Engineering* **34**, 426 (1995), and references therein.
5. Andreas K. Freund, *Optical Engineering* **34**, 432 (1995), and references therein.
6. R.C. Blasdell, L.A. Assoufid, and D.M. Mills, *Diamond Monochromators for APS Undulator-A Beamlines*, Argonne National Laboratory Report ANL/APS/TB-24 (September 1995).
7. G. Grübel, D. Abernathy, G. Vignaud, M. Sanchez del Rio, and A. Freund, *Rev. Sci. Instrum.* **67** (9) (1996).
8. Grown by De Beers and available in the USA through Harris Diamond Corporation, Mount Arlington, New Jersey.
9. G. Kowalski, M. Moore, G. Gledhill, and Z. Maricic, *J. Phys. D: Appl. Phys.* **29**, 793 (1996).
10. A.R. Lang, G. Pang, and A.P.W. Makepeace, *J. Synchrotron Rad.* **3**, 163 (1996).
11. C.S. Rogers, D.M. Mills, W.K. Lee, P.B. Fernandez, and T. Graber, in *High Heat Flux Engineering III,* A.M. Khounsary, ed., SPIE Proceedings **2855**, 170 (1996).

A Tunable Laue/ Bent-Laue Monochromator with Fixed Second Crystal for Synchrotron Radiation

Z. Zhong*, G. Le Duc*, D. Chapman[†] and W. Thomlinson*

National Synchrotron Light Source, Brookhaven National Laboratory, Upton, NY 11973
[†]*CSRRI, Illinois Institute of Technology, 3101 South Dearborn, Chicago, IL 60616*

Abstract. A Laue/ bent-Laue two crystal monochromator has been developed for producing a monochromatic x-ray fan beam with energy tunable over a wide range by adjusting the angle of only the first flat Laue crystal. Bending the second crystal increases the reflection bandwidth, making the monochromator very stable against vibrations. The monochromator was tuned to above and below the K edge energy of indium at the X12A beam line at the National Synchrotron Light Source for dual energy subtraction imaging.

INTRODUCTION

The x-ray and neutron diffraction characteristics of bent crystals have been studied by many authors [1–4]. Bent crystals have been used for sagittal focusing in Bragg mode [5], as analyzers in inelastic scattering experiments [6] and as monochromators in Cauchois geometry for producing areal monochromatic beam [7]. Multi-crystal x-ray optical systems involving bent crystals have been studied using, among other methods, phase-space analysis and Dumond diagram [8,9]. Despite these general and special treatments, the existence of a two-crystal Laue monochromator configuration allowing the beam energy to be tuned solely by adjusting the angle of the first crystal has not been studied. This configuration is implemented in a Laue/ bent-Laue double crystal monochromator by choosing the distance between the crystals to be a specific value. The purpose of this paper is to describe our investigation of the useful properties of such a system.

The monochromator was developed in the framework of an attempt to produce a tunable monochromatic fan beam. It was evaluated at the X12A beamline at the National Synchrotron Light Source. The goal was to apply the new monochromator in a feasibility study for digital subtraction imaging (above and below the indium K edge) to diagnose a tumor labeled with indium-containing porphyrin.

THE PRINCIPLE

The principle for the tunable Laue/ bent-Laue monochromator with fixed second crystal for synchrotron radiation is shown in Fig. 1. The two crystal monochromator system utilizes two crystals of the same crystal material and diffraction index in non-dispersive geometry. The first crystal can be a perfect or a slightly bent crystal reflecting in either Laue or Bragg mode. This crystal defines the energy and energy bandwidth of the x-rays reflected by the monochromator system. The second crystal is strongly bent cylindrically with the bending axis perpendicular to the plane of diffraction of both crystals. The bending radius ρ_2 of the second crystal is on the order of one meter. The distance s between the crystals is chosen to satisfy:

$$s = \frac{\rho_2}{2} \cos(\chi \pm \theta_{B0}) \tag{1}$$

where the asymmetry angle χ is defined as the angle between the crystal surface normal and the Bragg planes used, θ_{B0} is the Bragg angle for the middle energy E_0 in the desired tunable energy range. The upper sign corresponds to the case when the intersection of the beam and the first crystal and the center of bending are on different sides of

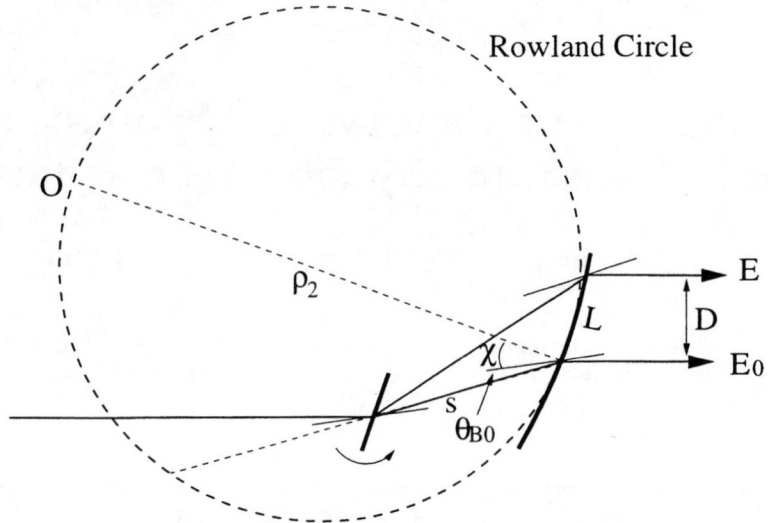

FIGURE 1. Design considerations

the diffraction planes, and the lower sign correspond to the case when they are on the same side of the diffraction planes. Fig. 1 shows the upper sign case. Since the Rowland focusing point for the second bent crystal is on the Rowland circle at a distance of $\rho_2 \cos(\chi \pm \theta_{B0})$ from the bent crystal, Eq. 1 shows that the first crystal is to be positioned half way between the Rowland focal point and the bent crystal.

If the first crystal angle is tuned to reflect a beam of energy E with corresponding Bragg angle θ_B, the monochromatic beam diffracted by this crystal lands on the second crystal at a distance L away from the corresponding landing point of the x-rays with energy E_0.

$$L \cong s \sin(2\Delta\theta)/\cos(\chi \pm \theta_{B0}) \qquad (2)$$

where $\Delta\theta \equiv \theta_B - \theta_{B0}$. Since the second crystal is bent cylindrically, the variation of the angle of incidence along the surface of the cylindrically bent second crystal is

$$d\theta = \frac{L}{\rho_2} \qquad (3)$$

Using Eq. 1, $d\theta \cong \Delta\theta$ for small $\Delta\theta$, thus the beam will be reflected by the second crystal for a wide range of beam energy defined by the Bragg angle of the first crystal.

The vertical motion of the reflected beam as the energy is changed from E_0 to $E_0 + \Delta E$ is

$$D \cong L \cos(\chi \mp \theta_{B0}) \cong \frac{\Delta E}{E} \rho_2 \cos(\chi \mp \theta_{B0}) \theta_{B0} \qquad (4)$$

In the present implementation, the first crystal is a perfect crystal which defines a beam of narrow energy and angular width. The second crystal is so strongly bent that the energy bandwidth far exceeds the bandwidth of the beam provided by the first crystal. Assuming parallel incident beam, for an incident beam with vertical height h, the change of angle of incidence across the beam height on the second bent crystal is

$$\Delta\theta(h) = \frac{h}{\rho \cos(\chi \pm \theta_B)} \qquad (5)$$

The flux rate per unit horizontal width F(ph/s/mm) of the beam diffracted by the second crystal is approximately

$$F(E, \theta_2) = \int_{-\frac{H}{2}}^{\frac{H}{2}} P_1(E, h) R_2(E, \Delta\theta(h) + \theta_2) \, dh \qquad (6)$$

where $\Delta\theta(h)$ is the change of angle of incidence on the second crystal as expressed by Eq. 5, P_1 is the intensity of the monochromatic beam diffracted by the first crystal and $R_2(E, \theta_2)$ is the reflectivity of the second crystal as function of θ_2 (the relative angle between the two crystals) for a given incident beam energy E.

According to the lamellar model [3], R_2 has a flat plateau for hard x-rays because the difference in absorption of the crystal due to beam path difference for x-rays diffracted by lamellae in the front and back of the Laue crystal becomes

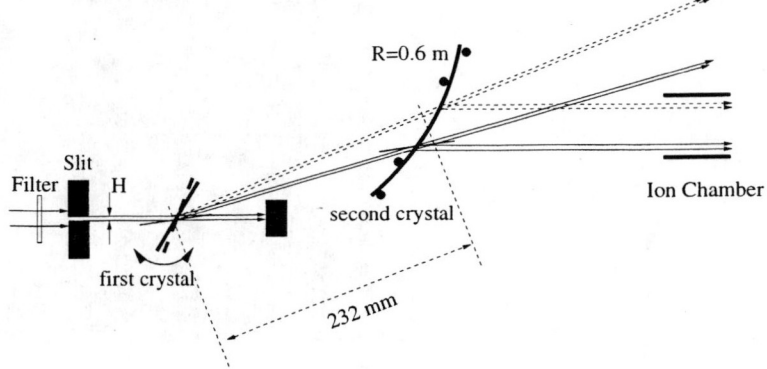

FIGURE 2. Experimental setup

negligible. For a given incident beam energy E, the FWHM (W) of the reflectivity curve $R_2(\theta_2)$ is approximately equal to the change in Bragg angle as a beam goes through the bent crystal [10],

$$W = \frac{T}{\rho}[\tan(\chi - \theta) + \frac{1+\nu}{2}\sin 2\chi + \tan\theta(\cos^2\chi - \nu\sin^2\chi)] \quad (7)$$

where ν and T are the Poisson ratio and thickness of the bent crystal, respectively.

Assuming a white beam intensity $G_0(E)$ (photons/s/keV/mm^2) with uniform vertical profile to be incident on the first crystal, P_1 is independent of h, and

$$P_1 = G_0(E)EI_1(E)/\tan\theta_B \quad (8)$$

where I_1 is the integrated reflecting power of the first crystal. The angular difference $\Delta\theta$ defined by Eq. 5 can be made smaller than the diffraction angular width of the bent crystal by limiting the vertical size of the incident beam. In this case, the maximum intensity of the beam produced by the monochromator is

$$P(E) = F(E,0)/H = R_2(E)G_0(E)EI_1(E)/\tan\theta_B \quad (9)$$

where $R_2(E)$ is the maximum reflectivity of the second crystal at energy E.

EXPERIMENTAL METHOD

The experiment was performed at the X12A beamline of NSLS which is a bending magnet beamline operated by the NSLS beamline R&D group. The experimental setup is shown in Fig. 2. The first crystal was on a Huber cradle which can adjust the Bragg angle of the first crystal. The crystal bender for the second crystal was put on a stage which has motorized adjustment over Bragg angle θ_2, height and the distance to the first crystal. The monochromator output flux was detected by an ion chamber with flowing argon at atmospheric pressure.

The crystals used were cut into rectangular shape from standard silicon [100] (the crystal surface normal corresponds to the [100] direction) wafers. The crystals were then etched for strain relief by a solution of 90% nitric acid and 10% hydrofluoric acid. The thickness of the crystals used were 0.46 mm. The [111] reflection used for the current experiment has an asymmetry angle $\chi = 35.3°$ for both crystals.

The second crystal was bent into a cylindrical shape with bending radius $\rho=600$ mm using a four bar bender [7]. The bending radius was measured by optical methods. The usable cylindrical region between the two inner bars was 50 mm wide by 25 mm high.

The monochromator system was aligned by first setting the first crystal to diffract x-rays with the indium K edge energy (27.94 keV), with the corresponding Bragg angle of 4.06° for the silicon [111] reflection. Then the second crystal was moved vertically so that the diffracted beam from the first crystal hit the middle of the second bent crystal. The center of the second crystal was put at a distance of $D = \rho\cos(\chi + \theta_B)/2 = 232$ mm from the center of the first crystal. The Bragg angle of the second crystal was then scanned so that the maximum ion chamber output was achieved.

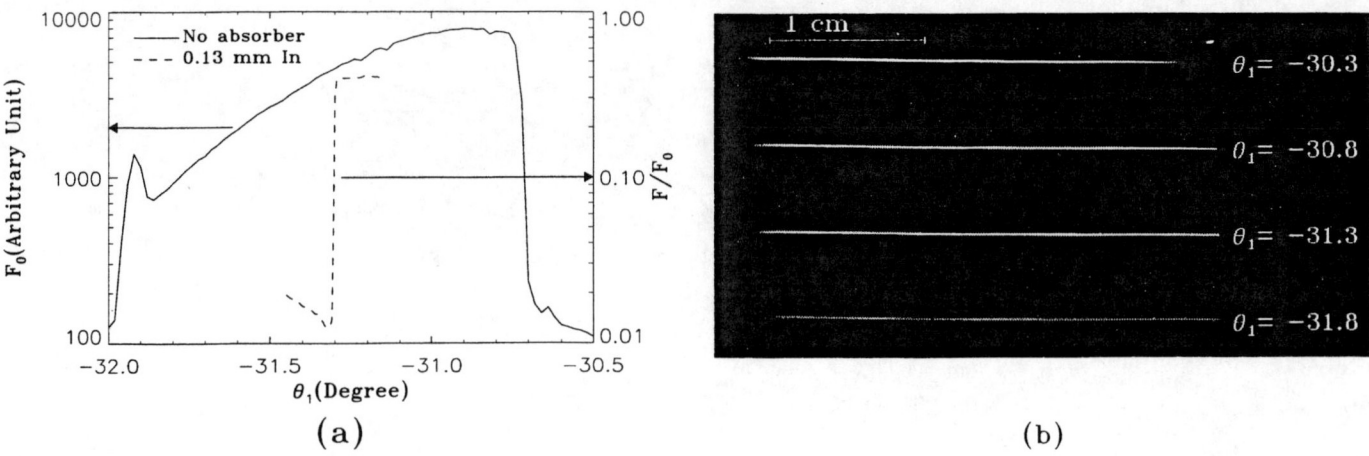

FIGURE 3. (a) Energy scan by changing the angle of the first crystal. The dashed line shows the absorption (F/F_0) of a 0.13 mm thick indium foil placed front of the ion chamber, the solid line is the intensity (F_0) of the ion chamber with no absorber. (b) Film taken at 155 mm from second crystal, with the first crystal at different angles.

RESULTS AND ANALYSIS

Tuning of Energy

Once the monochromator was aligned according to the procedure described above, tuning of the energy was achieved by rotating the first crystal in the plane of diffraction. The solid line in Fig. 3(a) shows the variation of the monochromator output intensity measured by the ion chamber with different first crystal angle θ_1 (corresponding to different energy E according to the Bragg's law), indicating that the energy is continuously tunable over a wide energy range. The shape of the curve is mainly due to the energy dependence of the flux rate of the bending magnet X12 (with a critical energy of 5 keV) and the filtering used (1 mm aluminum). The fall in intensity at $\theta_1 = -31.9°$ and $-30.7°$ was caused by the output beam of the first crystal hitting the edges of the lower and upper bending bars, respectively, for the second crystal.

A 0.13 mm thick indium foil was then placed in front of the ion chamber and the change of absorption (F/F_0) of the indium foil with different θ_1 settings was measured. The absorption is shown with the dashed line in Fig. 3(a). The slight change of ring current between the measurements with (F) and without (F_0) the indium foil was accounted for.

With a beam width of 25 mm (maximum available beam width at the NSLS X12A beam line) and a beam height of 0.2 mm, exposures were taken at several first crystal angle positions on a stationary film placed perpendicular to the beam at 155 mm down stream of the second crystal. The second crystal was stationary throughout the exposure process. Fig. 3(b) shows such a picture taken with the first crystal at -30.3, -30.8, -31.3 and -31.8 degrees. The energy of the diffracted beam was calibrated using the K edge of the indium. It is seen that uniform fan beams were produced at corresponding energies of 22.4, 24.9, 27.9 and 31.9 keV.

Vertical Beam Height and the Stability of the Monochromator

For a two-crystal monochromator, the stability is critical for applications such as monochromatic beam imaging (either radiography [11] or CT [12]) in which contrast on the order of 1% is typically sought and any beam intensity modulation on that order will degrade the image quality. With the first crystal tuned to reflect a beam of 27.94 keV, the angle of the second crystal was scanned. Fig. 4 shows the rocking curves obtained this way for different beam heights defined by a pair of tungsten slits in the white beam in front of the monochromator.

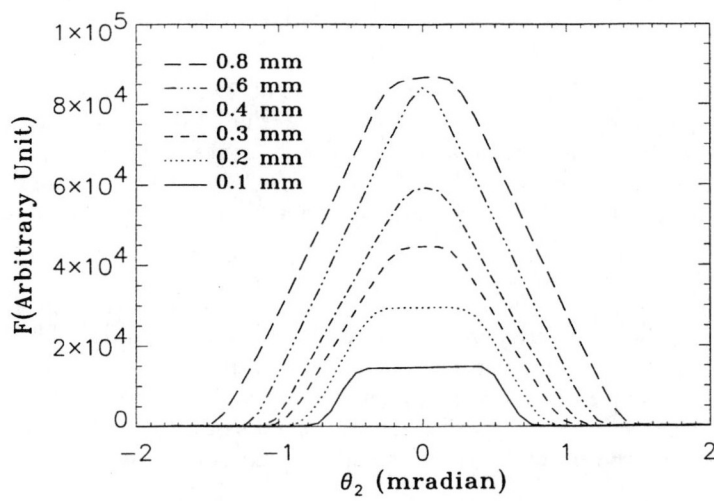

FIGURE 4. Rocking curves obtained with vertical beam heights of 0.1, 0.2, 0.3, 0.4, 0.6 and 0.8 mm

TABLE 1. Measured and theoretical W_p for different vertical beam heights

Beam Height H (mm)	0.1	0.2	0.3	0.4	0.6	0.8
Measured W_p (mradians)	0.84	0.66	0.42	0.31	0.21	0.54
Theoretical W_p (mradians)	0.87	0.67	0.48	0.28	0.11	0.49

It is seen from Fig. 4 that for beam vertical sizes less than 0.6 mm, there exists a flat plateau in the rocking curves. The width W_p of the flat plateau is a good indicator of how stable the monochromator is against vibration which causes oscillation of the relative angle (θ_2) between the two crystals. Using Eq. 6 and assumptions made in Eq. 8

$$W_p = |W - \Delta\theta| \qquad (10)$$

where W and $\Delta\theta$ are calculated using Eq. 7 and Eq. 5, respectively. Table 1 lists the W_p measured from the rocking curves shown in Fig. 4 and theoretical W_p for different vertical beam heights. For the experimental plateau widths listed, regions in each rocking curve with more than 95% of the peak intensity were used. The agreement between the theory and experiment is good despite many simplifications made in the theory.

For a vertical beam height of 0.2 mm, the width of the plateau is 0.04 degree (0.66 mradians) for the indium K edge energy. This is about two orders of magnitude larger than the Darwin width of a perfect crystal monochromator which is 10 μradians for the same energy.

Flux

The flux rate F of the diffracted beam which is measured by the ion chamber depends on the integrated reflecting power of the first crystal and the reflectivity of the second crystal (Eq. 6). As can be seen from Eq. 6, the maximum flux which can be achieved with optimal θ_2 tuning increases linearly with beam height H for small H and reaches maxima for

$$H_c = W\rho\cos(\chi \pm \theta_B) \qquad (11)$$

For the present system, H_c is around 0.5 mm. As can be seen from the rocking curves for the 0.6 and 0.8 mm beam heights, increasing the beam height beyond H_c does not increase the monochromator output flux.

At the indium K edge, with the ring operating at 2.58 GeV and 150 mA, with a incident beam of 0.2 mm high, the intensity P of the monochromator output measured by the ion chamber is about 2×10^7 ph/mm^2/s. This intensity agrees with the calculation of 2.2×10^7 ph/mm^2/s based on Eq. 9. The following parameters were used in the calculation: $G_0=3.1\times10^{10}$ ph/s/keV/mm^2, $I_1=6.2$ μradians, $R_2=0.28$. R_2 was calculated using the lamellar model for bent crystals.

DISCUSSIONS

The monochromator produces a fan beam with an intensity of 2×10^7 ph/mm^2/s and a uniformity which is adequate for dual energy imaging above and below the indium K edge. Bending the second crystal allows tuning of the monochromatic beam energy by adjusting the angle of the first crystal only. Using a bent second crystal instead of a conventional flat crystal also increases the bandwidth of the reflection of the second crystal. Since the rocking curve for a bent crystal has a flat plateau at the hard x-ray energy range used for imaging, the monochromator system is very stable.

Despite the system's excellent stability and easy tunability, which makes the monochromator a successful implementation for the production of beams above and below the indium K edge for digital subtraction imaging, two disadvantages were noted. First, the reflectivity of the second crystal is on the order of 0.3 due to the strong bending. For small beam height, the intensity of the output beam is about a quarter of the beam intensity produced by the first crystal depending on the bending radius, the asymmetry angle of the second crystal and the energy of the beam being reflected. Second, if the first crystal is a perfect crystal, there is an upper limit H_c (Eq. 11) for the height of the beam that can be reflected. In the present case, this height is around 0.5 mm.

ACKNOWLEDGMENTS

This work was supported by US DOE DE-AC02-76CH00016. The authors would like to thank Peter Siddons for about one month of ample beam time at X12A beamline, Avraham Dilmanian, Nicholas Gmür, and Peter Stephens for helpful discussions about the work and critical comments on the manuscript.

REFERENCES

1. G. Albertini, A. Boeuf, B. Klar, S. Lagomarsino, S. Mazkedian, S. Melone, P. Puliti, and F. Rustichelli. Dynamical neutron diffraction by curved crystals in the Laue geometry. *Phys. Stat. Sol.*, A44:127–136, 1977.
2. Z. H. Kalman and Z. Weissmann. One the x-ray reflectivity of elastically bent perfect crystals. *J. Appl. Cryst.*, 16:295–303, 1983.
3. E. Erola, V. Eteläniemi, P. Suortti, P. Pattison, and W. Thomlinson. X-ray reflectivity of bent perfect crystals in Bragg and Laue geometry. *J. Appl. Cryst.*, 23:35–42, 1990.
4. C. Schulze and D. Chapman. PEPO: A program for the calculation of the reflectivity of cylindrically bent Laue crystal monochromators. *Rev. Sci. Instrum.*, 66:2220–2223, 1995.
5. Peter W. Stephens, Peter J. Eng, and Teddy Tse. Construction and performance of a bent crystal x-ray monochromator. *Rev. Sci. Instrum.*, 64:374–378, 1993.
6. C. C. Kao, K. Hamalainen, M. Krisch, D. P. Siddons, T. Oversluizen, and J. B. Hastings. Optical design and performance of the inelastic scattering beamline at the national synchrotron light source. *Rev. Sci. Instrum.*, 66:1699–1702, 1995.
7. Z. Zhong, D. Chapman, R. Menk, J. Richardson, S. Theophanis, and W. Thomlinson. Monochromatic energy-subtraction radiography using a rotating anode source and a bent-Laue monochromator. *Phys. Med. Biol.*, 1997. Accepted for Publication.
8. P. Suortti and A. K. Freund. On the phase-space description of synchrotron x-ray beams. *Rev. Sci. Instrum.*, 60:2579–2585, 1989.
9. M. Popovici and A. D. Stoica. Interpretation of bent-crystal rocking curves. *J. Appl. Cryst.*, 21:258–265, 1988.
10. P. Suortti and W. Thomlinson. A bent Laue crystal monochromator for angiography at the NSLS. *Nucl. Instrum. Meth.*, A269:639, 1988.
11. R. E. Johnston, D. Washburn, P. Pisano, C. Burns W. C. Thomlinson, L. D. Chapman, F. Arfelli, N. F. Gmür, Z. Zhong, and D. Sayers. Mammography phantom studies with synchrotron radiation. *Radiology*, 200:659–663, 1996.
12. F. A. Dilmanian, X.Y. Wu, E. C. Parsons, B. Ren, J. Kress, T. M. Button, L.D. Chapman, J. A. Coderre, F. Giron, D. Greenberg, D. J. Krus, Z. Liang, S. Marcovici, M. J. Petersen, C.T. Roque, M. Shleifer, D. N. Slatkin, W. C. Thomlinson, K. Yamamoto, and Z. Zhong. Single- and dual- energy CT with monochromatic synchrotron x-rays. *Phys. Med. Biol.*, 42:371–387, 1997.

Beam Size Measurement of the Stored Electron Beam at the APS Storage Ring Using Zone Plate Optics and Undulator Radiation

Zhonghou Cai, Barry Lai, Wenbing Yun, Efim Gluskin, Dan Legnini,
Peter Ilinski, Emil Trakhtenberg, Shenglan Xu,
William Rodrigues, and Heung-Rae Lee

Advanced Photon Source, Argonne National Laboratory, Argonne, IL 60439

Abstract. Beam sizes of the stored electron beam at the Advanced Photon Source storage ring were measured using zone-plate optics and undulator radiation. A gold Fresnel zone plate (3.5 µm thick) located 33.9 meters from the x-ray source focused radiation of 18 keV, selected by a cryogenically cooled Si(111) crystal in horizontal deflection, and formed a source image in a transverse plane 2.41 m downstream. The sizes of the source image were determined from measured intensity profiles of x-ray fluorescence from a smooth nickel edge (1.5 µm thick), fabricated using a lithographic technique, while the nickel edge was scanned across over the beam in the transverse plane. The measured vertical and horizontal sizes of the electron beam were 60±4.3 µm and 300±13 µm, respectively, in reasonable agreement with the expected values.

INTRODUCTION

As the Advanced Photon Source (APS) begins operation, beam emittance measurements become an essential part of the commissioning process to diagnose the performance of the storage ring and insertion devices. In the center of the straight section, beam size and divergence are determined by the natural beam emittance, coupling between vertical and horizontal emittances, and the β functions characterizing the magnet lattice in the storage ring (1). Beam size measurements provide useful information about those values. When diagnosing undulator radiation, beam size is usually needed to convert the results obtained from absolute flux measurement to beam brilliance (2). Even after the APS enters the operational mode, beam size information will be desired for the experiments involving x-ray focusing.

Usually, the divergence of the photon beam can be determined accurately by measuring the photon beam size σ at a distance D from the source. That is

$$\sigma^2 = D^2 \sigma_{s'}^2 + \sigma_s^2, \tag{1}$$

where σ_s and $\sigma_{s'}$ refer to the size and the divergence of the photon source, respectively. When the measurement takes place at a large D, σ can be much larger than σ_s. The uncertainty of the measurement can be well managed and limited at a relatively small value. For undulator radiation, however, the inherent divergence of the photon beam, which is comparable with the divergence of the particle beam in APS storage ring, prevents us from accurately determining the divergence of the particle beam. As such, accurate measurements of beam size become an important issue in the diagnostics of beam emittance. Especially because our diagnostics were focused not only on the particle beam but also on the undulator photon beams, the technique selected has to accommodate the requirement for portability of the diagnostic equipment. In the next section, a comparison of pinhole optics and zone-plate optics is given to explain our choice of zone-plate optics.

ZONE-PLATE OPTICS AND PINHOLE OPTICS

A phase zone plate is a diffractive focusing device made of a number of concentric circular zones. The thicknesses of the zones are determined so that a phase shift of π is obtained for a given radiation wavelength. The focal length of the zone plate depends on the refractive index of the zone material, the wavelength of the radiation, and the Fresnel half-period radii. Zone plates work like thin lenses, so that the thin lens formula,

$$\frac{1}{Z_1} + \frac{1}{Z_2 - Z_1} = \frac{1}{f}, \qquad (2)$$

$$M = \frac{Z_2 - Z_1}{Z_1}, \qquad (3)$$

applies for the formation of the source image (3), where Z_1 is the source-zone-plate distance, Z_2 is the source-image-plane distance, f is the focal length of the zone plate, and M is the magnification factor of the imaging system. A zone plate has a diffraction limit to its transverse resolution similar to that calculated by Rayleigh for a thin lens, which turns out to be about 122% of the outermost zone width of the zone plates (3). With currently available manufacturing techniques for hard x-ray zone plates, the diffraction-limited resolution can be smaller than 0.2 µm.

Pinhole imaging has been used recently for source-size measurements during commissioning of several third-generation synchrotron radiation sources (4-7). The pinhole optics describing the relation between the measured size of the source image on the screen σ and the source size σ_s can be expressed as (7)

$$\sigma^2 = \left(\frac{Z_2 - Z_1}{Z_1}\right)^2 \sigma_s^2 + \left(\frac{Z_2}{Z_1}\right)^2 \sigma_p^2, \qquad (4)$$

where σ_p refers to the rms size of the pinhole, Z_1 and Z_2 are, respectively, the distances of the pinhole and the image screen from the source. The first term in Eq. (4) represents the size of the source image, and the second term reflects the size of the pinhole projected on the screen. Note that the image broadening given by the second term is evaluated with a point-source approximation. It is valid when $\sigma_s \ll Z_1 \sigma_s$. The Fraunhofer diffraction of x-rays from a pinhole results in an angular spreading of the x-ray beam and, thus, a broadening of the source image. The rms image broadening due to x-ray diffraction is about $(Z_2 - Z_1)\lambda / 7\sigma_p$, where λ is the wavelength of the x-rays (8).

For the APS, the radiation characteristics are dominated by the emittance of the stored particle beam. Given a natural beam emittance of 8.2 nm-rad, a horizontal β function of 14.2 m, and a vertical β function of 10 m, the horizontal and vertical sizes of the electron beam in the storage ring would be 325 µm and 86 µm for a coupling constant of 10%, and 333 µm and 62 µm for a coupling constant of 5% (9). The vertical beam size becomes smaller as the coupling constant decreases. Consider an undulator radiation source with rms size of 60 µm. For measurements using pinhole optics, if a pinhole of 5 µm (σ_p) is placed 10 meters from the source and the imaging screen is located 10 meters downstream, the rms sizes of the source image, the projection of the pinhole, and the broadening due to diffraction are, respectively, 60 µm, 10 µm, and 35 µm (for 10 keV x-rays). The size of the image on the screen is a convolution of the three quantities. Note that the diffraction broadening is about 60% of the size of the source image. The uncertainty in pinhole size will generate a significant error when deconvolution of the image size from the diffraction broadening and the finite size of the pinhole is carried out. This error becomes severe when the source size is small like that of the APS. In order to reduce the diffraction effect, one can either employ high energy radiation or make the distance $Z_2 - Z_1$ small. However, small pinholes for high energy x-rays are difficult to manufacture, and one would also reduce the source-pinhole distance Z_1 if $Z_2 - Z_1$ is reduced so that a magnification factor required for accurate measurement can be maintained. At the APS, this approach means that the measurement has to be conducted in the front end and a dedicated beamline is required for beam diagnostics. For the diagnostics of insertion devices, this approach is practically impossible because each device has to be installed in the section of the dedicated beamline once there is a need for the device to be diagnosed.

For measurements using zone-plate optics, if a zone plate with a focal length of 2.5 m is located 30 meters from the source, the rms size of the source image on the image plane will be 5 µm, and the diffraction-limited transverse resolution (0.2 µm) is only a small fraction of the source-image size. Because the diffraction-limited resolution is small, the accuracy of the source size measurement is not sensitive to the source-zone-plate distance. In other words, the measurements can take place in stations located at various distances from undulator sources. We have developed a technique for source size measurement using x-ray zone-plate optics, in which the source was imaged at the imaging plane of the zone plate, and the image size was measured by scanning a nickel thin edge across the beam in the imaging plane and measuring its fluorescence.

EXPERIMENTAL SETUP

The experiment was performed in the first optical enclosure (FOE) of the 2-ID beamline of the APS. The x-ray source was the synchrotron radiation generated from undulator A (9). Fig. 1 is a schematic of the experimental setup. In order to reduce the thermal load on the monochromator, the radiation extracted from the storage ring was, first, filtered with a set of filters consisting of 500 μm carbon, 150 μm diamond, and 500 μm beryllium. Then, the power of the undulator radiation was further reduced substantially by a water-cooled grazing incidence conical pinhole (copper) with the exit hole 800 μm in

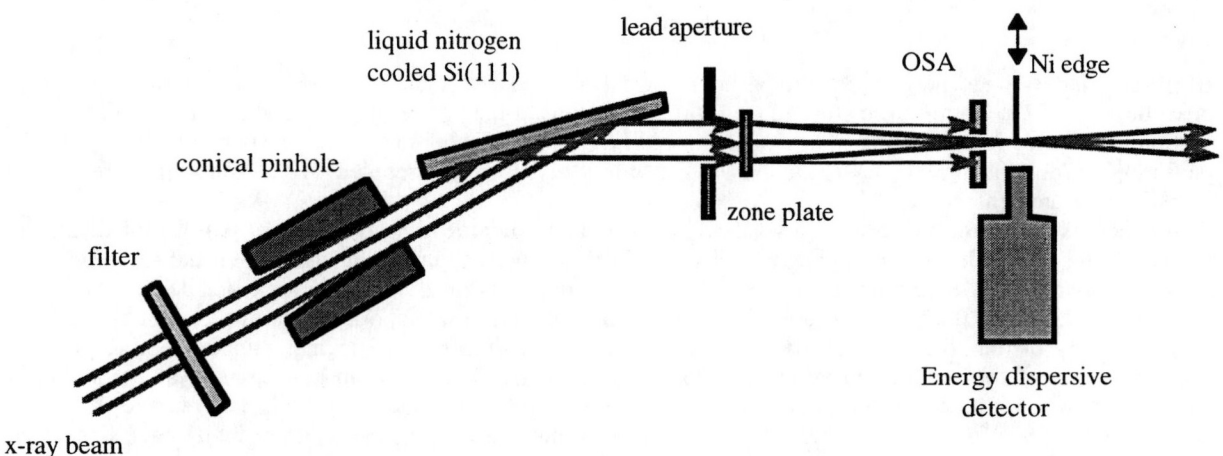

FIGURE 1. Schematic illustration of the experimental setup.

diameter. X-rays of 18 keV were selected by a cryogenically cooled Si(111) crystal in horizontal deflection. The vacuum chamber for the monochromator was sealed with two 150-μm-thick beryllium windows capable of tuning the x-ray energy from 7 to 30 keV. The monochromatic beam was apertured to a size of 600 μm before it illuminated the zone plate. A gold zone plate of 400 μm in diameter (3.5 μm thick) with a focal length of 2.25 m was located 33.89 m from the center of the straight section, and thus the x-ray source image was formed in a transverse plane 2.41 m downstream from the zone plate. A platinum order-sorting aperture (OSA) of size 30 μm was placed 100 mm upstream from the image plane to increase the contrast of the source image. An energy-dispersive detector (Si(Li), EG&G ORTEC SLP 06165PS) was located close to the imaging plane pointing in a direction normal to the x-ray beam to measure the x-ray fluorescence generated from a nickel thin

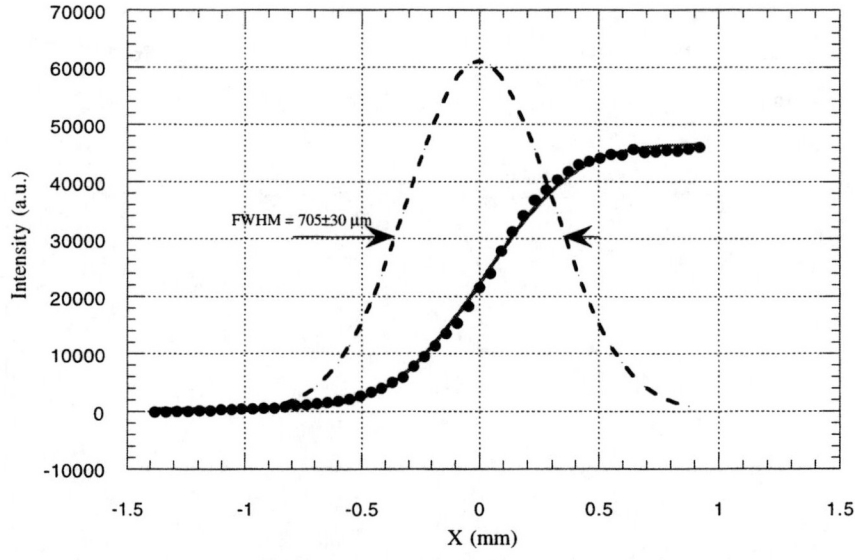

FIGURE 2. Fluorescence intensity profile of a horizontal knife-edge scan (dots). The solid line is the fitted curve to the experimental data, and the dashed line is the derivative of the fitted curve. The horizontal axis reflects the coordinates at the source location.

edge (Kα line) placed in the imaging plane. The beam sizes were determined from the fluorescence intensity profiles measured when the nickel edge was being scanned across the beam.

The FOE was chosen because the experimental stations downstream had not yet been commissioned when the measurements were made. The measurements were taken in the presence of a high radiation background because the detector had to be in the same station where radiation scattering components, such as the pinhole and the monochromator, were installed. Therefore, the conical pinhole was placed in a helium enclosure, and both the pinhole enclosure and the monochromator vacuum chamber were carefully shielded with lead and steel.

RESULTS

All measurements were performed with a beam of 7 GeV and less than 5 mA. For the natural emittance and the emittamce coupling of the beam in the storage ring, the β functions at the center of the straight section, and the experiment configuration, the source image on the image plane would be 6 μm vertically and 23 μm horizontally. The performance of the crystal monochromator and the energy bandwidth of the x-ray beam intercepted by the zone plate are critical to the accuracy of the measurement.

Main concerns about the performance of the crystal monochromator arise from the possibility of distortion in the atomic planes of the crystal due to the high power of the undulator radiation that would shift the virtual source away from the point of the real source. The undulator gap was set at 14.8 mm so that its third harmonic peaked at 18 keV. With a 800 μm aperture located at 33 m, the total incident power of the undulator radiation was about 5 W with a current of 5 mA. The high thermal conductivity of the silicon at liquid-nitrogen temperature should be able to maintain the atomic planes without distortion. The width of the rocking curve of the monochromator measured with an analyzing crystal (Si(111)) at beam current of 20 mA was found to be the same as the ideal value. Horizontal deflection takes advantage of less stringent requirements in slope error for avoiding image distortion in the vertical direction along which a small image size is expected.

Because of single bounce, the monochromatic nature of the x-ray beam is determined by the intrinsic energy bandwidth of the crystal and the angular acceptance of the zone plate. Given the distance of the zone plate from the source, we found that the energy bandwidth of the x-ray beam picked up by the zone plate was about 5 eV, and image broadening due to the bandwidth can be neglected.

We show the measured nickel (Kα) fluorescence intensity profile of a horizontal knife-edge scan in Figure 2. The experimental data were fitted with an error function and a uniform background was assumed. The horizontal intensity profile of the source image was then obtained by taking the derivative over the fitted fluorescence intensity profile. In the figure, the profiles are plotted versus the product of the dimension at the image plane with the demagnification factor of the zone-plate imaging system, representing the dimension at the source point. The beam intensity profile gives a horizontal source size of 705 μm at the full width half maximum, or 300 μm in rms size.

Figure 3 displays a similar fluorescence intensity profile for the vertical knife-edge scan. It provides a vertical source size of 141 μm at the full width half maximum, or 60 μm in rms size. The distance between the zone plate and the nickel edge was optimized so that the most narrow intensity profile of the source image was obtained. From the results of source-

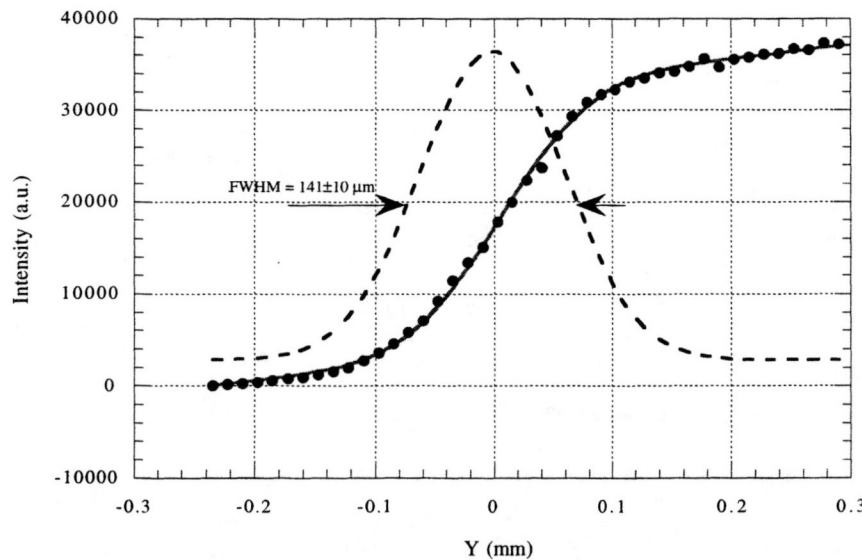

FIGURE 3. Fluorescence intensity profile of a vertical knife-edge scan (dots). The solid line is the fitted curve to the experimental data, and the dashed line is the derivative of the fitted curve. The horizontal axis uses the coordinates at the source location.

size measurements and beam-divergence measurements (2), we obtained a horizontal beam emittance of 7.5 nm-rad and a vertical beam emittance of 0.32 nm-rad, and, thus, the coupling constant is 0.043.

DISCUSSION

The accuracy of the measurement depends on the accuracies of the distances of the zone plate and the knife edge from the source, the quality of the Ni thin edge, and the mechanical stability of the imaging system. The Ni thin edge was manufactured using a lithographic technique and was checked by a scanning electron microscope (SEM) before the experiment. Figure 4 is a photograph of the SEM backscattering image of the nickel edge used in the experiment. The smoothness of the edge is better than 0.1 µm. In order to evaluate the performance of the thin edge and the stability of the mechanical system, the same nickel edge was used to measure the spot size of an x-ray beam focused with a 6-cm-focal-length zone plate using the same setup and the same source-zone-plate distance. A 0.2 µm beam spot size (rms) was obtained, indicating the roughness of the nickel edge and the instability of the system almost have no effect on the source size measurement.

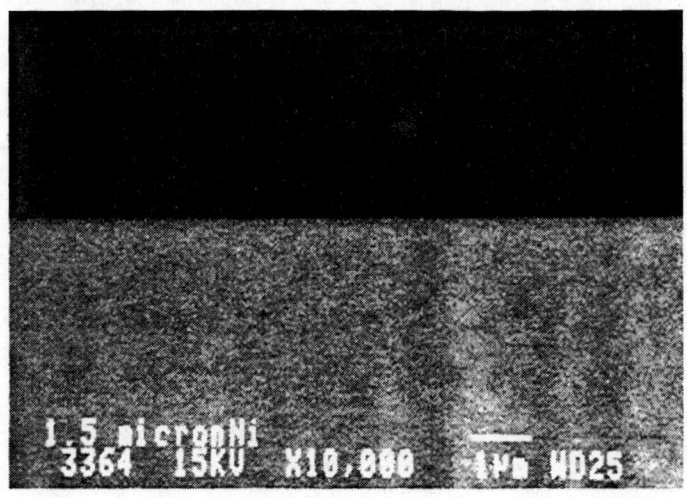

FIGURE 4. SEM backscattering image of the nickel thin edge.

In a summary, a zone-plate-optics-based, portable, and flexible beam diagnostics instrument for undulator source size (emittance) measurement has been developed. The instrument has been used for size measurement of the particle beam in the APS storage ring and for undulator photon beam diagnostics. Measured results were in reasonable agreement with expected values.

ACKNOWLEDGMENTS

Work was supported by the U.S. Department of Energy, BES. Material Sciences, under contract No. W-31-109-Eng-38.

REFERENCES

1. Sands, M., *The Physics of Electron Storage Ring*. Proceedings of the Int. School 'Enrico Fermi', 1971, Vol. **46**, p. 258.
2. Cai, Z., Dejus, R., Gluskin, E., Ilinski, P., Lai, B., Yun, W., and et al, *Rev. Sci. Instrum.*, **67** (1996) CD ROM.
3. Klein, M. V., *Optics*, Wiley, New York, 1970.
4. Mills, D., Viccaro, P., Merlini, A., and Shen, Q., *Nucl. Instrum. & Meth. Phys. Res.* **A291**, 481 (1990).
5. Elleaume, P., Fortgang, C., Penel, C., and Tarazona, E., *J. Synchrotron Rad.* **2**, 209 (1995).
6. Cai, Z., Lai, B., Yun, W., Gluskin, E., Dejus, R., and Ilinski, P., *Rev. Sci. Instrum.*, **66**, 1859(1995).
7. Cai, Z., Lai, B., Yun, W., Gluskin, E., Legnini, D., Illinski, P., and et al, *Rev. Sci. Instrum.*, **67** (1996) CD ROM.
8. Born M. and Wolf, E., *Principles of Optics*, Pergamon, Oxford, 1980.
9. Dejus, R., Lai, B., Moog, E., and Gluskin, E., Argonne National Laboratory Report, ANL/APS/TB-17 (1994).

Beam-Smiling in Bent-Laue Monochromators*

B. Ren[1], F.A. Dilmanian[1], L.D. Chapman[2], X.Y. Wu[1], Z. Zhong[3],
I. Ivanov[2], W.C. Thomlinson[3], and X. Huang[1]

[1] *Medical Department, Brookhaven National Laboratory (BNL), Upton, NY 11973*
[2] *Center for Synchrotron Radiation Research and Instrumentation (CSRRI),
Illinois Institute of Technology, Chicago, IL 60616*
[3] *National Synchrotron Light Source, BNL, Upton, NY 11973*

ABSTRACT

When a wide fan-shaped x-ray beam is diffracted by a bent crystal in the Laue geometry, the profile of the diffracted beam generally does not appear as a straight line, but as a line with its ends curved up or curved down. This effect, referred to as "beam-smiling", has been a major obstacle in developing bent-Laue crystal monochromators for medical applications of synchrotron x-ray. We modeled a cylindrically bent crystal using the Finite Element Analysis (FEA) method, and we carried out experiments at the National Synchrotron Light Source and Cornell High Energy Synchrotron Source. Our studies show that, while beam-smiling exists in most of the crystal's area because of anticlastic bending effects, there is a region parallel to the bending axis of the crystal where the diffracted beam is "smile-free". By applying asymmetrical bending, this smile-free region can be shifted vertically away from the geometric center of the crystal, as desired. This leads to a novel method of compensating for beam-smiling. We will discuss the method of "differential bending" for smile removal, beam-smiling in the Cauchios and the polychromatic geometry, and the implications of the method on developing single- and double-bent Laue monochromators. The experimental results will be discussed, concentrating on specific beam-smiling observation and removal as applied to the new monochromator of the Multiple Energy Computed Tomography [MECT] project of the Medical Department, Brookhaven National Laboratory.

I. INTRODUCTION

The x-ray reflectivity curve of a plate of perfect crystal bent from an unbent configuration becomes one with a flat top and large width. Typically, the integrated reflectivity increases by ten-folds, depending on the bending radius. These characteristics of bent crystals have led to an increased interest in the applications of the bent-Laue crystal monochromators for applications in which high beam flux and high beam stability in a two-crystal configuration are required. Single- and double-bent-Laue crystal monochromators have been constructed, or are being developed in several x-ray synchrotron laboratories around the world[2,3]. However, there is a persistent problem that hinders the application of these crystals for very wide beams, it is called the "beam-smiling" effect. "Beam-smiling" means that when a wide fan-shaped x-ray beam is diffracted by a bent crystal, the profile of the diffracted beam does not appear as a straight line, but as a line with its both ends curved up or curved down. For single bent-Laue crystal monochromator with a wide fan beam, the beam undergoing "smiling" will have a lateral energy dispersion, which appears as a curved shape in a photograph of the Laue diffraction taken far away from the crystal. For double bent-Laue crystal monochromator, the smiling effect causes the angular position at the center of the crystal in the "tuned configuration" to differ from that near the edge of the crystal. This will lead, for example, to an non-uniform intensity profile across the beam when the two crystals are tuned with respect to each other. In the extreme cases, the smiling causes partial disappearing of the beam after the second diffraction. Although the smiling effect has been a problem with many bent-Laue systems, there are no report in the literature on how to approach the problem.

In this paper we show, by simulation and by experimental results, that, while the beam smiling effect is the intrinsic characteristic of a bent Laue crystal due to anticlastic bending effect[4], there is a region vertically across the bent crystal from which the diffracted x-ray beam is almost flat. Furthermore, we show that by applying the "asymmetric bending" method, this region can be moved up and down, as desired. These two techniques provide us with a practical

---

*Research supported by the Office of Health and Environment Research, the U.S. Department of Energy under the Contract DE-AC02-76CH00016.

method of compensating the smiling effect. The experiments in developing a bent Laue-Laue monochromator for the Multiple Energy Computed Tomography MECT[1] program prove that our methods is effective and useful.

II. FINITE ELEMENT ANALYSIS OF BENT CRYSTAL

When a piece of rectangular shape, thin Silicon crystal layer is bent cylindrically with the axis parallel to one side of the rectangle, the shape of the bent crystal will show anticlastic bending[4]. The effect is due to the non-zero Poisson coefficient of Si. The analytical equation of the surface of the plate upon bending is only available for very simplified cases, such as when the crystal is bent by a torque which is constant along the edge of the crystal. In practice, it is difficult to provide a constant, uniform bending torque to the crystal. The more realistic case is to apply a constant displacement along the edge using a "four-rod" bending geometry. This problem is still difficult to solve analytically. Thus, we performed Finite Element Analysis (FEA) calculations for a cylindrically bent crystal.

A. Model description

We studied two types of crystals, one is a piece of Si wafer cut to a rectangular shape and another one is a bulk Si piece cut to have a thin layer in its center with two stiff ribs left at its edges. The bending mechanism was a four-rod-typed bender. It was composed of two rods fixed in a parallel geometry, and two other rods movable to adjust the bending torques. Although there are some differences in the shapes of the wafer and "ribbed" crystals, upon bending, the physical characteristics of bent crystal effects involved are the same. We, therefore, limit our discuss here to crystals with ribs.

B. Ribbed crystal geometry

Fig. 1 shows our modeling and defines the coordinates. We note that in the following discussion the motion of the two movable rods are always parallel to the fixed rods, i.e., no twisting of the crystals are involved. Our FEA model took the advantage of the crystal's symmetry, and simulated only one half of the crystal's area. Linear stress analysis was performed over a crystal of the pre-defined shape, using the commercially available Finite Element Analysis package, ABAQUS[6]. The crystal was assumed to be isotropic in the simulation (although Si crystals are orthotropic, as we know in practice, simulations using a single Young's modules and Poisson coefficient provide a good approximation of the real crystal[4]). The two ribs were modeled by 3-D brick elements and the central thin layer was modeled by 2-D shell elements. The boundary condition was approximated by fixing the crystal's vertical movement, as is practically imposed by the two fixed rods, and applying uniform displacement to the movable rods in Z direction. The parameters we used in FEA were: overall crystal size: 17.3 x 10.2 x 1.3 cm; rib size: 17.3 x 2.5 x 1.3 cm; thin layer size: 17.3 x 5.1 x 0.08 cm.

C. FEA Results

Fig. 2 shows the shape of the central (thin) part of the crystal after bending. The displacements of the two movable rods were set to 50 μm for both of them; the bending radius at the center of crystal is 6.5 m. The shape of the crystal is quite non-uniform after bending; the anticlastic bending effects show along the x direction, and the two edges bow out; therefore both the slopes and curvatures are non-uniform along the x direction.

The change in the slope in the vertical direction, which is of most relevance to the beam intensity, is ploted in Fig. 3. The plot shows the slope as a function of x at different y positions. From Fig. 3 we see that there is only one region along the x direction in which the slope is constant. This region is located in the center of the crystal vertically (i.e. $y = 0$). For $y > 0$, the slope at the edge curves up, while at $y < 0$ it curves down.

The above results is for symmetric bending of the crystal when the displacement of the two rods are the same. If these two displacements are different, one rib of the crystal will be bent more than the other. The shape of the crystal will then deviate from the cylindrical one. We define the D_1 as displacement for the upper movable rod, and D_2 for the lower movable rod. They can be expressed as: $D_1 = D - \delta$, $D_2 = D + \delta$ where D is the symmetric displacement and δ is the differential displacement. In our example, D is 50 μm. For FEA simulation at the asymmetric bending, we varied δ from 0 μm to 50 μm. The results for $\delta = 20$ μm are showed in Fig. 4. We note that while the anticlastic bending introduces the same slope-profile variations as in symmetric bending, the uniform slope region has moved down to y slightly less than -0.635 cm.

Fig. 5A shows the relationship between the y position of the slope-uniform region vs. the differential displacement δ; Fig. 5B shows the resulting bending radius R at that slope-uniform region as a function of δ. The results indicate that asymmetric bending at the differential displacement of $\delta = \pm 50$ μm shifts the slope-uniform region vertically to $\pm 0.9 cm$ from the geometric center line. Correspondingly, the bending radius at the slope-uniform region changes from 6.5 m at $\delta = 0$ μm to 2.8 m at $\delta = \pm 50$ μm.

D. Implications of the FEA results

The anticlastic bending causes the shape of the cylindrically bent crystal to be non-uniform in its surface slopes and surface curvatures, both of which will affect the x-ray diffraction pattern. In this paper, we will concentrate on the surface slope non-uniformity that contributes only to the beam-smiling effect. Discussion of the variation in the bending radius that affects the shape of the rocking curve is out of the scope of this paper.

Depending on the specific surface-slope variations, the profile of x-ray may appear as a straight line, or a line with its both ends curved up or curved down. Fig. 6 shows a series of Laue spot photographs taken far away behind a single crystal for the x-ray beam diffracted from different location on the bent crystal.

When a cylindrically bent crystal is used to diffract x-ray beam, two possible setup geometrics can be used: Cauchois geometry, in which the concave side of the bent crystal is toward the x-ray source, and polychromatic geometry in which the convex side is toward the x-ray source. The distributions of the surface-slope non-uniformity are different in these two geometries. If we take a series of Laue spot photographs far away behind the single bent crystal for the x-ray diffracted at different parts of the bent crystal, the Laue spots will show the following pattern illustrated in Fig. 6.

In making a bent-Laue crystal monochromator for wide fan-shaped x-ray beam, it is very important to let the x-ray beam diffract from the slope-uniform region of the crystal to have a smile-free monochromatic beam. Otherwise, any residual beam smiling can be compensated either by moving the crystal up or down relative to the beam, or by adjusting the differential displacement parameter δ. The latter is referred as differential bending[5]. Since the anticlastic bending happens, obviously, near the two edges of the crystal along the bending axis direction, the width of the crystal should be made much larger than the length of the fan beam. The height of the crystal should be comparable to its width.

In double-bent-Laue crystals monochromator, the beam smiling may cause regional disappearing of the monochromatic x-ray beam. This is because the Bragg diffraction angles in the two crystals may not match along the beam's length. Given the small reflection widths of x-ray diffraction in crystal ($1 \sim 100$ $\mu radian$), small surface-slope variations in the crystal can shift the angular position for Bragg diffraction outside this range. To have a uniform intensity along the length of the crystal, the two crystals should be either smile free or to have smiles that compensate each other. To align the monochromator for the latter effect is difficult, as it involves identifying the smile in the two crystals and aligning them accordingly.

III. EXPERIMENTAL METHODS AND RESULTS

A. MECT's new bent Laue-Laue monochromator: general

The Multiple Energy Computed Tomography program at BNL takes advantage of the wide-energy-band, high intensity synchrotron radiation from the X17B superconducting wiggler beam line of the National Synchrotron Light Source (NSLS) to develop a monochromatic CT system for imaging human head and neck[1]. Earlier, MECT images were obtained with a flat Laue-Laue monochromator[7]. The aim in developing this new monochromator is to further improve the quality of the monochromatic x-ray beam toward possible future MECT clinical applications. The x-ray beam at NSLS's X17B beam line has 5 $mrad$ (horizontal) and 0.2 $mrad$ (vertical) opening angles. MECT's new monochromator is designed to be positioned 22.5 meters from the source, where the beam has 113 mm width. The preliminary testing of the monochromator was carried out at the X17B1 hutch, 29.5 meters from the source. The beam width in the center of the hutch was 145 mm. To comply with the beam-line restrictions, the monochromatic beam was chosen to be 15 mm above the white beam. The two bent crystals were either in the lower-lower setup or in the higher-higher one, in an attempt to avoid the laue spot contamination. The details of the monochromator designs will be published in a separate paper. The testing of the new monochromator is described below.

B. Crystals cut with ribs

The silicon crystals were cut from a five-inch-diametered Float-Zone (FZ) boul. The plane $<331>$ was used as the diffraction plane. The crystals in the monochromator are in the upper-upper setup, as showed in Fig. 7. Due to the limitations in the vertical size of the crystals in this test, the monochromatic beam could not be positioned in the center (vertically) of the second crystal. This limitation was imposed to allow the white beam to pass above the lower rib. The parameters for the crystals were: the overall size of the crystal was 12.7 x 6.5 x 1.0 cm; the size of the rib was 12.7 x 3.5 x 1.0 cm; the size of the central region was: 12.7 x 3.0 x 0.09 (and 0.11) cm.

In practice, the observation of beam-smiling by taking a photograph of Laue diffraction is not very informative. For most cases, the bending radius of the crystal is so large that, at limited distance ($\approx 2m$), the small shape-variation is unobservable. Instead, we used a method we call Rocking Graph. As we rocked the second crystal relative to first crystal around the Bragg diffraction peak, we acquired the intensity profile of a vertically narrow monochromatic beam from the two-crystal diffraction using the linear array detector of the MECT program. Fig. 8. presents three such plots. In these plots, each vertical line is equivalent to a rocking curve and each horizontal line is the beam profile at a certain Bragg angle. The angular range of the plots in Fig. 8 is 300 μrad. The beam's vertical width in the test was 0.4mm, and the beam energy was around 40keV.

These pictures were acquired using the feature of our bending mechanism that allowed easy unbenting of each crystal by using spacers that nentralized the bending forces on each bar. In this way, we could set one of the crystal flat and let it act as an analyzer, while we observed the pure smiling from the other bent crystal (the flat analyzer does not contribute to smiling). The first rocking graph, Fig. 8A is produced by a configuration in which the first crystal is bent and the second is flat. Since the vertical height of the beam on the first crystal is close to its center, very little beam smiling should be produced unless the first crystal is under asymmetric bending. The wavy shape of the rocking graph suggests that the first (bent) crystal have contributed to the slope variations. (Our rocking graph for flat-flat configuration - not shown - indicates that the second crystal when flat has very small distoration.) The first bent crystal is probably bent asymmetrically, and the small distoration of the second (i.e. flat) crystal is probably due to the weight from the upper rib. For Fig. 8B, the first crystal is flat and the second crystal is bent, and the second crystal's rocking graph shows a large smile with little distortion: the two ends curved symmetrically. This probably comes from the non-uniform slopes at the two edges of the upper part of the second crystal where the monochromatic beam hits. The above two rocking graphs indicate the beam-smile in individual bent crystals. However, the beam of interest is that from the bent-bent system. Such a rocking graph is showed in Fig. 8C. It is flat for nearly 95% of the horizontal width of the middle part of the beam, with the other 5% is close to the edges and is curved up. The rocking graph indicates that while we get maximum peak diffraction in the middle, we can only get reduced intensity at the edges.

C. Wafer crystals

In an attempt to increase the crystal size, we turned to large crystal wafers of Czochralski (CZ) type, available from semiconductor industry. The silicon wafers used were $15.5 \times 11.5 \times 0.7$ cm^3. The surface normal of the wafer was in $<001>$ direction, the diffraction plane was $<111>$, and the asymmetric angle for $<111>$ plane was 35.26°. The lower-lower setup in the Cauchios geometry was used. The first crystal was bent to $R = 8m$ and the second to $R = 12m$. The crystals were bent using a He-Ne laser and the reflections from the mirror-type surface of silicon wafer for guidance. The dimensions of the incident beam were 12.5×0.05 cm, and the beam energy was at 40keV. Differential bending was applied for smile-removal. The vertical distances of the beam from the centers of the crystals 1 and 2 are denoted in these figures by $Z1$ and $Z2$ respectively.

Fig. 9A shows that when the x-ray beam goes through the middle of the crystal the smile disappears, except for some small wiggles that will be discussed later. Besides the differential bending method, we can also move the crystal up or down to find the best region for x-ray diffraction. Our FEA simulations has shown that the smile exists in the bent crystal nearly symmetrically, and that by shifting the crystal we can bring the smile-free region into the x-ray diffraction. A test of our FEA simulations was done by moving the first or second crystal up and down respectively and record the rocking graph. We found that whenever the crystal is moved away from the centered position the smile will show up in the rocking graph and the direction of smile flips as one crosses the center of the crystal. From direction of the smile in the rocking graph in Fig. 9B..9E, we can predict how to adjust the height of the crystal to eliminate the beam smile.

We estimate that the remaining wiggles in the rocking graphs come from the imperfection of the silicon wafers, probably from stress frozen in the crystals during the cutting process or from the imperfect surface of the wafers or the rods. Although the surface of the wafer has been finished to a very high quality, once we hold and bend the wafer by the four rods, the shape of the wafer may start to show irregularity. Given the fact that over a bending radius of 10 to 20 meter, the displacement of the rod is only around 50 μm, any variation in the order of a few micrometers in either the wafer surface or the straightness of rods will change the crystal's shape considerably.

When the crystal is bent cylindrically, commonly some twisting of the crystals exist together with bending, due to non-balanced torque. The twisting generally causes a linear slope-change across the crystal, which is independent of the beam smiling. Experimental rocking graphs often consist of the contributions from both effects. Only after the twisting has been removed can the pure smiling effect be seen, as demonstrated in Fig. 8 and Fig. 9. Without the removal of the twisting, the rocking graph may appear as an tilted distribution in which the tilting does not depend on the crystal's vertical height. Both the twisting and the smiling effects influence the performance of the bent Laue monochromators. Furthermore, the weight of the crystal and the heat-loading may also influence the twisting and the smiling since they introduce additional strain to the bent crystal. Experimentally, both effects can be compensated for by the differential bending.

IV. CONCLUSIONS

Our results demonstrate the magnitude of the beam-smiling problem in bent Laue-Laue crystal monochromator. The following conclusions can be drawn from our experiments: 1) the crystal width should be much larger than the beam's width to avoid the edge effects from anticlastic bending; 2) the crystal height should be comparable to the crystal width; 3) the beam height should always be aligned to land in the center of the crystal to minimize the beam smiling effect (it is possible to let the beam go through the upper or lower part of the two crystals, as long as the beam smilings match in the two crystals and compensate each other, which is difficulty to achieve); 4) the bending mechanism should allow differential bending and twist removal in order to compensate for the non-uniform surface slope of the crystals; 5) ribbed crystals seem to be advantageous over crystal wafers because of the wafer's extraordinary senitivity to the bending mechanism and crystal-surface treatment. The final design of the MECT's monochromator has taken all these effects into consideration: the crystals are ribbed and the benders employ four independent, motorized adjustments through springs. Preliminary tests of the new device carried out at CHESS with a 5-cm-wide x-ray beam showed very encouraging results. The final test will be done at X17B beam line in the next two months.

In summary, we studied the beam-smiling effect, which is a limiting factor in the development of bent-Laue crystal monochromators for wide beam, in the course of developing a bent Laue-Laue monochromator for MECT program. We developed a method of producing rocking graphs, and a differential-bending method for the removal of beam-smiling. The results indicate that the beam-smiling effect can be controlled to a large extent.

V. ACKNOWLEDGEMENT

The authors would like to thank G. Rosenbaum of APS, and B. Dowd, J. Gatz, N. Gmür, J. Hastings, A. Lenhard, N. Satterley, P. Siddons, G. Van Derlaske and M. Woodle of BNL, B. Blank, E. Fonts, K. Finkelstein of CHESS, E. Black, S. Cimpose, A. King and W. Lavender of CSRRI, C. Schulze of ESRF, and J. Quintana of Northwestern University for their help during the design and the execution of the experiments.

VI. REFERENCES

1. F.A. Dilmanian, X.Y. Wu, E.C. Parsons, B. Ren, J. Kress, T.M. Button, L.D. Chapman, J.A. Coderre, F. Giron, D. Greenberg, D.J. Krus, Z. Liang, S. Marcovici, M.J. Petersen, C.T. Roque, M. Shleifer, D.N. Slatkin, W.C. Thomlinson, K. Yamamoto, and Z. Zhong. *Phys. Med. Biol.*, **42** 371-387, 1997.

2. P. Suortti, W.C. Thomlinson, D. Chapman, N. Gümr, D.P. Siddons and C. Schulze, *Nucl. Instr. and Meth.*, **A336** (1993) 304-309.

3. B. Ren, F.A. Dilmanian, X.Y. Wu, L.D. Chapman, I. Ivanov, Z. Zhong. *1996 Activity Report of NSLS.* **BNL 52517**, p. B-239.

4. V.I. Kushnir, J.P. Quintana and P. Georgopoulos, *Nucl. Instr. and Meth.*, **A328** (1993) 588-591.

5. Z. Zhong, D. Chapman, R. Menk, J. Richardson, S. Theophanis, and W.C. Thomlinson. *Phys. Med. Biol.*, 1997 (in press).

6. Program ABAQUS was developed by Hibbitt, Karlsson & Sorensen, INC., 1080 Main St., Pawtucket, RI 02860-4847, USA.

7. M. Shleifer, F.A. Dilmanian, F.A. Staicu, M.H. Woodle. *Nucl. Instr. and Meth.*, **A347** (1994) 356-359.

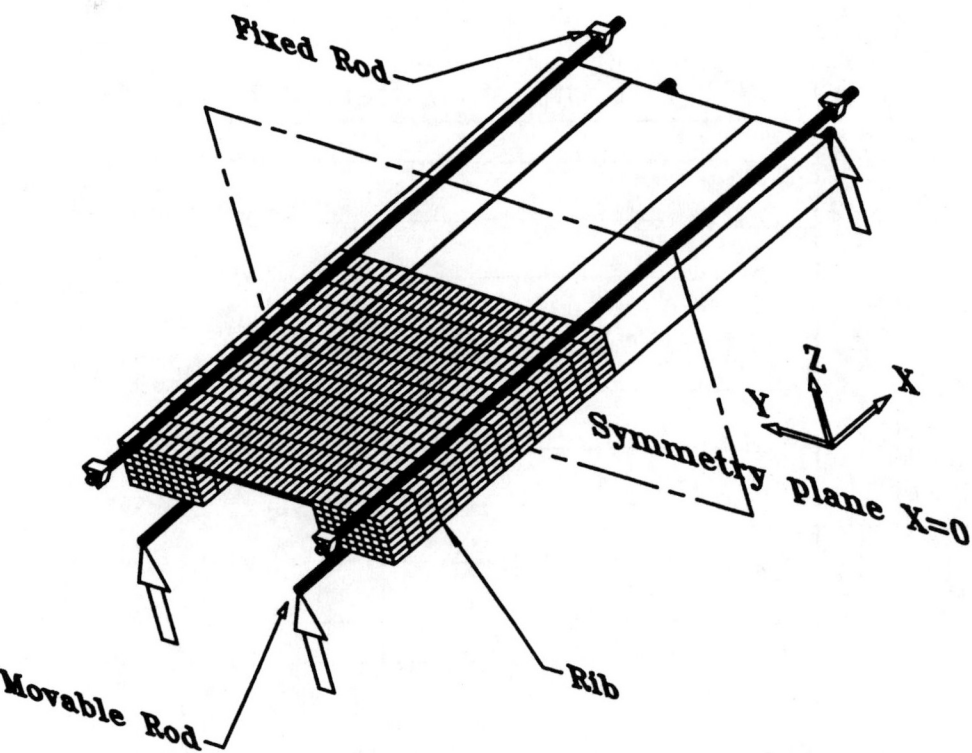

Fig. 1 FEA model for the crystal bent by the four-rod bender

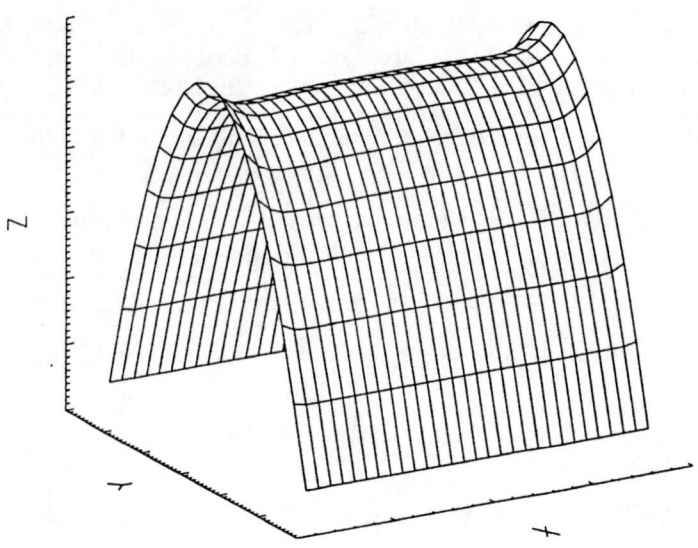

Fig. 2 Shape of the thin crystal layer upon bending

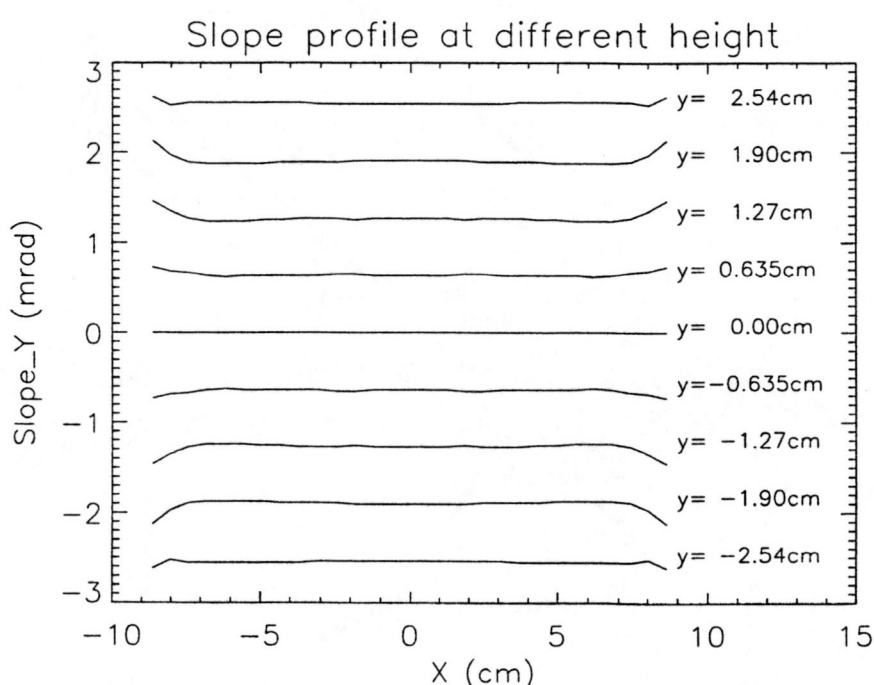

Fig. 3 Slope profiles under symmetric bending

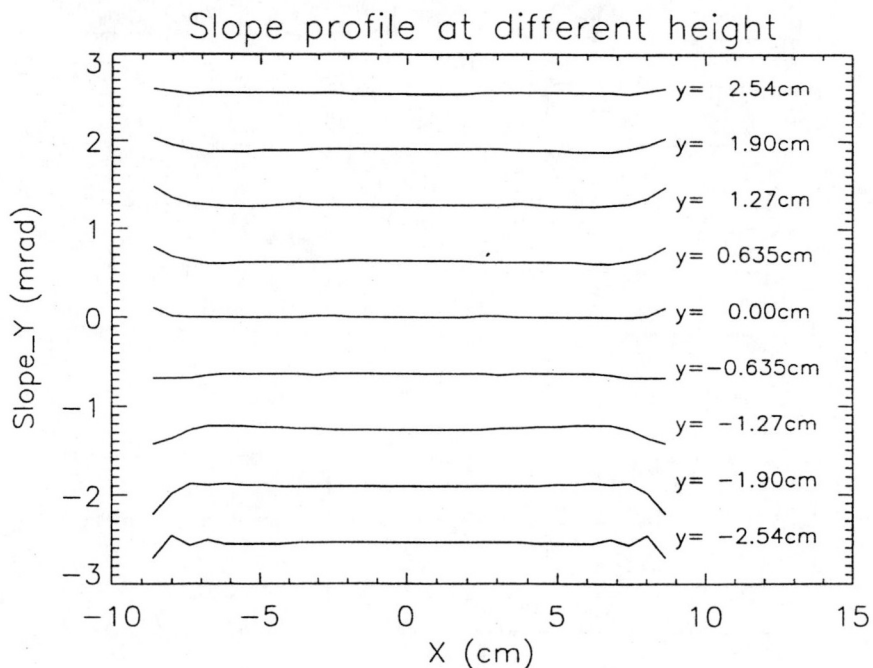

Fig. 4 Slope profile for asymmetric bending

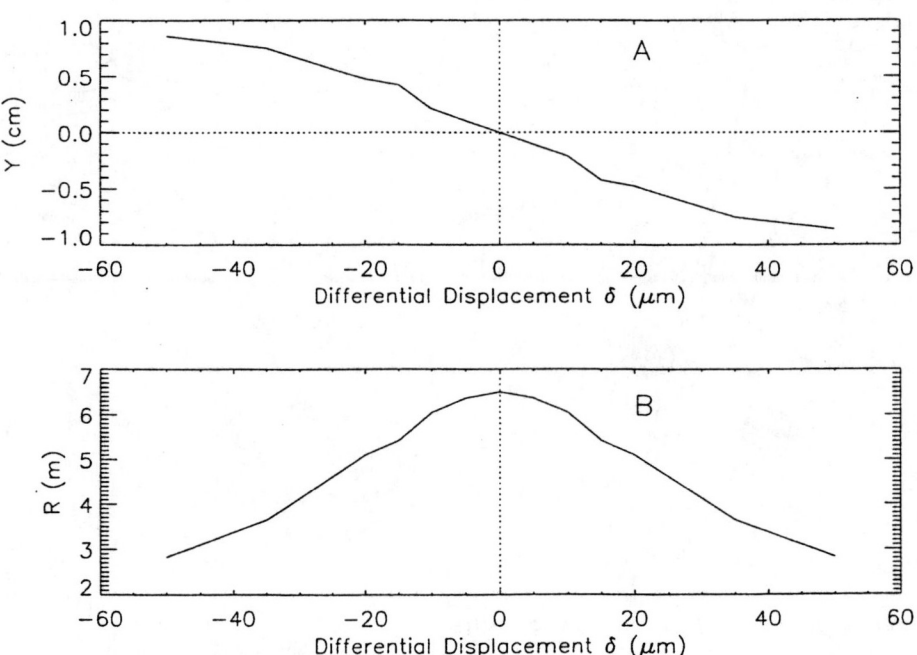

Fig. 5 Position(A) and Radius(B) change of the slope-uniform region vs. differential displacement δ

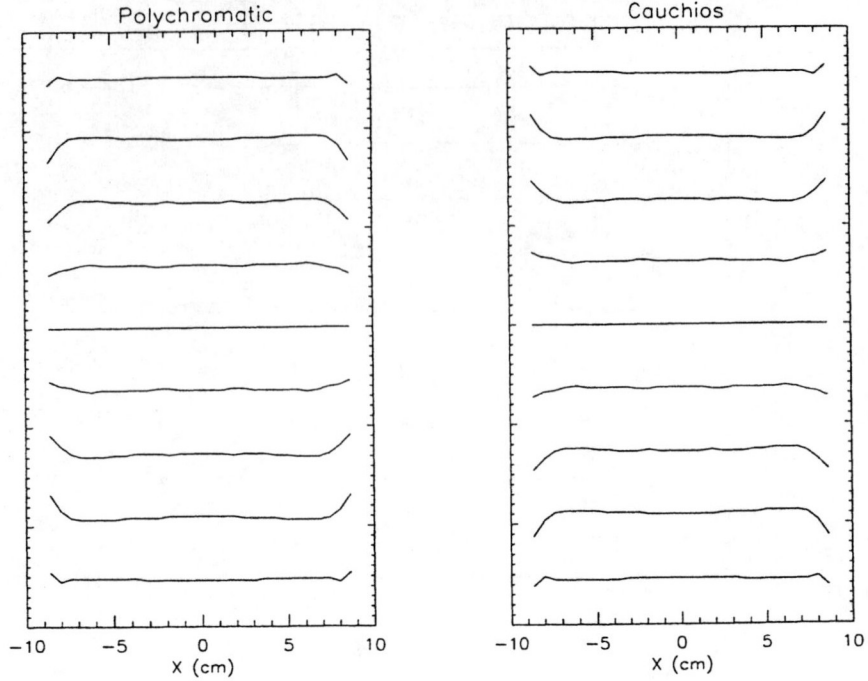

Fig. 6 Laue spot pattern in polychromatic and Cauchois geometry

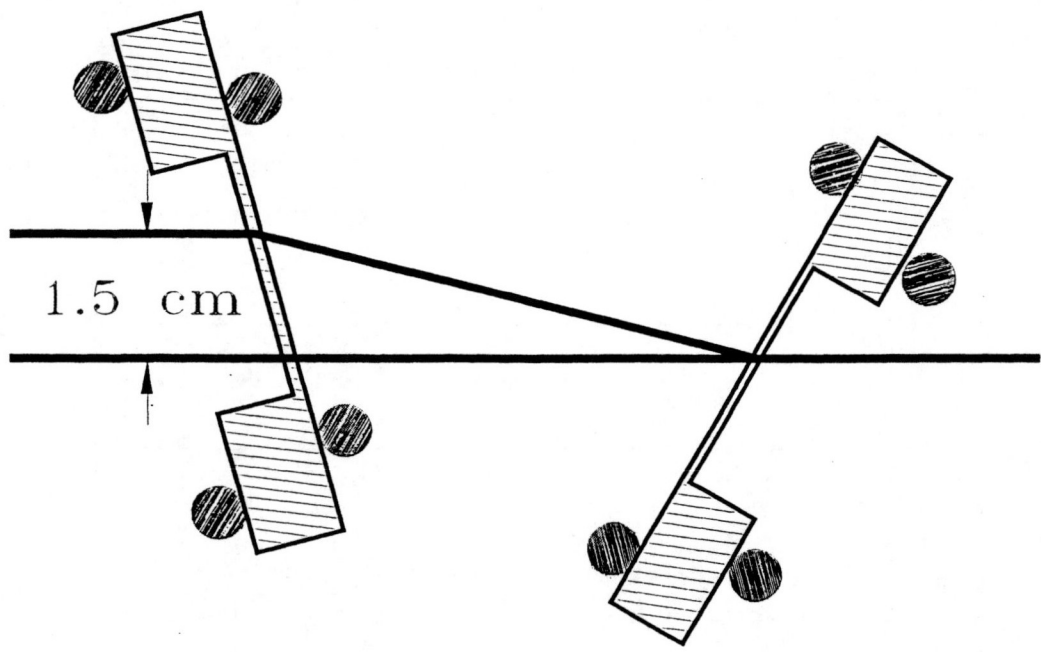

Fig. 7 Bent Laue-Laue crystals in the upper-upper setup.

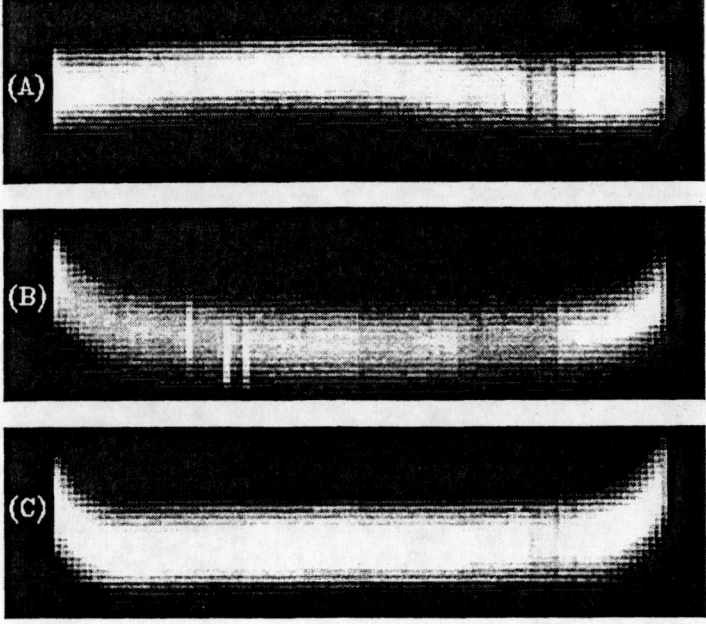

Fig. 8 Rocking graph for Laue-Laue monochromator   A: $R_1 \approx 12.3m$, 2nd crystal flat; B: 1st crystal flat, $R_2 \approx 15.6m$; C: $R_1 \approx 12.3m$, $R_2 \approx 15.6m$.

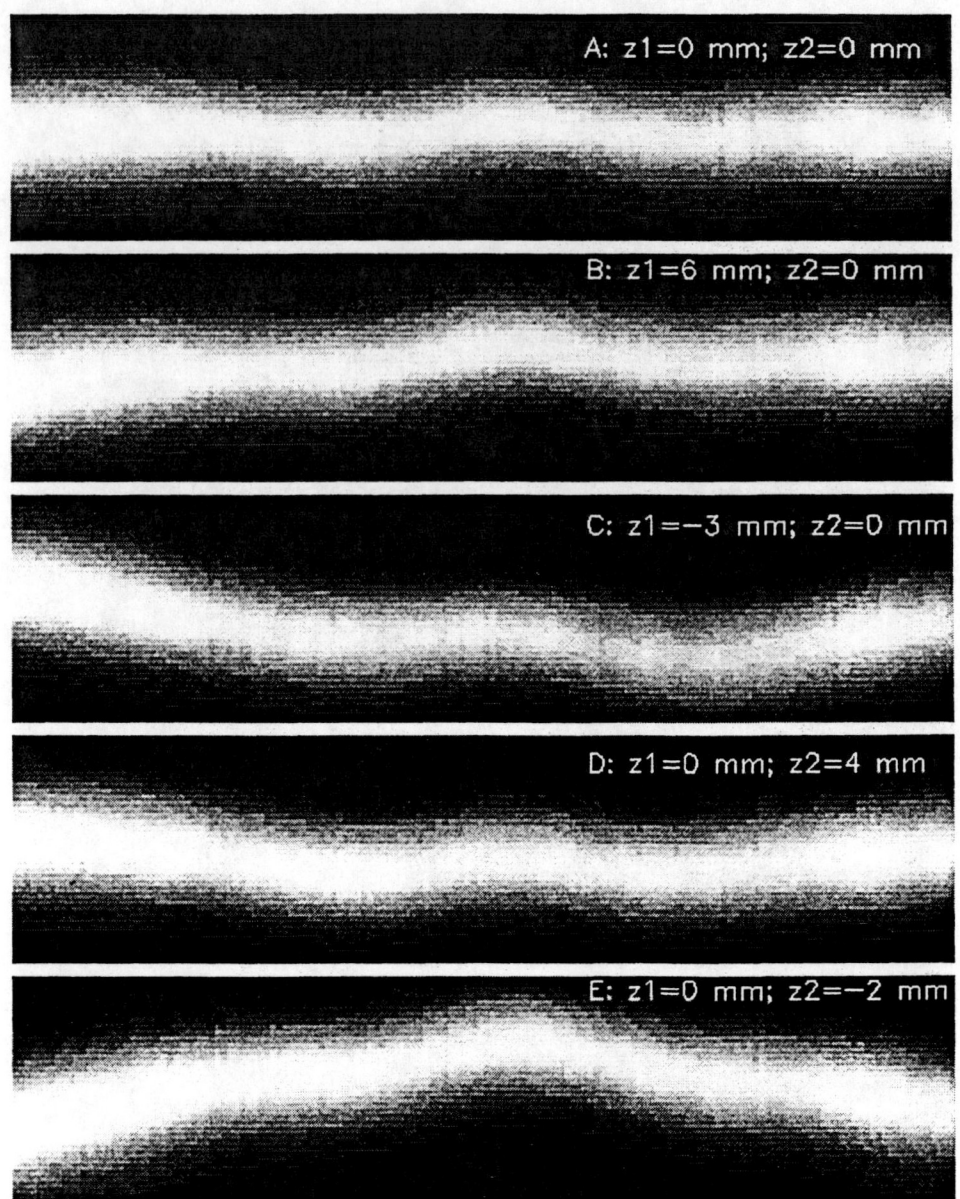

Fig. 9 Rocking graphs for the Laue-Laue system with wafers

The performance of a wide band x-ray Bragg polarizer grown by molecular beam epitaxy

J. O. Cross, B. R. Bennett and M. I. Bell

Naval Research Laboratory
4555 Overlook Avenue Southwest
Washington DC 20375

K. J. Kuhn

Intel Corporation
Beaverton OR 97005

A wide band Bragg polarizer for photon energies around the L3 absorption edge of erbium has been grown from a semiconductor alloy using molecular beam epitaxy. The active optical element is an 8 micron single crystal film of In(.51)-Ga(.49)-Sb grown on a GaAs(001) substrate. The composition of the ternary alloy was calculated using Vegard's rule to give maximum linear polarization for the target energy of 8358 eV at the (006) Bragg reflection. The full width at half maximum of the polarizer's energy acceptance was found to be 27 eV centered at 8359 eV using synchrotron radiation by scanning the incident photon energy and maintaining a fixed angle of 90 degrees between the incident and diffracted beams. The results illustrate a general approach to preparing linear x-ray polarizers for hard x-rays.

Measurement of Diffraction Gratings with a Long Trace Profiler with Application for Synchrotron Beamline Gratings

S. C. Irick and W. R. McKinney

University of California,
Lawrence Berkeley National Laboratory
Advanced Light Source
Berkeley, CA 94720, USA

Abstract. The Long Trace Profiler (LTP) is used primarily for measuring the figure of long synchrotron beamline mirrors. The LTP has also been used for measuring the figure of the substrate of beamline gratings. We propose a method for measuring the effective figure that comes from the grating groove pattern on the substrate of long beamline gratings. Analysis of grating groove patterns can be useful in determining cause of poor imaging of the diffracted light, but requires investigation of small changes of the groove frequency over the entire clear aperture of the grating.

A diffraction grating that is small enough to be measured by a general purpose six inch aperture interferometer is measured by both this interferometer and the LTP, so that results for two different instruments may be compared. The height profile of the substrate light (m = 0) measurement is subtracted from the height profile of the diffracted light (m = 1) measurement, and the result is the effect of only the diffraction from the grooves along the entire surface. This procedure is also used for a diffraction grating that is too long to be measured by the general purpose interferometer, but is easily measured by the LTP.

INTRODUCTION

Focusing X-rays onto a small spot has placed difficult requirements on production of optical components in synchrotron beamlines. These requirements have spawned new instrumentation over the last few years. One of these instruments is the Long Trace Profiler (1,2,3,4), which is able to characterize the slope profile over the length of a mirror. The slope profile $s(x)$ may be integrated with respect to the measurement direction x, in order to get a height profile $h(x)$ which can be compared to results from other height-measuring instruments. However, the slope profile is often retained for a more direct analysis of mirror imaging performance. In other words, if the light source is a simple geometric shape (e.g. a point), then the image will be a distribution of light that is directly related to the histogram of slopes of the imaging mirror.

The Long Trace Profiler (LTP) has been used to measure the substrates of X-ray beamline diffraction gratings, which is equivalent to measuring the specular order of a grating (m = 0). Recently there has been a need to determine the effect of the grating grooves on imaging quality of the main diffracted order (m = +1). The main diffracted order is used in X-ray beamline monochromators, so it is this imaging characteristic that needs to be analyzed for quality assurance of the component.

The technique of measuring grating groove effect would be similar to measuring the profile of a mirror. First the grating substrate (m = 0) would be measured to give a slope profile $s_0(x)$ over the entire length of the grating clear aperture (CA). Then the grating would be tilted by an angle to give a Littrow configuration (the diffracted beam is coincident with the incident LTP probe beam), and this measurement would yield a slope profile $s_1(x)$ of the diffracting surface (m = 1). The diffraction effect of the grating grooves alone would be obtained by subtracting $s_0(x)$ from $s_1(x)$.

In order to verify this technique, the same procedure is done with a large aperture height-measuring interferometer. A small diffraction grating (50 mm square) is measured both on the LTP and on a Zygo Model GPI interferometer, which has a 150 mm diameter aperture and a resolution of 640 x 480 pixels. Both diffraction orders (m = 0 and m = 1) are measured on both instruments, and the results are compared.

MEASUREMENT OF A SMALL GRATING

A straight-ruled plane diffraction grating with a nominal groove density f of 1200 grooves/mm and outside dimensions of 50 mm x 50 mm x 2 mm was measured on the LTP for m = 0. In order to ignore irregularities near the edges, measurements were actually made 5 mm from each edge. Thus the LTP measurement is 40 mm long (see Figure 1a). The grating was then tilted at an angle α to satisfy the Littrow condition (Figure 1b), and the diffracting surface was measured for m = 1. Angle α is determined from the grating equation:

$$m \lambda f = \sin(\alpha) + \sin(\beta). \tag{1}$$

In Littrow $\alpha = \beta$. Also, m = 1 and λ = 670 nm (the LTP uses a laser diode). Thus

$$\alpha = \arcsin(m \lambda f / 2) = 23.7 \text{ degrees}. \tag{2}$$

The diffracted beam is compressed laterally by a factor $\cos \alpha = 0.9157$, which means that the LTP measurement interval must be 0.9157 times as great for m = 1 as it is for m = 0. Since the measurement interval was $\Delta x_0 = 1.0$ mm for m = 0, then for m = 1, $\Delta x_1 = 0.9157$ mm. This ensures that there will be the same number of data points in each measurement, and the measurements may easily be compared. Figure 2 gives results of those LTP measurements as slope profiles.

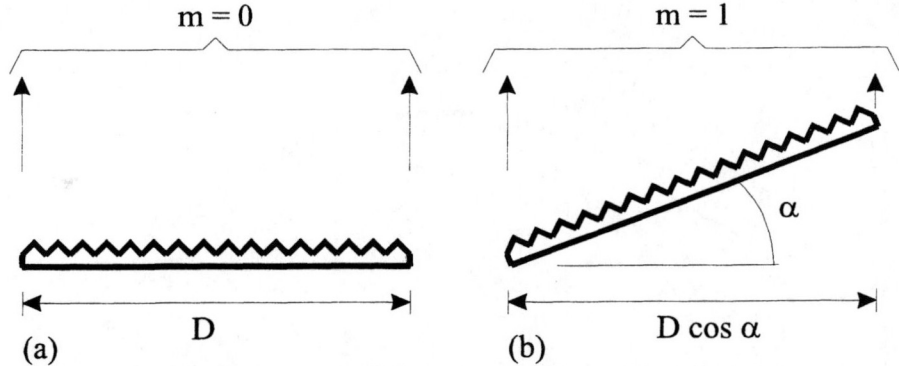

FIGURE 1. Rays of specular reflection (a) and rays of first order diffraction (b) returning to the measuring instrument.

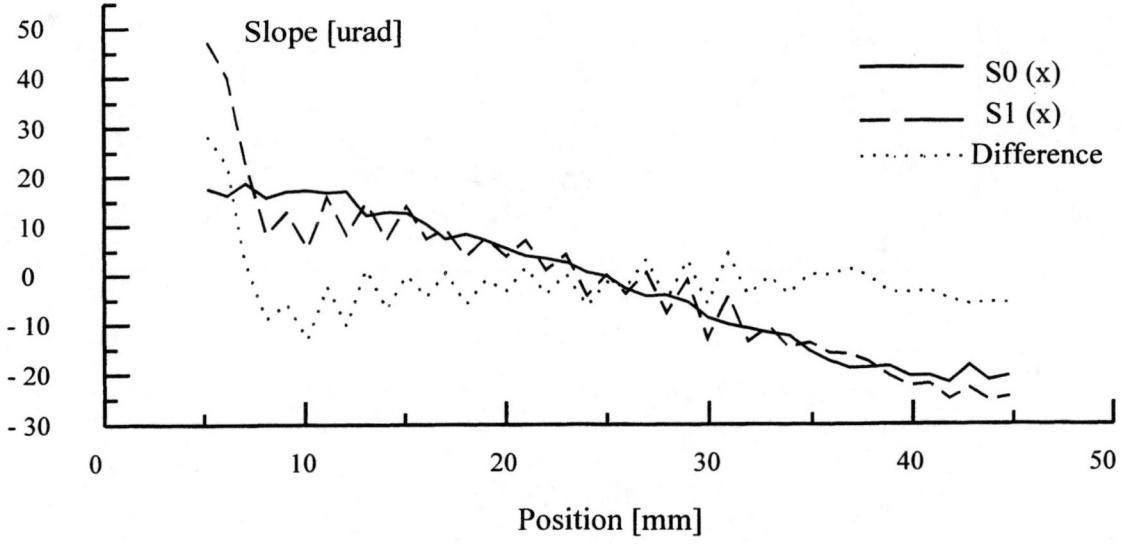

FIGURE 2. Slope measurements from the LTP.

Likewise, care must be used when measuring the grating for both orders with the Zygo GPI. Just before the grating is measured for m = 0, the Zygo system is calibrated for lateral length. The outside dimensions of the grating are known accurately, so this may be used as a calibration standard. Just before the grating is measured for m = 1, the Zygo system is zoomed to a magnification of M = 1 / cos α, and the calibration is not changed. This ensures a constant correspondence between the measurement points for m = 0 and m = 1.

The center 40 mm line trace was taken from each Zygo height measurement and plotted in Figure 3. For comparison to Zygo results, the LTP slope functions were integrated to give height profiles for a center 40 mm trace, also shown in Figure 3. There is good agreement with both Zygo and LTP measurements.

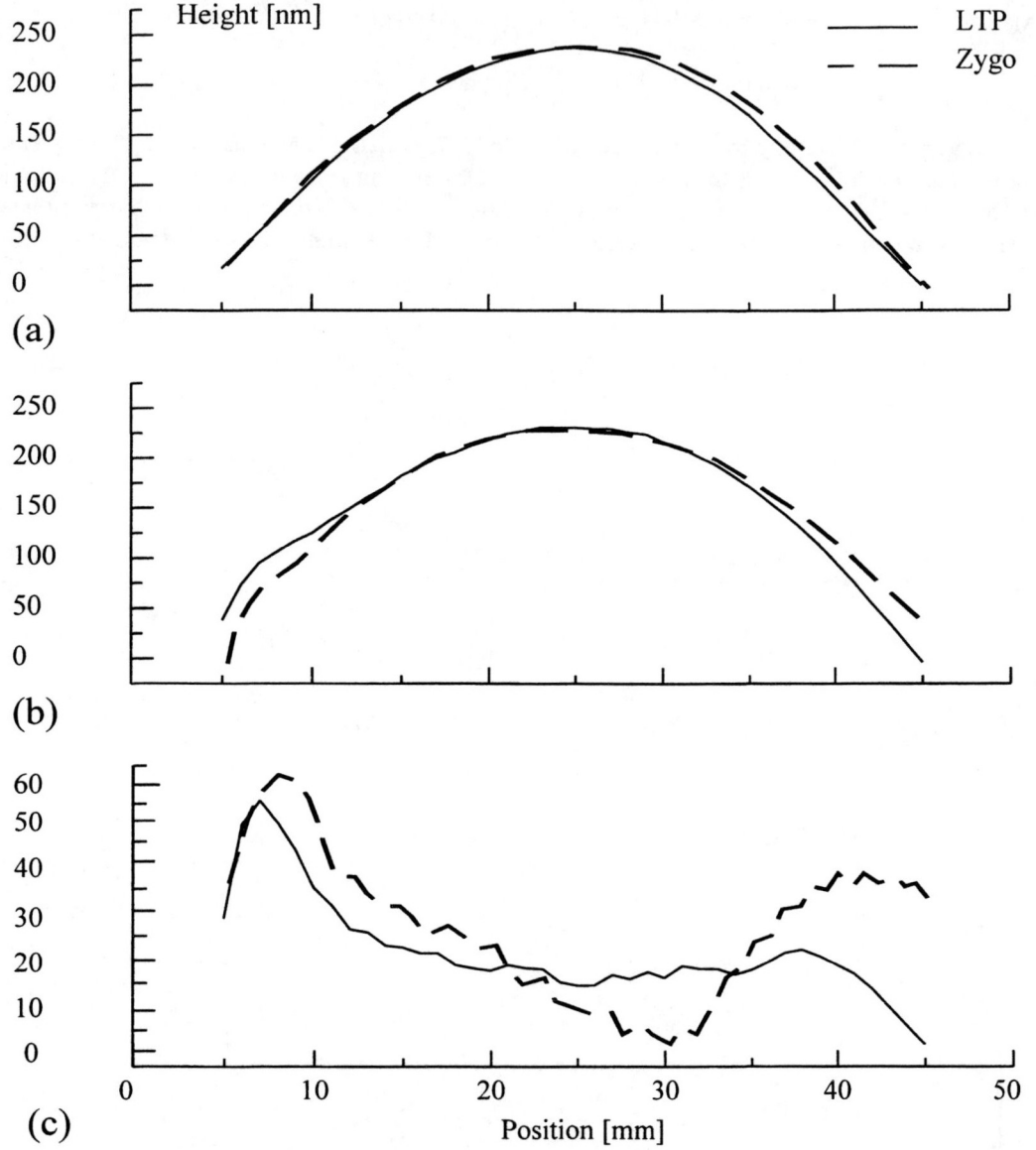

FIGURE 3. Height profiles for a center line trace on the small grating.
(a) Profile for m = 0; (b) profile for m = 1; (c) profile for the difference (m = 1) − (m = 0).

MEASUREMENT OF A LONG BEAMLINE GRATING

Many commercial profiling instruments are inappropriate for measuring X-ray beamline components, not only because the instrument aperture is too small but also because the fringes of high-curvature surfaces cannot be resolved. This is the case for the diffraction grating from the ALS beamline 7.3.1. The grating clear aperture is 217 mm long, and the grating's substrate has a specified radius of curvature of R = 250 meters. For the LTP, however, this is a routine measurement.

Again, to measure the m = 1 diffracted order the grating is tilted by an angle α in order to give a Littrow condition. Figure 4 shows height profiles of a center trace of the grating. Although the two measurements are almost on top of each other (dashed and dotted lines in Figure 4), their difference gives the effect of the grating grooves alone. This grating groove effect (solid line in Figure 4) has a convex height profile. The radius of curvature of this convex profile is 16000 m, so the grating's substrate curvature (R = 250 m) is effectively reduced by a small amount.

As mentioned earlier, it is often more instructive to retain the measurement results as slope profiles. Figure 5 shows the slope function $s(x) = s_1(x) - s_0(x)$. The maximum and minimum slope values can then be used to determine the amount of groove spacing change. The optical arrangement is the Littrow condition, but the incident angle is constant; only the diffracted angle β changes with groove frequency f change.

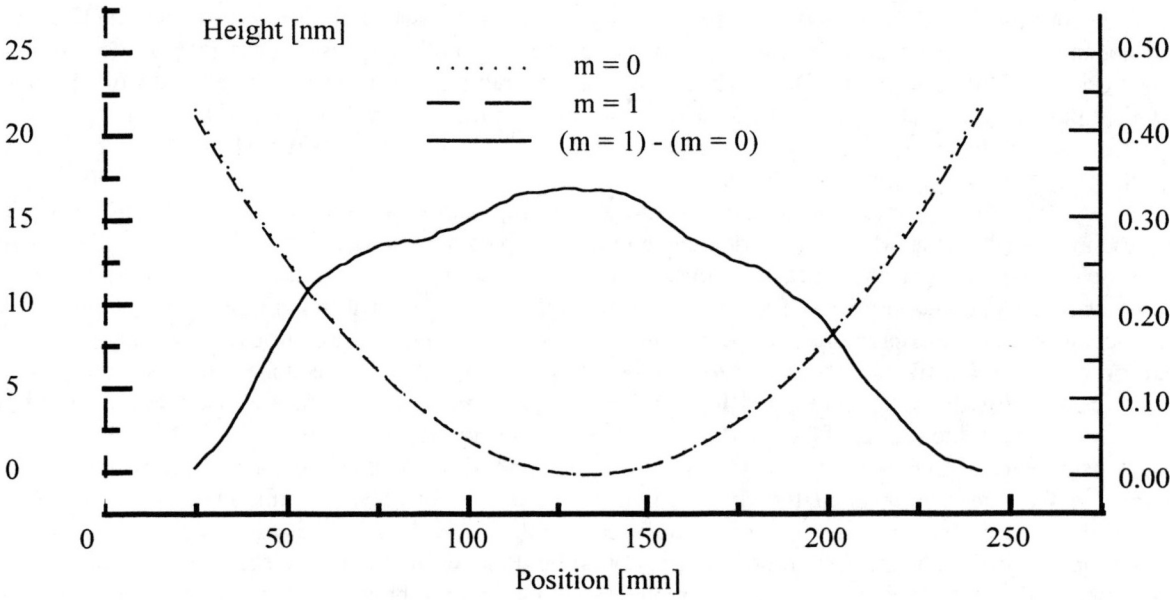

Figure 4. Measurements of the beamline 7.3.1 grating.

Therefore, we take the derivative of the grating equation, Equation (1), with m = 1:

$$\lambda f = \sin(\alpha) + \sin(\beta) \tag{3}$$

$$\lambda \Delta f = \cos(\beta) \Delta \beta \tag{4}$$

The relative groove frequency change will then be

$$\Delta f / f = \Delta \beta \cos(\beta) / (\lambda f). \tag{5}$$

For this grating and measurement, λ = 0.670 μm, f = 0.200 grooves/μm. Therefore, $\beta = \alpha$ = 3.84 degrees. When the LTP measures the slope profile of a mirror, it takes into account the factor of 2 between surface and optical deflection angles.

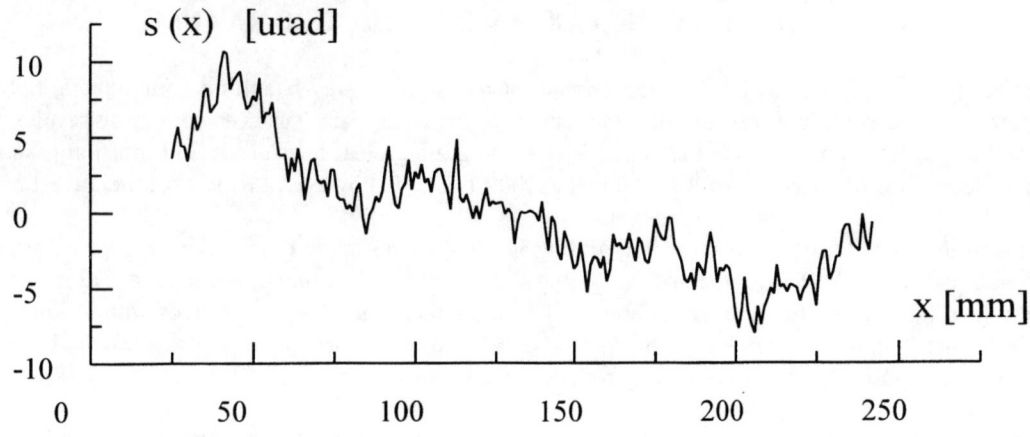

FIGURE 5. Slope profile of the diffracting surface of the BL 7.3.1 grating. The peak-to-valley slope variation here is 18.7 μrad.

Figure 5 reflects this same interpretation, so $\Delta\beta$ is actually twice the above peak-to-valley slope variation; $\Delta\beta = 37.4$ μrad. Then $\Delta f / f = 0.000278$.

Estimated error for this value $\Delta f / f$ would come from the uncertainty of $\Delta\beta$, which in turn comes from the LTP accuracy. LTP accuracy depends on 1) the noise floor of the instrument, 2) the length of the measurement, and 3) the nonlinearity of the LTP optical system. The noise floor of the LTP is less than 0.5 μrad rms for a measurement length of 250 mm. This produces a typical peak-to-valley error of 1.4 μrad. The nonlinearity of the LTP produces less than 0.7 μrad error for R = 250 m and a measurement length of 250 mm. The typical combined error is then 2.1 μrad. With this conservative error figure for $\Delta\beta$, the error bar for $\Delta f / f$ becomes 11.2%.

To check this value of $\Delta f / f$, it was suggested to measure the grating with an atomic force microscope (AFM). The BL 7.3.1 grating was measured by the AFM at the positions corresponding to approximately x = 40 and x = 220 in Figure 5, where the largest slope variations occur. At each of those positions an area of 75 μm x 75 μm was scanned with a resolution of 256 x 256 points. A rough measurement of groove frequency was obtained by determining the distance between cursors in the analysis section of the AFM program. A much more accurate way of determining groove frequency was to take the Fourier transform of a height profile along a line perpendicular to the grooves. This was done for 1) several orders of the transform at the center of each measurement and for 2) the +1 and -1 orders of the transform at several places in each measurement. Table 1 shows the results of the AFM measurement of the grating.

If both methods of determining groove frequency in Table 1 are valid, then a huge error bar of 1000% must be placed on this measurement. If the measurement of the first method (averaging many orders) is influenced too much by higher frequency noise, then the second method should be more believable. Clearly this analysis points out the difficulty in quantifying AFM lateral measurements. Of more importance is the direction of frequency change. For both methods the frequency is greater at x = 220. In Figure 1(b) the diffracted ray at the left will be inclined away (clockwise) from a ray of m = 0. This corresponds to a lower slope value, which agrees with a lower slope value at x = 220 in Figure 5.

TABLE 1. AFM measurements on the BL 7.3.1 grating. Values are for f [grooves/mm].

Method	x = 40	x = 220	$\Delta f / f$
Average 12 orders of center of area	200.70347	201.06363	0.0018
Average +1, -1 orders at five places in area	193.424	193.461	0.000185

CONCLUSION

Using a LTP has shown to be a useful and valid method for measuring the diffracting surface of a long beamline grating. Measurements of a small grating show good agreement between the LTP and Zygo large aperture interferometer. Many long beamline gratings can be measured only on the LTP.

For the BL 7.3.1 grating the value $\Delta f / f$ of 0.000278 falls within the large error bar placed on the AFM measurement, and is closer to the measurement of the better of the two AFM methods. While the AFM would be valuable for revealing groove shape and microroughness at high spatial frequencies, the AFM is not suited for determining the quality of the entire diffracting surface.

In addition to diagnosing the diffraction quality of gratings, this LTP measurement technique will also be useful for measuring long variable line spacing gratings.

ACKNOWLEDGMENTS

The authors thank Tim Renner for supplying the X-ray synchrotron diffraction grating from the ALS beamline 7.3.1. We thank Malcolm Howells for discussion of measurement methods of the grating. Thanks also to Steve Lindaas and Dan Pinkas for helping with AFM measurements.

This work was supported by the Director, Office of Energy Research, Office of Basic Energy Sciences, Materials Sciences Division of the U. S. Department of Energy, Under contract No. DE-AC03-76SF00098.

REFERENCES

1. Takacs, P. Z., and Qian, S, *United States Patent* 4884697, 1989.

2. Takacs, P. Z., Feng, S. K., Church, E. L., Qian, S., and Liu, W., "Long trace profile measurements on cylindrical Aspheres," *Proceedings of the SPIE,* **Vol. 966**, pp. 354-364, (1988).

3. Irick, S. C., McKinney, W. R., Lunt, D. L. J., and Takacs, P. Z., "Using a straightness reference in obtaining more accurate surface profiles from a long trace profiler," *Review of Scientific Instruments,* **Vol. 63**, No.1 (Part IIB), pp. 1436-1438, (January, 1992).

4. Irick, S. C., "Improved measurement accuracy in a long trace profiler: Compensation for laser pointing instability", *Nuclear Instruments and Methods in Physics Research,* **Vol. A347**, pp. 226-230 (1994).

Design and Performance of Two New Double Crystal Monochromators

D. J. Holly, W.P. Mason, F. H. Middleton, T. Sailor, and R. E. Smith

University of Wisconsin Physical Sciences Laboratory, Stoughton, Wisconsin 53589

Abstract. The University of Wisconsin-Madison Physical Sciences Laboratory (PSL) has designed and built two new types of double crystal monochromator (DCM): the Compact DCM and the Broomstick DCM.

The Compact DCM has a single rotational stage and features a frame footprint only 0.74 m by 0.84 m, with a distance of only 0.35 m from beam centerline to the back of the instrument. It features full frontal access to the crystals without the need for a track system. The instrument is UHV-compatible by virtue of its all-metal construction and features motorized in-vacuum adjustments using bellows-sealed drive motors.

The Broomstick DCM uses a patented mechanism which combines a cotangent generator with a pair of unique half-angle mechanisms. An external linear drive is the only input required to scan the monochromator; fine angular adjustments are made via PZT or manually.

We present the final design of both DCMs and results from their performance tests.

DOUBLE CRYSTAL SCAN MOTIONS

In a parallel crystal double crystal monochromator, the diffracting planes of the two crystals must remain parallel as the monochromator is scanned through different Bragg angles. In addition, at least one of the crystals must translate if the beam offset is to remain fixed as the monochromator is scanned. The two monochromator designs described here achieve this motion in different ways.

The basic design of the Compact DCM is similar to that described by Mills and King (1). Both crystals are mounted on a large scan plate so that they rotate together as the plate is scanned, maintaining the diffracting planes of the two crystals parallel to each other. The center of rotation of the main scan plate is located on the surface of the exit crystal, at a point such that a line drawn from this point perpendicular to the surface of the exit crystal intersects the entrance crystal at the location of the incoming beam. To keep the exit beam at a constant vertical offset h from the entrance beam, the entrance crystal must be translated perpendicular to its surface so that the spacing between the crystal surfaces is $y_1 = h/(2 \cos \theta)$, where θ is the Bragg angle. To keep the beam impinging in the same location on the second crystal, the second crystal must be translated by a distance parallel to its surface $z_2 = h/(2 \sin \theta)$. All required motions are mechanically independent and can be individually controlled by computer to achieve constant exit beam height and fixed beam spot on the crystal surface.

In the Broomstick DCM, the motion of the entrance crystal during scanning is a pure rotation about a point where the entrance beam intersects the crystal surface. The mechanism (U.S. patent No. 5,268,954) maintains the two crystals parallel while the angle is scanned and causes the center of rotation of the second crystal to translate along a horizontal path by an amount $x = h \cot(2\theta)$, where h is the height difference between entrance and exit beams and θ is the Bragg angle of the crystal. The basic mechanism (Fig. 1) consists of a pair of half angle mechanisms combined with a parallelogram "broomstick" cotangent generator. The half-angle generator is a simple mechanism with two arms which constrains the angle of the second arm to be precisely half that of the first arm. An external linear drive stage provides the only mechanical input needed to scan the monochromator, apart from fine adjustments of one crystal using a piezoelectric element.

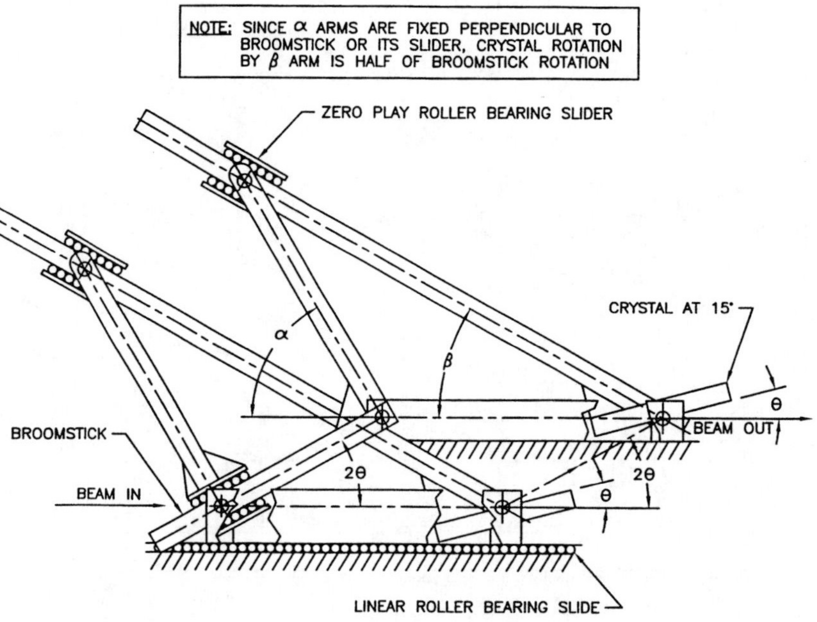

FIGURE 1. Basic Mechanism of the broomstick monochromator.

COMPACT DCM DESIGN

Figure 2 shows the main mechanism of PSL's Compact DCM. Two such instruments are in use at the Advanced Photon Source (by SRI-CAT and UNI-CAT). While they are very similar, some of the features described below are present on only one or the other of the Compact DCMs.

The heart of the Compact DCM (2) is the main θ scan plate, which is mounted by a single large four-point-contact ball bearing to the 50 mm thick stainless steel backplate of the vacuum chamber. The main scan plate is driven by an arm which is bellows-coupled to a linear scan stage mounted (in air) to the frame beneath the vacuum chamber. A counterweight mechanism is used to balance part of the load on the main scan drive. The range of motion for the main scan plate is -5° to 30° (SRI-CAT) and 0° to 45° (UNI-CAT).

The linear scan stage which drives the main scan plate features a linear optical encoder to encode the stage position; the encoded value is accurately mapped to scan plate angle by laser interferometer calibration. The scan stage is mounted below the chamber for compactness, while the linear encoder is mounted at scan arm height to minimize the Abbé offset. The scan stage travels on four hardened steel wheels which rotate on precision ball bearings and ride in commercial vee-rails. The linear scan stage is driven by a stepper motor using microstepping at 50,800 microsteps per revolution, giving a motion resolution of 0.009 arcseconds. The encoder resolution corresponds to a scan angle resolution of 0.037 arcseconds. The motor rotation is converted to translation using a precision leadscrew having 40 threads per inch and a custom-made silicon bronze nut which is adjustable for preload.

In addition to the main scan plate rotational motion, the entrance crystal requires translation perpendicular to the crystal surface (y1) and the exit crystal requires translation parallel to the crystal surface (z2). These motions are accomplished using in-vacuum linear stages driven by hermetically-sealed linear actuators. The y1 stage accomodates a motion of 20 mm and uses four recirculating crossed roller trucks riding on vee rails. The z2 stage accomodates a total motion of 200 mm and uses either four recirculating crossed roller trucks or six hardened steel wheels which rotate on ball bearings and ride in commercial vee-rails.

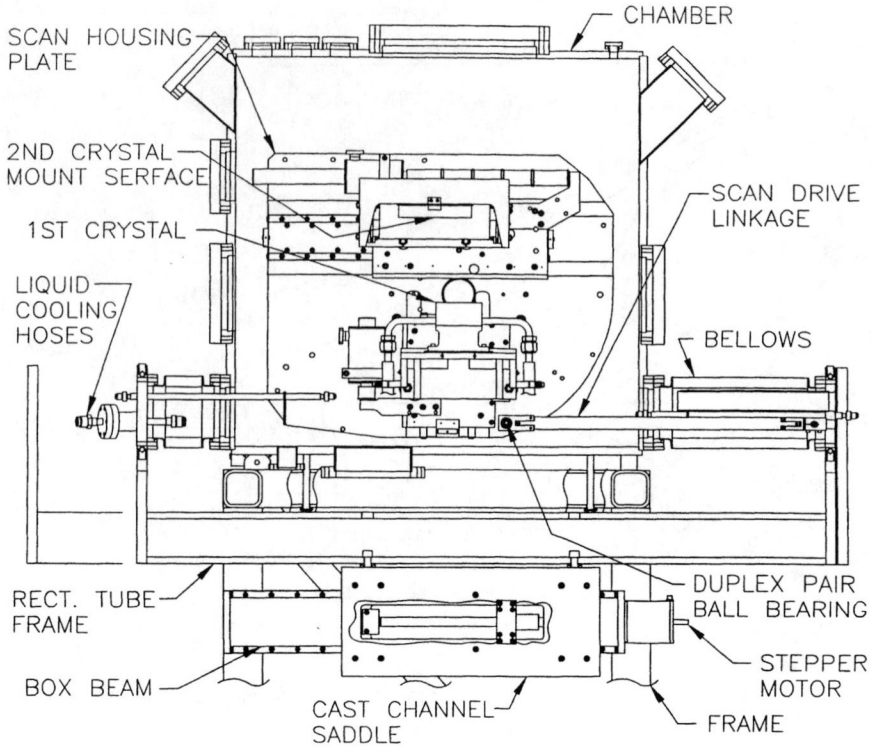

FIGURE 2. Main mechanism of the Compact DCM.

These linear motions are driven by modified commercial linear actuators using a hollow-rotor stepping motor and a lead screw with 40 threads per inch. The motor and leadscrews are enclosed in stainless steel cans and communicate the linear motion via stainless steel bellows which are attached to the leadscrews. Another set of flexible stainless steel bellows is used to carry the wires from the motor (and rotary encoder, in some cases) outside the vacuum chamber and to allow the motor to run in air.

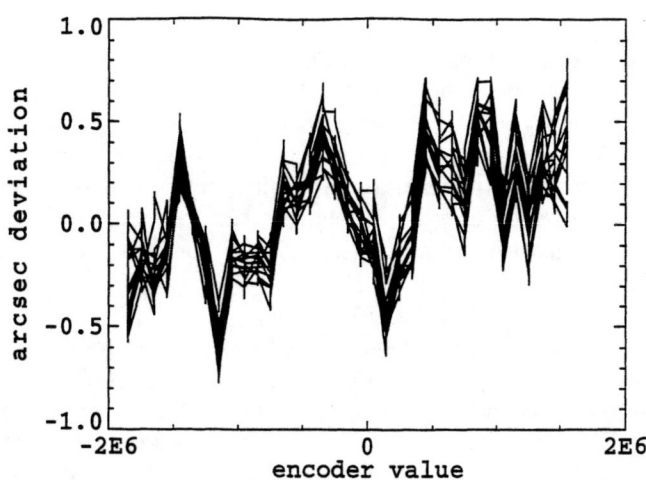

FIGURE 3. Overlay of 13 full scans of the Compact DCM. Each scan is taken with the second crystal stage at a different z position. The angle shown is the deviation from a polynomial fit (the same polynomial is used for all scans) using a simple linear correction for the z position.

In addition to the main motions θ, y1, z2 described above, the Compact DCM provides a number of remotely-actuated fine crystal adjustments. These include adjustments of first and second crystal pitch (θ1 and θ2), second crystal translation parallel to scan axis, second crystal yaw (rotation about y), and second crystal roll (rotation about z). The actuators for all these motions are based on hermetically-sealed, bellows-coupled linear actuators similar to those which move y1 and z2, and in some cases include rotary encoders on the hollow rotor shaft. Microstep drives with 50,800 microsteps per revolution provide fast adjustment response and good resolution.

In addition, one of the monochromators has a manually-actuated first crystal yaw adjustment and a piezo-actuated exit crystal pitch dither which can move 1 arcsec at 100 Hz (+/- 45 arcsec total range). The piezo actuator is sealed in a stainless housing and actuates through a bellows similar to the other adjustments. A strain gauge monitor provides direct measurement of the piezo transducer motion.

Flexible stainless steel hoses with braid reinforcement are used to cool the first crystal stage, using water or liquid nitrogen as the cooling medium.

COMPACT DCM PERFORMANCE

The main (θ) scan mechanism is similar to that used in PSL spherical grating monochromators (3) and exhibits similar performance with one exception: in the Compact DCM the loading on the scan mechanism changes significantly as z2 is scanned. If the scan stage linear encoder is maintained at a fixed position as z2 is translated over its full 200 mm range, this loading change causes a shift of about 6 arcsec in the main scan plate angle with the crystal faces near horizontal (less if the crystal faces are not near horizontal). This overall scan plate shift can readily be compensated by a slight adjustment of the main scan stage position, based on the known z2 position and main scan stage encoder position. Figure 3 shows a superposition of thirteen full scans, with z2 stepped between scans to cover the full 200 mm z2 motion. The raw data has been corrected by a simple linear equation using the z2 position and main scan stage encoder value to simulate a simple correction that could be made using the main scan drive. Note that with the correction, most of the points deviate by less that 0.5 arcsec total range over thirteen scans, even though the measurements were done in a location with no special temperature control.

Since z2 motion and y1 motion are performed simultaneously with the scan, pitch variation during these motions is an important parameter. The measured total pitch excursion during 20 mm y1 motion for the two Compact DCMs was 3.5 arcsec and 2.3 arcsec; the measured total pitch excursion during 200 mm z2 motion for the two monochromators was 2.2 arcsec and 4.3 arcsec. An example of the pitch excursions during this motion is plotted in Fig. 4, which shows the pitch excursions for the stage with hardened steel wheels. The stage using crossed roller bearings shows a similar range of pitch variation, but with several reproducible excursions of about 2 arcsec each having a z2 width of about 10 mm.

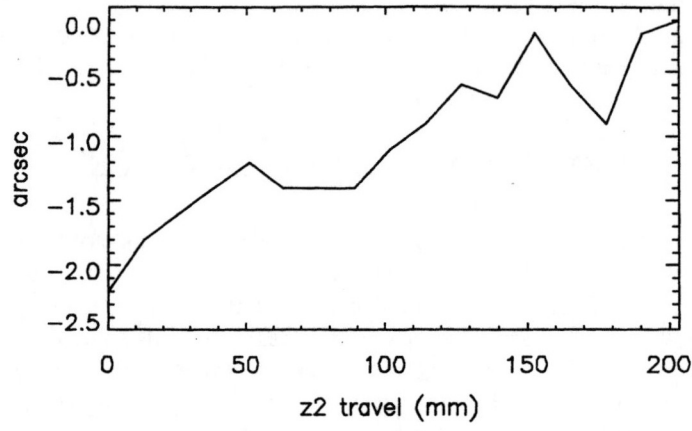

FIGURE 4. Compact DCM second crystal pitch variation over the full z range.

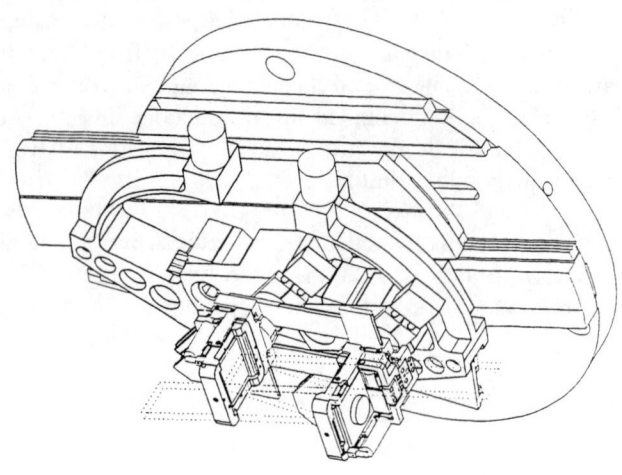

FIGURE 5. Broomstick monochromator mechanism.

The hermetically-sealed linear actuators must move in very small steps to secure the high resolution needed in some of the fine adjustments. Autocollimator measurements show that these actuators are capable of these fine, sub-arcsecond adjustments. In order to test motor temperature rise, we scanned the main scan stage to 30° and ran the z2 stage continuously at high speed with the chamber evacuated. A thermocouple mounted to the motor indicated that even after several hours of this continuous high-speed motion the motor temperature would not rise above 150 F. All motors are equipped with thermocouple sensors so their temperature can be readily monitored under vacuum conditions.

Initial tests suggest that the parallelism between the two crystals during a scan can be improved by making the first stage stiffer and thus better able to resist forces from the cooling hoses. We are presently modifying the first stage to decrease its compliance. The change in parallelism during a scan is reproducible and exhibits very little hysteresis, so that the remaining change should be correctible via the fine pitch and roll adjustments.

BROOMSTICK DCM DESIGN

Figure 5 shows the broomstick DCM mechanism which PSL designed and built for the Pohang Light Source in South Korea. The mechanism is designed for an input beam – exit beam separation of 50 mm and accomodates a Bragg angle range of 8° to 80°. Both crystal assemblies are balanced with counterweights. All linear bearings use silicon nitride balls; all rotary bearings use hollow rollers. Further details of the design and construction are given in reference 4. In addition to the scan motions which are driven by the broomstick mechanism, one of the crystal mounts provides pitch and roll adjustment using piezoelectric transducers, and both crystal holders provide manual (in-vacuum) pitch, roll, yaw, and elevation adjustments using a special rotary, linear gimbaled UHV feedthrough.

BROOMSTICK DCM PERFORMANCE

Crystal parallelism was tested using an autocollimator with mirrors mounted in place of the crystals. Figure 6 shows the deviation from parallelism over the full scan range. Note that the total deviation from parallelism is less than 6 arcsec over the full range, and varies much less over any selected small portion of the full range, except around $\theta = 45$ where one crystal holder is directly above the other. An expanded view of this region reveals a highly reproducible shift of about 10 arcsec which occurs over a travel of 1 cm. This shift corresponds to the position at which the linear slides reverse direction. We believe this shift can be significantly reduced in future designs by providing an improved seating surface for the linear slide bearings, reducing the frictional force on the linear slides.

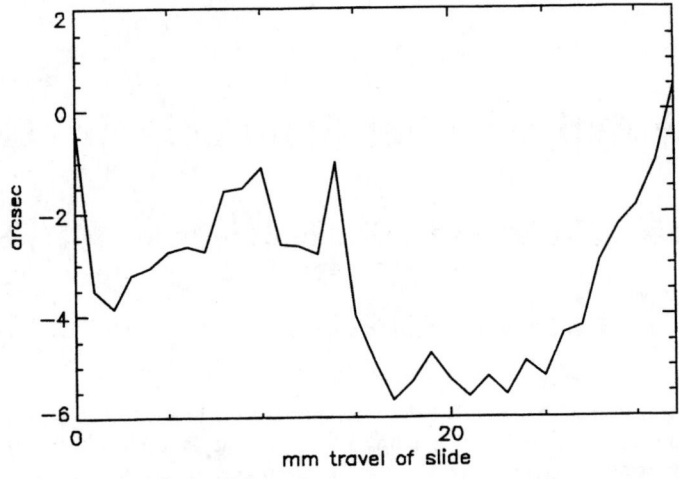

FIGURE 6. Autocollimator measurement of crystal parallelism over full range of Broomstick monochromator

CONCLUSION

PSL's Compact DCM mounts both crystals on a large scan plate which is supported by a single large precision bearing. The necessary linear motions and fine angle adjustments are made via independent sealed motor assemblies.

The Broomstick DCM uses a patented mechanism which combines a half-angle mechanism with a cotangent generator to provide the required crystal motions with a single linear drive.

While the two DCMs are based upon different mechanisms and are configured for different energy ranges, both are compact designs and feature true UHV compatibility to the 10^{-10} torr range. Innovations incorporated into their design and improvements made as a result of their performance measurements will be incorporated into future instruments as well.

REFERENCES

1. D. M. Mills and M. T. King, Nucl Instrum and Meth **208**, 341 (1983)
2. F. H. Middleton, G. R. Emmel, F. Feyzi, and G. M. Gregerson, Rev Sci Instrum **67** (9), 3351 (1996)
3. S. L. Hulbert, D. J. Holly, F. H. Middleton, and D. J. Wallace, Nucl Instrum and Meth **A291**, 343 (1990)
4. W. P. Mason, G. R. Emmel, F. Feyzi, D. J. Holly, F. H. Middleton, and A. J. Pagac, Rev Sci Instrum **67** (9), 3350 (1996)

New Actively Bent Plane Mirror at SRC

Tim Kubala, Mike Fisher, Ruben Reininger, Mary Severson

University of Wisconsin Synchrotron Radiation Center, 3731 Schneider Drive, Stoughton, WI 53589

Abstract. A dynamically tunable mirror is currently being developed as the first (M0), horizontally focusing optical element in a new 4 meter NIM beamline at the SRC. Balanced bending moments at the ends of the 375 x 60 x 10 mm single crystal silicon optic are generated by compressing a system of highly compliant springs over a relatively long translation, allowing for active beam focusing during monochromator scanning. A coaxial flexural pivot and hinge arrangement provides positional constraint of the water cooled optic, while controlling the small torque rotations (< 3.5 mrad) in tuning the radius of curvature (49.5 to 55.5 meters) and accommodating thermal expansion due to a bulk temperature rise. This mechanism is due to be commissioned in late 1997.

1. INTRODUCTION

It is possible to achieve a stigmatic image at the exit slit of a vertically dispersing normal incident monochromator by incorporating an active cylindrically bent plane mirror focusing horizontally before the entrance slit [1]. This mirror is the first optical element (M0) in our beamline design [2] and absorbs 23 of the 31 watts of power incident to it when the undulator is tuned to 6 eV. The balance of the power is absorbed by a subsequent water cooled plane mirror (M1). A spherical optic (M2) focuses the source vertically onto the entrance slit, and contributes to the horizontal focusing. The radius of curvature of the cylindrical mirror can be easily calculated such that, in combination with the M2 and spherical grating, the correct wavelength is focused onto the exit slit both vertically and horizontally. By dynamically varying a force applied as a couple to the ends of the M0 optic, the calculated radius of curvature can be simultaneously tuned with the monochromator. Utilizing a commonly accepted scheme [3] of deflecting compliant elements to generate these forces, a mechanism was developed which reflects a design philosophy emphasizing simplicity and durability while remaining sufficiently close to optimal. The mechanical aspects of this scheme are described below.

2. M0 MIRROR

The M0 optic will be a continuously adjustable cylindrically bent plane mirror whose radius of curvature is to vary between 49.53 and 55.52 meters at a 5 degree grazing angle of incidence. This mirror is placed approximately 14 meters from an electromagnetic undulator source in a 4 meter NIM beamline [2]. An optimal beamline layout compatible with the available floor space was developed after numerous calculations and ray traces using SHADOW [4]. Ray tracing also provided beam profiles at discrete locations. An example of the improvement in the image at the sample position due to an optimized radius of curvature of the M0 is shown in Fig. 1. Both ray traces were obtained with the monochromator and the undulator tuned to 30 eV and using 30 μm slits. The left ray trace shows the image when the mirror has a radius of 54.6 m, the radius required for 30 eV. The right ray trace is the case when the mirror has a radius of curvature of 49.4 m, corresponding to a radius appropriate for 12 eV. Significant improvement in the spot is evident from the figure.

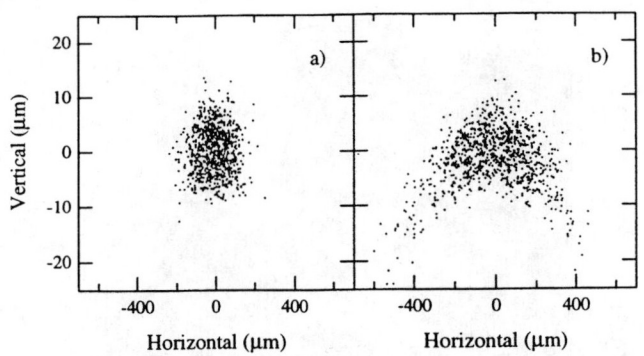

FIGURE 1. Ray traces at the sample position for a photon energy of 30 eV. Left: mirror radius tuned for 30 eV. Right: mirror radius tuned for 12 eV.

A mirror blank with dimensions of 375 x 60 x 10 mm (L x W x H) was determined as sufficient and appropriate for optimum filling with a 300 x 20 mm beam footprint. A single crystal of silicon was selected as an excellent material for the optic substrate on the merits of its homogeneity, its ease of polishing and its high thermal conductivity to thermal expansion ratio. A maximum slope error of 1.0 arcsec in both the meridian and sagittal directions was specified for the manufacturing of this optic.

3. PHYSICAL AND THERMAL LOADS

Reactions to physical and thermal loads were calculated for the single crystal silicon optic utilizing the physical properties listed in Table 1.

Table 1: Material properties of single crystal silicon.

coefficient of expansion	2.3×10^{-6} /K
conductivity	0.168 W/mm-K
density	2.3×10^{-3} g/mm^3
specific heat	0.75 J/g -C
Young's modulus	16.7×10^4 N/mm^2

Because deflections are small and radius of curvature large for mirrors of grazing incidence, such mirrors can be satisfactorily characterized by the formula for a beam in pure bending [5]:

$$R = EI/M. \qquad (1)$$

Here, R = radius of curvature, E = modulus of elasticity, I = bh^3/12 = moment of inertia of a rectangular cross section of the beam, (where b = width and h = thickness) and M = applied moment. Given a uniform rectangular cross section (10 x 60 mm) subjected to equal and opposite couples, a bending moment of 16.86 Nm (149 in lb) is required to yield the smallest radius of curvature of the M0 silicon mirror. Maximum deflection of the mirror surface is at the pole and given by:

$$\delta = L^2/(8R) \qquad (2)$$

with L = mirror length and R = radius of curvature. Gross pole deflections from flat (R = infinity) to maximum (R = 49.53 m) curvature amounts to 300 µm, with a total pole motion of 24 µm within the range of adjustment. This motion is insignificant when compared with the horizontal FWHM of the focused image, 400 µm. By producing balanced moment forces and assuring symmetric mechanical geometry, the mirror pole remains on a line normal to the optic containing the center of curvature.

The maximum absorbed power on this optic of 23 Watts corresponding to an undulator setting of 6 eV was calculated using a version of URGENT [6], which allows one to determine the thermal load on optical components. A three dimensional plot of the absorbed power density on the mirror surface obtained with this code is shown in Fig. 2.

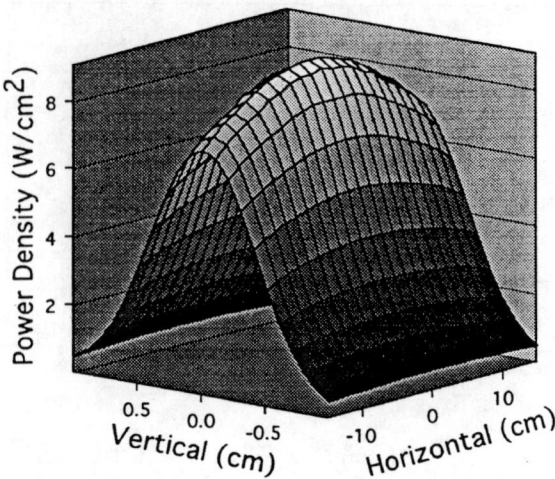

FIGURE 2. Absorbed power density on M0

A strong variation of power occurs along the mirror width. The mirror is cooled by conduction along the mirror's length and through an indium film interface to ground copper blocks that are clamped to the mirror end. The indium film interface reduces the contact resistance to heat transfer (h = 3000 W/m²K) [7]. The copper blocks themselves are attached to a water cooled continuous copper tubing loop (no coolant-to-vacuum joints) that penetrates the UHV chamber through a pair of strain-relieved bellows feedthrough assemblies. Thermal analysis of the maximum absorbed power loading indicates a peak temperature rise at the center of the mirror of 15°C within 15 minutes (Fig. 3).

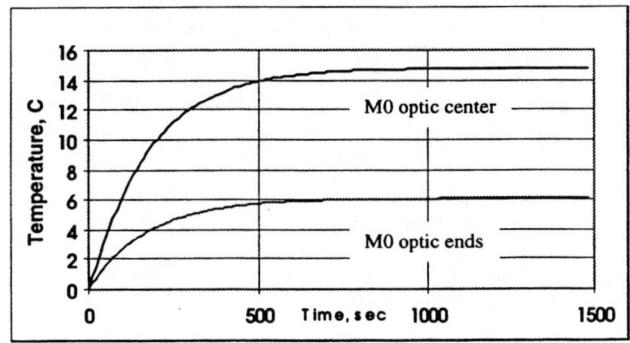

FIGURE 3. Time to Equilibrium for M0 optic

The relatively small temperature rise in the optic minimizes radiative losses (only 2% of the total heat transfer) and allows the majority of the absorbed power to be channeled directly to the cooling loop, eliminating concerns over thermally induced expansions in the mechanism and support structure. The bending mechanism is designed to be insensitive to the thermal expansion of the mirror substrate (maximum change of 17 μm in length) which changes as a function of both undulator energy setting and beam decay.

4. BENDING MECHANISM

Mechanical symmetry in a longitudinal direction is maintained and utilized in both the bending and driving mechanisms. The deflection of the 375 x 60 x 10 mm silicon substrate ① to a desired cylindrical shape of prescribed radius is facilitated by applying couples of an opposite sense at the ends of the mirror about parallel axes in the neutral plane of the optic (fig. 4). A flexural hinge and flexural pivot pair combination allows rotational freedom about, and is coaxial to, these moment centerlines. The flexural hinge ② provides compliance in a direction parallel to the optic length while constraining four degrees of freedom. This single axis flexural hinge was designed to provide maximum stiffness [8], yet remain compliant about its input axis to small changes in mirror length due to thermal expansion. The flexural hinge and flexural pivot clamps were EDM wire cut to assure a common centerline of rotation. The flexural pivot pairs ③ [9] experience maximum rotations of 3.5 mrad, are always loaded in the same sense, and inherently eliminate the liabilities of sliding friction, backlash and rolling contact in a UHV environment.

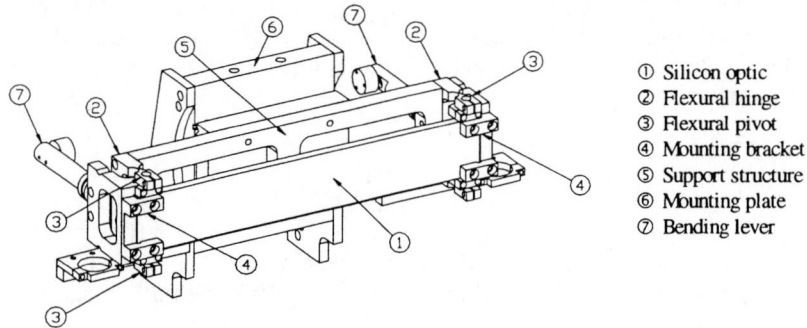

① Silicon optic
② Flexural hinge
③ Flexural pivot
④ Mounting bracket
⑤ Support structure
⑥ Mounting plate
⑦ Bending lever

FIGURE 4. Bending Mechanism

Use of face mounted brackets ④ to transmit the bending torque at the mirror ends eliminates contact stress near the pole and potential distortions due to compressive strain, while allowing dissipation of the clamping forces outside of the acceptance area of the optic. Additionally, these torque brackets function as mounting plates for the copper cooling blocks on the back of the optic and when phosphorous coated, as first order beam position diagnostics. A single rigid support structure ⑤ reacts the bending moments generated by the driving mechanism and bellows feedthroughs, and maintains the relationship between the bending levers at each end of the mirror. An intermediate mounting plate ⑥ functions as a section of the vacuum chamber and supplies a three ball kinematic mounting scheme for the optic support structure on its internal (UHV) side. The external side of this plate is directly bolted to a fully adjustable kinematic mount system used to accurately position the mirror along the beamline. This effectively isolates the bending mechanism from chamber stress and vacuum loading. Bending levers ⑦ complete the link from a completely external driving mechanism to the bending mechanism within the UHV chamber.

5. DRIVING MECHANISM

Small variations in curvature of the bendable mirror are realized by compressing a system of highly compliant springs over a relatively long translation, generating the required force (in contrast to a discrete displacement) and applying that force to the mirror as a couple. This easily controlled applied force is insensitive to functional variables such as dimensional changes due to thermal expansion (very small mirror length changes relative to driver translations) and eliminates the need to produce and control very small (submicron) displacements.

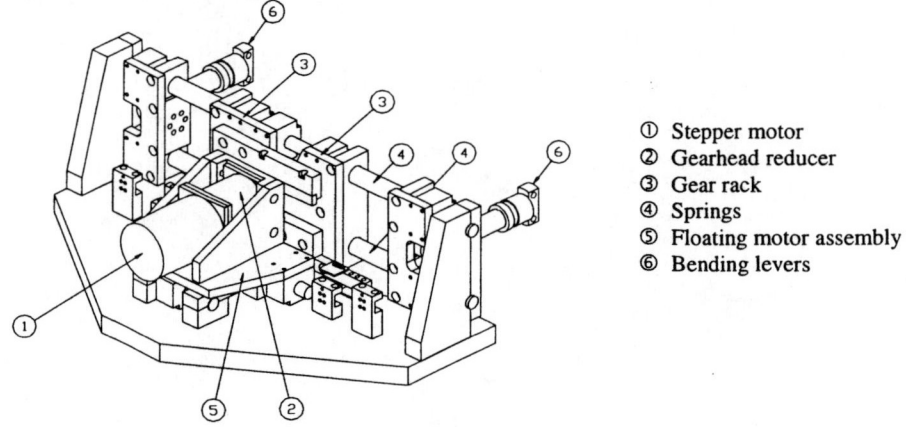

① Stepper motor
② Gearhead reducer
③ Gear rack
④ Springs
⑤ Floating motor assembly
⑥ Bending levers

Figure 5. Driver Mechanism

The driving mechanism (figure 5) was developed as a discrete entity external to the UHV environment, allowing use of lubricated stock precision components. A 4000 microstep/rev stepper motor ① is coupled to a 20:1 gearhead reducer ② which, through a single pinion, drives separate rack sections mounted to slider assemblies ③ in opposite directions. These assemblies compress pairs of compression springs ④ (~3 inches total translation) to produce and maintain the necessary force for a given radius of curvature. The mirror is subjected to a bending torque at its ends that originate as equal and opposite forces generated by the driving unit and transmitted to bending levers across ball on flat contacts in an attempt to minimize undesirable moments and deformation of the mirror due to misalignments. The floating stepper motor assembly ⑤ assures balanced forces at both lever arms which penetrate the vacuum chamber through welded bellows assemblies. A closed loop control system makes use of an integrated compression load cell directly measuring the applied force at the bending lever end ⑥. The combination of load cell resolution and stepper motor/reducer ratio yields a high sensitivity of control to the magnitude of the bending forces.

6. SUMMARY

All the mechanical components of the bending mechanism have been manufactured and delivered. The bender is presently being assembled with a test mirror. The characterization of the image quality in the full tuning range of the mechanism will be performed in the near future.

ACKNOWLEDGMENTS

Helpful consultation with D. Wallace and R. Hansen, as well as the efforts of T. Nelson and W. Thatcher is greatly appreciated. This work is based upon research conducted at the Synchrotron Radiation Center, Univ. of Wisconsin, which is supported by the NSF under award No. DMR-9531009.

REFERENCES

1. Koike, M., Heimann, P.A., Kung, A.H., Namioka, T., DiGennaro, R., Gee, B., and Yu, N., *Nucl. Instr. Methods* A **347**, 282 (1994).
2. Reininger, R., Severson, M.C., Hansen, R.W.C., Winter, W.R., Green, M.A., and Trzeciak, W.S., *Rev. Sci. Instrum.* **66**, 2194 (1995).
3. Padmore, H. A., Howells, M. R., Irick, S., Renner, T., Sandler, R., and Koo, Y-M., *SPIE Proceedings* **2856**, 145 (1996).
4. Lai, B. and Cerrina, F., *Nucl. Instr. Methods* A **246**, 337 (1986).
5. Roark, R.J. and Young, W.C., McGraw-Hill, New York (1975)
6. Walker, R.P. and Diviacco, B., *Rev. Sci. Instrum.* **63**, 392 (1992).
7. Khounsary, A., personal communication, (June 1996)
8. Paros, J.M. and Weisbord, L., *Machine Design*, 151-156, (1965)
9. Lucas Aerospace Corporation, Utica, New York

Preliminary Results From A New Plane Grating Monochromator at SRC

Mary Severson, Mark Bissen, Ruben Reininger, Mike V. Fisher, Greg Rogers, Dave Eisert, Tim Kubala, William Wood

University of Wisconsin Synchrotron Radiation Center, 3731 Schneider Drive, Stoughton, WI 53589

Abstract. A plane grating monochromator (PGM) has recently been completed and installed on an undulator port at the Synchrotron Radiation Center (SRC). The beamline is designed to be a high resolution, high throughput beamline with an energy range of 8 to 245 eV. Initial results show the undulator is performing very well and agreeing closely with theoretical predictions. The undulator and monochromator can be scanned together throughout the energy range using software on the beamline computer. Undulator scans are routinely performed during the morning shift and the small effects this induces on the storage ring are being further minimized. Monochromator scanning is achieved by rotating the plane mirror and rotating and translating the grating. An advanced control system maintains sub-arcsecond resolution for the rotations. Flux and resolution tests have shown that the grating performs poorly. A new laminar profile grating will be installed soon.

1. INTRODUCTION

A plane grating monochromator (PGM) has recently been completed at the Synchrotron Radiation Center (SRC) (1). An overhead view of the beamline is shown in Fig. 1. The main components of the beamline are a cylindrical mirror (M0) deflecting the beam horizontally and focusing sagittally into the entrance slit, the entrance slit, a plane mirror (M1) that rotates as the energy is scanned, a plane 1200 l/mm grating that rotates and translates, an ellipsoidal mirror (M2) that focuses the light horizontally and vertically onto the plane of the exit slit, the exit slit, and an ellipsoidal refocus mirror (M3). A plane mirror (M2a) can be lowered into place to divert the beam into the branch line. A second exit slit and a toroidal refocus mirror (M3a) complete the branch line. Also shown are the two sets of beam position monitors (BPM #1 and BPM #2) (2). Details of the optical (1,3) and mechanical design have been published (4). The SRC PGM design is expected to provide high resolution and high flux throughout the energy range of 8 to 245 eV using a single grating.

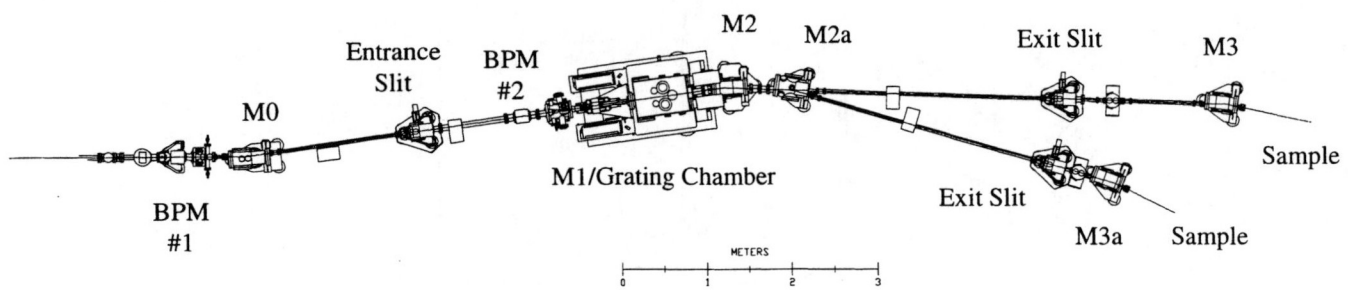

FIGURE 1. Overhead view of the PGM.

2. BEAMLINE SYSTEMS

Almost all beamline functions are computer controlled. A VME computer controls all the motors and feedback loops and interfaces with two PCs. One PC is dedicated to the two sets of beam position monitors. The beam position monitors can be scanned to map the beam horizontally and vertically or can remain stationary and monitor beam position. Another PC handles the main software for beamline scanning, undulator scanning, slit motions, vertical blade motions for aperturing and the positioning of the M2a diverting mirror for the branch line. The vertical blades before the M1 move as the monochromator is scanned to prevent overfilling of the M1 which would create scattered light in the system.

Monochromator scanning is achieved by rotating the plane mirror and rotating and translating the grating. Laser interferometers measure the rotations of the plane mirror and grating directly and are used in a feedback loop to control the stepping motor-piezoelectric actuator scan drive for each optic. Sub-arcsecond resolution is obtained for these rotations. The grating translates approximately 200 mm with a positional accuracy and vertical stability of a few microns. Step sizes as small as 0.1 meV have been used in several regions throughout the energy range and no energy calibration instabilities have been observed.

Undulator scans are routinely performed during the morning beam. The global feedback system on the storage ring prevents undulator scanning from producing significant changes in the orbit position. There are some residual beam size variations that occur during scanning and these are being further minimized.

3. FLUX MEASUREMENTS

The undulator is a 3.5 meter long permanent magnet device (5) located in long straight section 3 (LSS3) of the Aladdin storage ring. It has a 7 cm period (50 periods) and is designed to work in the energy range of 8 to 245 eV at 800 MeV using the first, third, and fifth harmonics.

A representative scan of the monochromator over its full energy range with the undulator set, in this case, at a fixed gap of 50.9 mm (40 eV) is shown by the solid line in Fig. 2. The scan was recorded using a gold diode, 300 micron entrance and exit slits and converted to "standard" synchrotron radiation units using the photoyield of gold. A large entrance slit was used in the measurement to ensure that no photons were lost at the entrance slit. A simulated spectrum for the same gap was calculated using the URGENT (6) code. This spectrum was then multiplied by the reflectivity of the optics, the acceptance of the beamline, and the efficiency of the ideal grating. The latter was calculated using the laminar theory for grating efficiency including shadowing (7). The acceptance of the beamline is determined by the M1 angle and therefore changes with energy. URGENT calculates using only one acceptance value in a given run, so several runs were done using different acceptance values corresponding to the peak energies of the harmonics. These runs were combined to produce the simulated curve. The simulated curve *multiplied by 0.05* is represented by the dotted line in Fig. 2. As seen in the figure, this factor brings the simulated and experimental fluxes at the first harmonic to roughly the same value. This is also the case for the second harmonic. For the higher harmonics displayed in the figure, this factor does not bring the peak values for the two curves into agreement indicating a difference in grating behavior for that portion of the spectrum. The experimental spectrum also shows

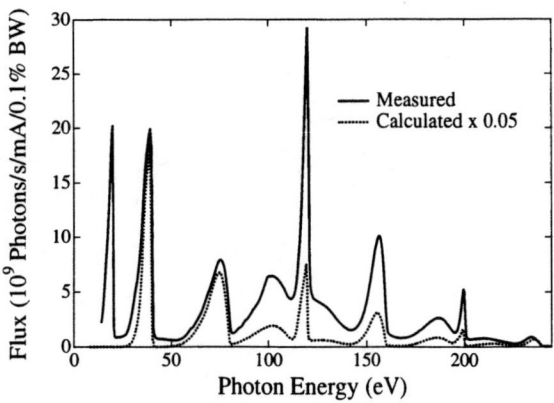

FIGURE 2. Comparison of the measured and simulated (multiplied by 0.05) undulator spectrum for a fixed undulator gap of 50.9 mm (40 eV). The entrance and exit slits were set at 300 microns for the experimental data. Details of the simulated curve calculation are given in the text.

a significant second order peak for the first harmonic despite the fact that the grating has a laminar profile. It is also worth noting that the overall shape of the calculated harmonics is very similar to those measured. Similar flux losses were noticed for other undulator settings and for scans in which the undulator and monochromator were scanned.

To get a better understanding of where the flux losses were occurring, several measurements were performed for different undulator gaps using BPM #1, BPM #2, and a gold diode after the exit slit. The measurements using the BPM's were done by scanning one of the molybdenum blades over the whole beam footprint and integrating the current. The integrated current at each BPM was estimated in the following manner. Undulator spectra were generated using URGENT and converted to units of photons/sec/eV/mA. They were subsequently multiplied by the yield of Mo and integrated as a function of the energy. For BPM #2, the spectra were also multiplied by the reflectivity of the M0 and then integrated. No losses were found at the position of BPM #1. The measurements with BPM #2, done with the entrance slit set to 300 microns, yielded one half of the estimated values.

The flux at the gold diode after the exit slit was measured using the monochromator in a zero order configuration. In this situation, the grating acts as a mirror thus eliminating any losses due to grating efficiency. With the exit slit set at 300 microns, the plane mirror was set to a given angle of incidence and the grating was scanned around the same angle of incidence using steps corresponding to the slit width. The gold photodiode current was then integrated and compared with the URGENT calculations which included, in these cases, the gold photoyield, the reflectivities of M0, M1, the grating, and M2, and assumed the efficiency of the grating for zero order is 50%. The measured values were found to be approximately 40% of the calculated values. These results and those of BPM #2 led us to conclude that the flux losses shown in Fig. 2 can be attributed to a very poor grating efficiency in first order.

4. RESOLUTION TESTS

A gas cell was placed on the end of the main branch of the PGM to verify the energy calibration and resolution. Using this cell several photoionization spectra of neon, krypton and xenon were recorded. Preliminary tests showed a non-linear behavior of the energy calibration. This was solved by introducing an offset on the M1 angle.

Figure 3 shows the Rydberg series $3d_{5/2} \rightarrow np$ and $3d_{3/2} \rightarrow np$ of krypton (8,9) recorded with entrance slit and exit slit set at 71 and 42 microns, respectively. These slit settings have the same contributions to the resolution. The equivalent transitions in xenon (8,9) recorded with the same slit settings are shown in Fig. 4.

A stringent test for the monochromator resolution is the spectrum of the high n Rydberg transitions of neon $2s \rightarrow np$ above 48 eV (10). Figure 5 (a) shows this spectrum recorded with an entrance slit of 17 microns and an exit slit of 10 microns. Obviously, the results demonstrate that the theoretical resolving power greater than 20000 in this region is not achieved. A careful examination of the lower energy part of the spectrum reveals that it actually contains a ghost spectrum shifted 4 meV to higher energies. The same spectrum, recorded by masking 86% of the grating is shown in Fig. 5 (b). The energy separation between the maximum and minimum on the $2s \rightarrow 11p$ transition is 3 meV which corresponds to the best resolution measured with this grating.

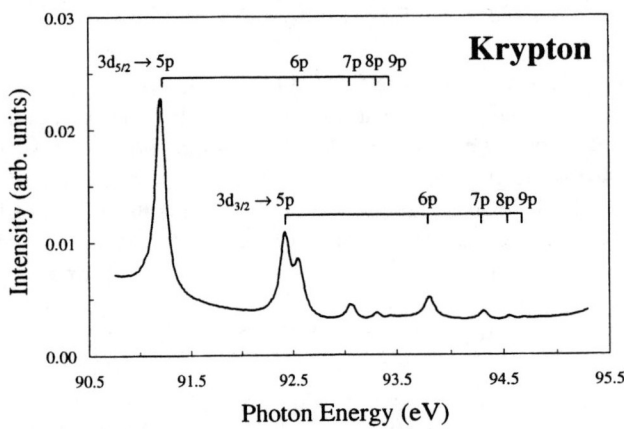

FIGURE 3. The Rydberg series $3d_{5/2} \rightarrow np$ and $3d_{3/2} \rightarrow np$ of krypton. The entrance slit is at 71 microns and the entrance slit is at 42 microns.

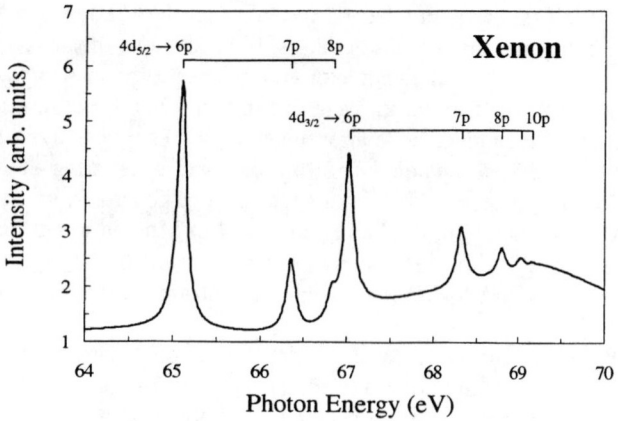

FIGURE 4. The Rydberg series $4d_{5/2} \rightarrow np$ and $4d_{3/2} \rightarrow np$ of xenon. The entrance slit is at 71 microns and the entrance slit is at 42 microns.

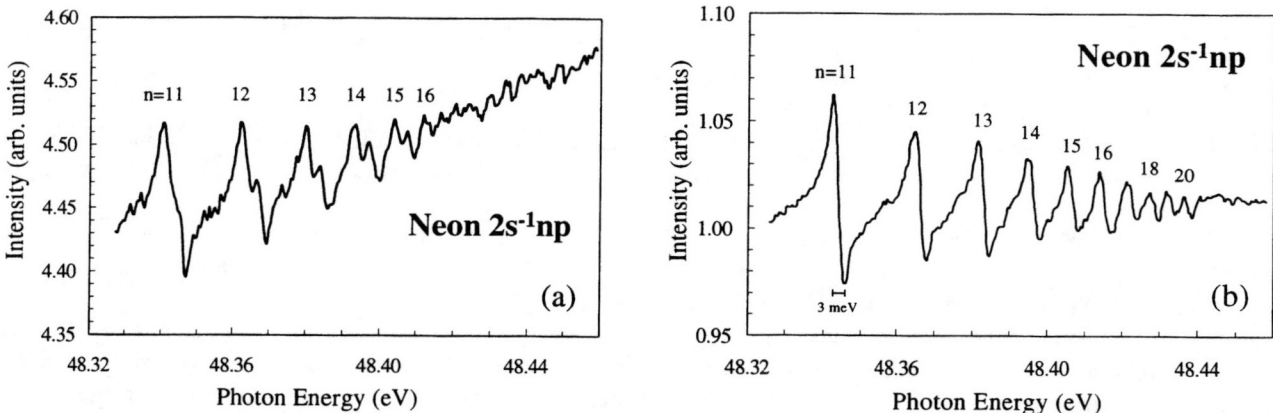

FIGURE 5. Neon Rydberg transition $2s \rightarrow np$ above 48 eV. Comparison of spectra taken with the entire grating illuminated (a) and with approximately 86% of the grating masked (b). Entrance and exit slits are at 17 and 10 microns, respectively.

5. SUMMARY

All the systems of the new PGM beamline at SRC are fully operational. At present the flux is low by more than one order of magnitude and the resolution using full grating illumination is far from design specifications. Several observations point to the installed grating as the reason for this poor performance. It is known that the groove profiles and depths on this grating are not correct. The grating efficiency drops off quickly below 80 eV and also produces a large amount of second order light. A new ion etched holographic grating has been manufactured by Hyperfine, Inc. of Boulder, CO. Atomic force microscope measurements on the new grating show that the groove profile and depth fulfill the specifications. The grating will be installed in the very near future and we expect the beamline will then fulfill its performance specifications.

ACKNOWLEDGMENTS

The authors wish to acknowledge the contributions to this work by Tom Nelson, and Wayne Thatcher (precision machining), Menghort Thikim (vacuum assembly), Al Riley (mechanical design), John McMurray (electronics), Dan Wallace (alignment) and Rick Keil (cabling and power distribution).

This work is based upon research conducted at the Synchrotron Radiation Center, University of Wisconsin - Madison, which is supported by the NSF under Award No. DMR-95-31009.

REFERENCES

1. Reininger, R., Crossley, S. L., Lagergren, M. A., Severson, M. C., and Hansen, R. W. C., *Nucl. Instrum. Methods* A **347**, 304-307 (1994).
2. Mossessian, D. A., Rogers, G. C., Bissen, M., Severson, M. C., and Reininger, R., *Rev. Sci. Instrum.* **67** (9), 1-4 (1996).
3. Jark, W., *Rev. Sci. Instrum.* **63**, 1241-1246 (1992).
4. Fisher, M. V., Bissen, M., Bourgeois, F., Eisert, D., Kubala, T., Reininger, R., and Severson, M., *SPIE Proceedings* **2856**, 212-219 (1996).
5. Younger, F. C., Pearce, W. J., and Ng, B., *Nucl. Instrum. Methods A* **347**, 98-101 (1994).
6. Walker, R. P., and Diviacco, B., *Rev. Sci. Instrum.* **63**, 392-395 (1992).
7. Franks, A., Lindsey, K., Bennett, J. M., Speer, R. J., Turner, D., and Hunt, D. J., *Phil Trans. R. Soc. Lond A* **277**, 503-543 (1975).
8. King, G. C., Tronc, M., Read, F. H., and Bradford, R. C., *J. Phys. B: Atom. Molec. Phys.* **10**, 2479-2494 (1977).
9. Haensel, R., Keitel, G., Schreiber, P., and Kunz, C., *Phys. Rev.* **188**, 1375-1380 (1969).
10. Schulz, K., Domke, M., Püttner, R., Gutiérrez, A., and Kaindl, G., *Phys. Rev. A* **54**, 3095-3111 (1996).

Variation of Q with Energy in Mosaic Analyzers for Inelastic X-ray Measurements

J.Z. Tischler*, B.C. Larson*, and Paul Zschack†

*Solid State Division, Oak Ridge National Laboratory, Oak Ridge, TN 37831-6030
†Univ. of Illinois, UNICAT, Bldg. 438D, Argonne National Lab., Argonne, IL 60439-4863

Abstract. Curved mosaic graphite analyzers have been used for many years for inelastic scattering measurements with both conventional x-ray sources as well as synchrotron sources (1). The trend in recent years has been to use spherically bent perfect crystal analyzers to collect large solid angles with high energy resolution. Although, these spherical analyzers achieve excellent energy resolution, the large solid angle limits the Q resolution. For cylindrically bent mosaic graphite, it is possible to obtain good energy and Q resolution simultaneously, while maintaining a large solid angle by collecting a range of energies dispersed along a linear position detector. However, if the mosaic spread of the crystal is less than the acceptance angle subtended in the scattering plane, the energy spectrum from a mosaic analyzer as collected in a linear detector will have Q varying with energy. The resolution and the variation in Q with energy along a linear detector are discussed in relation to inelastic x-ray scattering measurements.

Introduction

Inelastic x-ray scattering measurements of the dynamical structure factor $S(\vec{q},\omega)$ for the investigation of electronic structure and electron correlations in condensed materials require moderate Q resolution with scattered energy resolution of less than a few electron volts. Unfortunately, even using second and third generation synchrotron sources, the scattered intensities are very low (2). As a result, considerable effort has been spent to make detector/analyzer systems that collect the scattering over large solid angles with high energy resolution.

Two rather different methods have been used to achieve large solid angles with good energy resolution. One method utilizes large spherically bent Si and Ge single crystals, and the other uses a cylindrically bent mosaic graphite analyzer to collect a range of energies simultaneously. Spherically bent analyzers result in rather coarse Q resolution, but are acceptable because the electronic inelastic scattering, in general, varies slowly with Q. Mosaic analyzers provide higher Q resolution, but are complicated by a variation in Q with scattered energy and the requirement of calibrating the analyzer efficiency.

In this paper, we describe the use of a cylindrically bent graphite analyzer for inelastic x-ray scattering measurements, and discuss this measurement system in relation to the variation of Q in the energy loss spectrum.

Scattering Geometry

Most measurements of inelastic x-ray scattering are performed using thin spherically bent Si (or Ge) crystal analyzers (3). These analyzers focus diffusely scattered x-rays to a single channel detector, and provide a large Q coverage for a single scattered energy, as shown in Fig. 1a. Thus, measurements of the inelastic x-ray scattering spectrum are performed by placing the analyzer at the desired Q, and scanning the incident energy. Since the detector is a single channel device, and the area of the analyzer crystal is large, the Q resolution depends solely upon the solid angle subtended by the analyzer.

The cylindrically bent graphite mosaic analyzer method works differently; it disperses the scattered beam onto a linear detector, and simultaneously measures the energy of all the scattered rays. This is commonly referred to as the para-focusing geometry, and is shown in Fig. 1b. This type of analyzer, used for short range order diffuse scattering measurements, has been well described by Ice and Sparks (1). For the separation of resonant-Raman scattering, energy resolution of ~15 eV is adequate; for inelastic x-ray measurements investigating electronic structure, a resolution ≤ 2 eV is needed. Such energy resolution requires the graphite analyzer to operate near back scattering. The main advantage of a mosaic analyzer is the ability to measure all energies simultaneously using the entire face of the analyzer. The desired condition is for the mosaic spread of the analyzer to equal the angular size of the analyzer as seen from the sample. When this condition is met, each point of the analyzer crystal contributes to every point of the energy spectrum, and the maximum count rate is achieved.

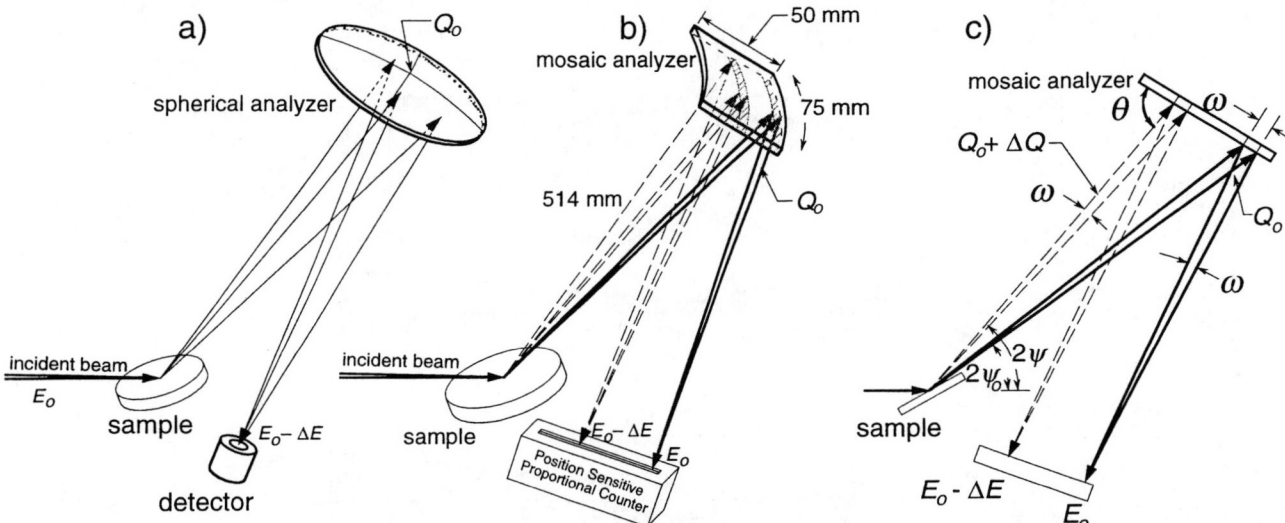

FIGURE 1. a) Ray diagram showing the layout for a spherical analyzer. All rays of one energy scatter into the detector. b) Ray diagram for a cylindrically bent mosaic analyzer showing para-focusing, and the sagittal focusing. The hatched bands on the analyzer show the acceptance angle for a single energy as limited by the mosaic spread ω. c) Ray diagram for a cylindrically bent mosaic analyzer; this cut is in the plane of diffraction to illustrate the angles clearly. Acceptance regions of analyzer for two energies are shown; the size of these regions is determined by the mosaic width ω. Note that ψ is used to indicate angles from the sample, and θ for the analyzer.

For an analyzer subtending an angular width (viewed from the sample) that is greater than the mosaic width, the angular acceptance width on the analyzer that contributes to each point in the energy-loss spectrum is limited by the mosaic width. This leads to a variation in the scattering angle as a function of energy, and, hence, to a variation of Q over the energy loss spectrum. For inelastic scattering measurements, the scattered intensity over an energy range of 0 to ~100 eV is desired in many cases.

Energy and Q Resolution

Fig. 1c shows the elastically and inelastically scattered rays as analyzed by a mosaic crystal. For a ray which has lost an energy ΔE, the expression for the Q transferred to the sample is,

$$Q(E_o, \Delta E, \psi) = \frac{4\pi \sin(\psi)}{\lambda_o} \sqrt{1 + \frac{\Delta E}{E_o} + \left(\frac{\Delta E}{E_o} \frac{1}{2\sin(\psi)}\right)^2} . \quad (1)$$

Where E_O is the energy of the incident beam, ΔE is the energy loss relative to E_O, and ψ is half the angle between the incident and scattered beams, the term under the square root is usually small compared to the Q resolution, and can be ignored. This equation reduces to the familiar $Q_o = 4\pi \sin(\psi_o)/\lambda_o$ when $\Delta E=0$.

For a mosaic width of the analyzer larger than the angular size of the analyzer (as seen from the sample) a scattered x-ray incident upon any part of the analyzer will penetrate until it encounters a crystallite satisfying the Bragg condition. And if the detector and sample are at equal distances from the analyzer (the para-focusing condition), then all rays of one energy will focus at one point on the linear detector. Thus, the position that rays strike the detector becomes a measure of the energy loss relative to the incident energy. The energy $E = E_O - \Delta E$ along the linear detector is given by $\Delta E = E_O \cot(\theta_O)\Delta\theta$, where $\Delta\theta = 2\Delta\psi$, ΔE, ψ, and θ are defined in Fig. 1c.

However, in most cases, the angular size of the analyzer is considerably larger than the mosaic ω. Therefore, only a limited region of the analyzer will contribute to the scattering at each energy, as depicted by the bands in Figs. 1b and 1c. For this situation, there is a simple relation based on Bragg's Law between ψ, the energies E_O and ΔE; and $\psi = \psi(E_O, \Delta E)$. For small changes in energy ΔE, the change, dQ, in Q as a function of energy can be approximated as,

$$dQ(E_o, \Delta E, \psi_o) = Q_o(E_o, \psi_o) \cot(\psi_o) d\psi \sqrt{1 + \frac{\Delta E}{E_o} + \left(\frac{\Delta E}{E_o} \frac{1}{2\sin(\psi_o)}\right)^2} , \quad (2)$$

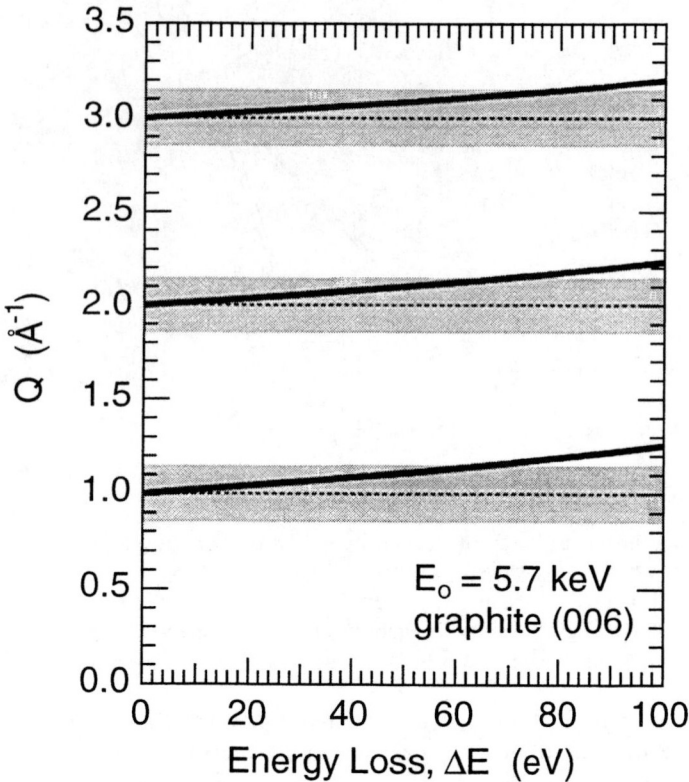

FIGURE 2. The actual Q for a mosaic analyzer for three values of Q_o across a 100 eV spectrum at 5.7 keV. The gray bands show a typical Q resolution for a spherically bent perfect crystal analyzer. The Q resolution at each point comes from the mosaic, and is approximately the thickness of the line.

where Q_o is the Q at energy of E_o, and ψ_o is the half-scattering angle of the $\Delta E=0$ ray. For most purposes, it is possible to set the expression within the square root to 1, since $\Delta E/E$ is ~1%. However, near back scattering, the change in energy ΔE creates significant changes in ψ, so it is necessary to use Eq. 1.

The change in Q as a function of ΔE for three values of Q_o is plotted in Fig. 2 for conditions used in recent experiments (4,5) on the X-14 beamline at the National Synchrotron Light Source. These conditions correspond to $E_o = 5.7$ keV, a Bragg angle of 76.6° for the 006 graphite reflection (d=1.118 Å), and a 50 mm graphite analyzer (with a ~0.5° mosaic width ω) located 510 mm from the sample. For Q_o ranging from 1 to 3 Å^{-1}, the change in Q for a ΔE range of 100 eV is about 0.2 Å^{-1}.

The energy resolution for these conditions is shown in Fig. 3 for a 1 mm high spot on the sample. For a mosaic analyzer, Ice and Sparks (1) show the energy resolution (for negligible particle size broadening or Darwin width) to be,

$$\delta E_m = E \cot(\theta_o) \, \delta\theta / 2 \quad , \tag{3}$$

where $\delta\theta$ is the angle subtended by the spot on the sample as seen from the analyzer. This expression is specifically for mosaic crystals, and it differs by a factor of 2 from the expression for perfect crystals, which is $\delta E_D = E \cot(\theta_o) \, \delta\theta$. Eq. 3 is displayed in Fig. 3, where we note that at higher energies the variation in Q is reduced, but the energy resolution becomes poor. The overall energy resolution of an inelastic measurement is given by δE from Eq. 3 convolved with the resolution of the monochromator (Si 111). The resolution of the system at X-14 was 2.4 eV, with 2 eV from the Si monochromator and 1.35 eV from the graphite analyzer.

Also shown in Fig. 2 are shaded bands representing the range in Q expected for a typical spherical analyzer, 50 mm diameter located 1 m from the sample. For a spherical analyzer, the Q resolution is given by the variation in Q across the analyzer. Although Q varies with energy for mosaic analyzers, the Q resolution at each energy is quite good, and is given by the width of the line in Fig. 2. We note that the Q resolution (~0.02 Å^{-1}) is approximately 10 times smaller than the variation of Q with energy.

Given the size of the Q variation, it is important to include the Q variation in the data analysis process. This is illustrated in Fig. 4, which shows the effect of the Q variation on $S(\vec{q},\omega)$ for jellium with non-interacting electrons (6). The dotted line is the jellium response calculated for a fixed Q of $Q_o = 3$ Å^{-1}. The solid line is jellium calculated for the same Q_o, but

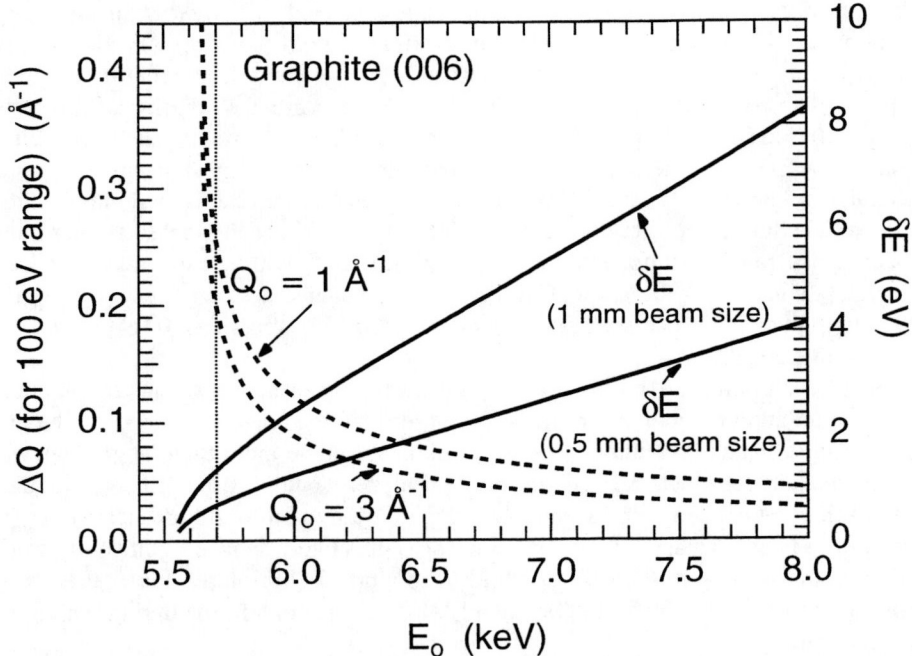

FIGURE 3. For the Graphite 006 reflection, the energy resolution is determined from a spot size of 1 mm at a distance of 500 mm. The solid curves show the range in Q for a 100 eV energy range at two different Q's.

allowing Q to vary with energy loss. The difference is clearly quite important; for instance, the normalization procedure to put experimental measurements on an absolute basis is based upon a computation of the first moment of the scattered intensity using the f sum rule (6),

$$\int_0^\infty \Delta E \cdot S(Q, \Delta E) \, dE = \hbar^2 Q^2 / 2m \, . \tag{4}$$

For the two curves shown in Fig. 4, the first moments differ by 10%; this difference is significant and necessitates interpolation of measurements made at two closely spaced Q_o positions to obtain a proper normalization factor.

Another important aspect associated with the use of mosaic analyzers is that the detector/analyzer combination, in

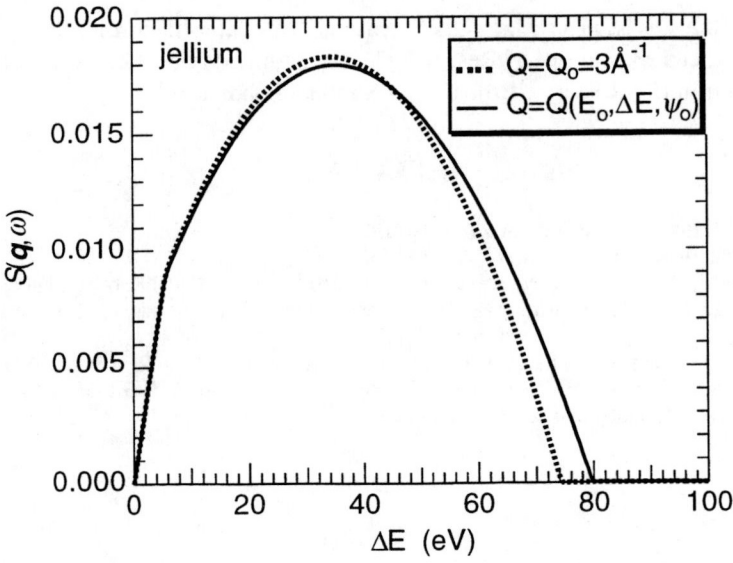

FIGURE 4. $S(\vec{q}, \omega)$ for jellium with non-interacting electrons showing the effect of including the Q variation for $Q_o = 3$ Å^{-1}. In this case, the first moment of the two curves differ by 10%.

general, will not have a uniform response across the energy range collected. This variation must be removed from the measured spectra by a calibration of the response of the system as a function of energy loss ΔE.

The procedure for calibrating the response of the detector/analyzer system as a function of energy loss ΔE uses quasi-elastic diffuse scattering (i.e. thermal scattering) at different energies to measure the response of the system. Step scanning the monochromator moves the quasi-elastic peak across the entire energy range of the analyzer. Since only the energy changes, the position on the sample and other angles remain fixed, the quasi-elastic peak intensity measures the correct part of both the analyzer crystal and detector. Of course, only the narrow quasi-elastic peak is used in the calibration; the inelastic scattering component is excluded using the energy resolution of the system. Since thermal scattering increases quadratically with Q, and tends to peak near Bragg reflections, the inherent variation in Q with energy makes it necessary to change the scattering angle ψ to keep Q constant as the monochromator is step scanned across the energy range. Variations in the mosaic width of pyrolytic graphite analyzer crystals may be as large as ~±15% across the 100 eV range, so it is important that these corrections be performed carefully.

Considering that high Q resolution is not required for most inelastic scattering measurements, an alternative procedure is possible that is analogous to an analyzer with a mosaic width larger than the angle subtended by the analyzer crystal. By masking the edges of the analyzer and continuously rocking the analyzer by more than its angular width, the integrated reflectivity (rather than the mosaic dependent reflectivity) of the analyzer would be used. This would allow all points on the analyzer to contribute to each point of the energy spectrum, and the Q resolution would broaden to the Q range described in Fig. 2. While this procedure has not been applied (to our knowledge), it is likely to have significantly smoother response as a function of energy, and it would result in a constant Q (albeit with a broad Q resolution) over the entire energy spectrum. The efficiency of the mosaic analyzer system would be reduced, by the fraction of time during which parts of analyzer are outside the diffraction condition.

Conclusions

Cylindrically bent mosaic analyzers represent an alternative to spherically bent perfect crystal analyzers for inelastic x-ray scattering measurements. Although mosaic analyzers provide higher effective count rates due to their ability to disperse multiple energies onto linear detectors, they introduce a Q variation into the energy loss spectrum. The Q variation can be used to obtain an improved Q resolution relative to large spherical analyzers; however, the need for accurate calibrations and interpolation between measurements to obtain constant Q results suggest that rocking of the analyzer crystal to achieve a constant average Q to reduce the impact of calibration may be an attractive alternative for future work.

ACKNOWLEDGMENTS

Research sponsored by the Division of Materials Sciences, U.S. Department of Energy under contract number DE-AC05-96OR22464 with Lockheed Martin EnergyResearch Corp. performed at the Oak Ridge National Laboratory Beamline X-14 at the National Synchrotron Light Source, Brookhaven National Laboratory.

REFERENCES

1. G.E. Ice, C.J. Sparks Jr., Nucl. Instrum. Methods, A **291**, 110 (1990).
2. E.D. Isaacs and P.M. Platzman, Rev. Sci. Instrum., **67**, 3381 (1996).
3. E. Burkel, B. Dorner, Th. Illini, and J. Peisl, Rev. Sci. Instrum. **60**, 1671 (1989), C.-C. Kao, K. Hamalainen, M. Krisch, D.P. Siddons, T. Oversluizen, and J.B. Hastings, Rev. Sci. Instrum. **66**, 1699 (1995), and A.T. Macrander, V.I. Kushnir, and R.C. Blasdell, Rev. Sci, Instrum. **66**, 1546 (1995).
4. B.C. Larson, J.Z. Tischler, E.D. Isaacs, P. Zschack, A. Fleszar, and A.G. Eguiluz, Phys. Rev. Lett., **77**, 1346(1996).
5. P.M. Platzman, E.D. Isaacs, H. Williams, P. Zschack, and G.E. Ice, Phys. Rev. B **46**, 12943 (1992).
6. Gerald D. Mahan, *Many-Particle Physics*, 2nd ed., Plenum Press 1993.

SMALL STUFF: MAKING AND IMAGING MICROSTRUCTURES

Glass capillary optics for making x-ray beams of 0.1 to 50 microns diameter

Donald H. Bilderback and Ernest Fontes

*Cornell High Energy Synchrotron Source (CHESS) and
School of Applied and Engineering Physics*

*Cornell University, Ithaca, New York 14853, USA
email: dhb2@cornell.edu, ef11@cornell.edu*

ABSTRACT

We have fabricated a unique computerized glass puller that can make parabolic or elliptically tapered glass capillaries for microbeam x-ray experiments from hollow glass tubing. We have produced optics that work in a single-bounce imaging mode or in a multi-bounce condensing mode. The imaging-mode capillaries have been used to create 20 to 50 micron diameter x-ray beams at 12 keV that are quite useful for imaging diffraction patterns from tiny bundles of carbon and Kevlar fibers. The condensing-mode capillaries are useful for creating submicron diameter beams and show great promise in x-ray fluorescence applications with femtogram sensitivity for patterned Er and Ti dopants diffused into an optically-active lithium niobate wafer.

INTRODUCTION

Although the variety of systems studied with synchrotron radiation is growing daily, it is increasingly common for light source visitors to share one need: they wish to study smaller specimens than ever before - or they wish to examine very small regions of larger samples. An example of the first might be the biochemist who can only grow a macromolecular crystal to 20 microns in size; for the later a material scientist who needs to understand the sub-micron crystallite formation in a 0.25 micron patterned conductor on an integrated circuit. Growing from this need, there is considerable technical interest in making microbeams on the scale of microns and smaller to explore, with hard x-rays, the structure of small objects such as buried conductors in integrated circuits, crystalline and amorphous regions of tiny fibers, fluid inclusions in geological minerals, etc.

Small diameter x-ray beams can be produced by a variety of focussing schemes. The properties of the focussed beam depend sensitively on both the nature of the x-ray source and the focussing elements in the beamline. Properties of the focussed beam that must be considered include spot size, intensity profile, energy spread (bandwidth), and angular spread. Each focussing system produces a beam with characteristic properties, and because of the diverse applications, no focussing scheme produces beams suitable to all measurements.

Although a full review of all focussing systems is beyond the scope of this paper, we mention several to provide a framework for appreciating the distinct features and applications of tapered glass capillaries (TGCs). Among the "classical" methods are using fixed-curvature reflecting mirrors (the KB geometry[1] utilizes two orthogonal mirrors to focus both vertically and horizontally), or by active mechanical bending of either mirrors[2] or diffracting crystals in order to cause the reflecting or diffracting beams to converge onto a focal plane. Acting as geometric lenses, mirrors and sagittal focusing crystals produce an image of the source. Since modern third generation storage rings have x-ray source sizes between 50-100 microns, these classical systems can produce an image of the source having equal size. In some cases, sagittal focussing has acheived 3:1 or 4:1 demagnification. Small, highly curved KB mirrors have produces a 100:1 focus at the ALS, providing a beam with a near gaussian profile with 1 micron full-width at half maximum (FWHM)[3]. And although imperfections due to surface roughness, figure error, or strain could reduce the quality of the focussed spots, recent advances in fabricating metal-coated total reflection mirrors have resulted in extremely good optical components; surface roughness and slope errors can be specified to the unprecedented levels of several Angstroms and sub-arcsecond.

"Recent" methods of focussing x-ray beams have arisen from advances in fabricating patterned materials with extremely good tolerances. For instance, Fresnel phase zone plates[4] have yielded 1 micron beams at NSLS and 0.25 micron beams at the Advanced Photon Source. And most recently, Snigirev has realized[5] the first refractive hard x-ray lense using a series of patterned hollow cavities of 300 micron dimension. Both the Fresnel and refractive lenses can have short focal lengths, and can therefore produce micron-sized beams.

In this paper we discuss fabricating TGCs to produce small diameter hard x-ray beams. We delineate between two extreme case: **single-bounce imaging mode** and **multi-bounce condensing**. This paper will not discuss using multi-capillary bundles as TGCs; see reference[6] for further details. We will point out three features of TGCs that distinguish them from other optical elements: 1) with sufficient care, glass can be made to assume an almost arbitrary shape. Thus we can create custom shapes to achieve specialized focussing geometries, 2) based on the total external reflection of x-ray from the glass surface, TGCs can be used over an extremely wide energy range - even with white beam under suitable conditions, and 3) the output tip of the TGC behaves like a strict aperture, so that very small illuminated volumes (and high spatial resolution) can be realized for condensing optics.

In the first case (see figure 1a), a hollow tapered tube is used in a single-bounce geometry that resembles a parabolic mirror with cylindrical symmetry about the long axis of x-ray propagation. A typical TGC uses a large-bore tube, of order 1 mm input diameter or larger, with an output diameter of 0.3 mm and length of 25 cm. The emerging rays meet in a so-called focal plane that can be 20-30 mm from the tip. We have fabricated a large-bore TGC that produce focussed spot sizes ranging from 20-50 microns FWHM. In this mode, a single-bounce from the glass surface produces a demagnified image of the source in the image plane. The real size and divergence of the x-rays in the image plane depend on the source size, the shape of the glass tube, the slope errors from manufacture. Also, because of the spread in focal lengths of rays reflecting from different distances from the focal point, there is no single image of the source. Instead, the single-bounce TGC acts like a series of extended demagnifying lenses, the rays closest to the output tip have the strongest focussing and those from the input end have the smallest angular deflection. This attribute can be exploited to customized the angular spread of the focussed beam by varying the parts of the TGC that are illuminated by the incident x-ray beam.

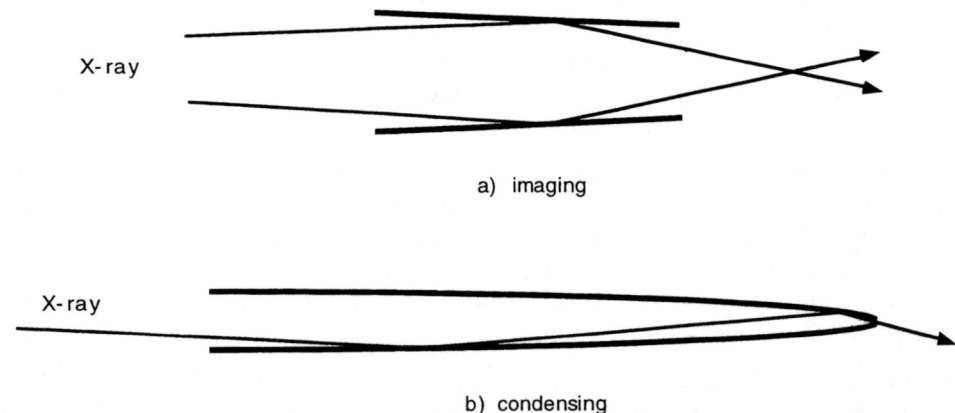

Figure 1. Glass capillaries working in a) imaging and b) condensing modes. An extended image of the source appears in the focal plane when the optical surface intercepts the x-ray and reflects it once. In contrast, the condensing optic supports a ray bouncing many times between its inner surfaces. In the condensing geometry, the emerging x-ray beam is smallest at the tip of the optic.

The second geometry, shown in figure 1b, uses the TGC as a **multi-bounce condensing** or concentrating element. In a sense like a "funnel", x-rays may bounce repeated until they emerge from the tip. Condensing TGC tips are very efficient apertures, producing very sharply defined beams. A beam as small as 0.05 microns has been demonstrated[7]. In the condensing mode, x-rays undergo multiple bounces in the same glass tube before they exit the optic. The size and divergence of the x-ray beam again depend on the glass shape, but differently than for imaging optics. The exit beamsize, though, depends on the exit hole size in the capillary and not on aberrations of the capillary shape, source size, or roughness. This is an advantage when making the smallest diameter x-ray beams. Because of the divergence in the exiting beam, the highest spatial resolution is realized by using the condensing TGC as a proximity probe. Fortunately, the glass tips are small and can be easily moved to within microns of the specimen.

In both cases, glass capillaries are pulled on glass pullers to create a long hollow glass tube whose outlet size is smaller in diameter than the inlet size. X-rays will efficiently pass down this long hollow tapered "needle" if at each point the angle of incidence on the smooth glass surface is kept below the criitical angle, typically about 3 milliradians for 10 keV x-rays and glass surfaces. Like any kind of optical component, a capillary takes a relatively parallel beam of x-rays and makes it smaller at the expense of creating divergence. Tradeoffs can be made between beamsize and divergence by adjusting the design shape of the capillary.

The efficiency of both imaging and condensing TGCs depends sensitively on a high reflectivity of the x-ray from the inner glass surface. To keep the x-rays from penetrating the TGC wall, the angle of incidence must be kept below the critical angle. The reflectivity and critical angle are critically important in the condensing systems, since x-rays may undergo many bounces and with each reflection the incidence angle increases. This is not a stringent condition for the single-bounce system. For instance, the reflectivity of the inner surface for a particular large-bore TGC was measured at 75% for 12 keV photons. While this decreases the throughput of the entire TGC by roughly 25% for one-bounce, a five bounce efficiency would be reduced on order $(0.75)^5 = 0.24$, and a much larger loss could result from the increasing angle of the successive bounces.

In the remainder of the paper we discuss how capillaries are drawn on a computerized glass puller and recent results with imaging and condensing capillaries.

COMPUTERIZED GLASS PULLER

The first attemps at pulling glass at Cornell University were based on a simple gravity drawing process that was achieved by hanging a piece of tubing vertically in a furnace and letting a weight draw the parts until they separated into fine needle-like shapes[8]. Though the resultant capillaries had irregular and non-reproducible shapes, they nonetheless produced usable, but not optimized beams for x-ray experiments. It was with this generation of capillaries that a 50 nm beam diameter of 5 to 8 keV x-rays was created and employed in an x-ray imaging demonstration[7].

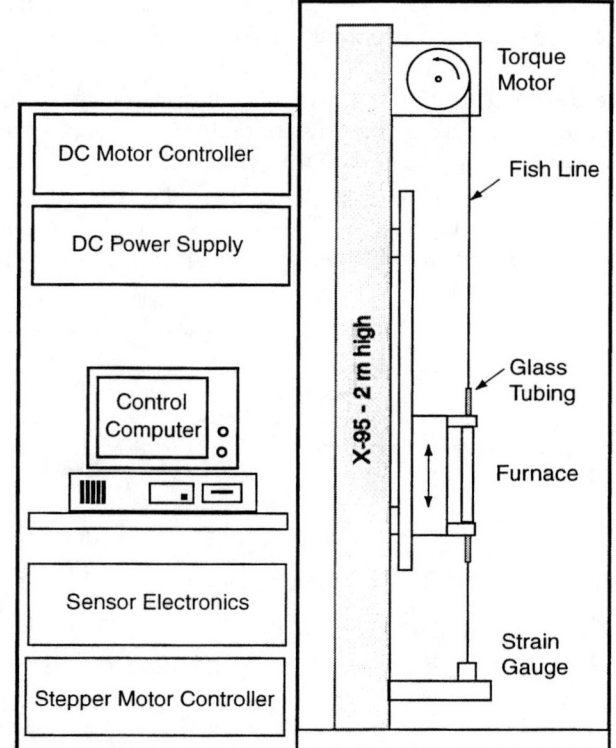

Figure 2: Schematic of the DB-1 tapered glass capillary drawing tower. The computer controls the position of the furnace as the glass elongates. A torque motor and strain gauge keep the tension on the glass capillary fixed.

Subsequently, a computerized glass capillary puller was fabricated in order to obtain a more reliable and reproducible source of capillary optics. A CHESS Microscience group was formed to build a custom glass puller that would provide the needed control to make to ideally-shaped parts. The project is still underway. A number of other groups have had similar ideas[9].

The group designed a computerized instrument that controls all the relevant parameters during the pulling process, including control of the furnace heating profile, furnace motion to generate the different tapers needed, tension control to simulate a variable weight to pull the glass, etc. All of these ideas are imbeded in the design of the DB-1 puller shown in Figure 2. With this equipment, we can make parts of about 7 to 40 cm in length starting with tubing diameter up to about 2 mm in diameter. The nichrome wire in the furnace limits the peak temperature to about 900 C, so the low temperature glasses such as leaded glass, borosilicate glass, soda-lime glass, etc. can be pulled with the present equipment[10,11].

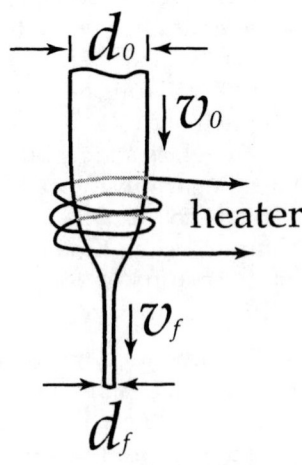

Figure 3: Schematic of the drawing process. Through the heat zone the velocity and diameter of the glass tube change.

The design procedure for working with the puller is to first determine the starting and ending diameters over some capillary length. The test shape is often analyzed with ray tracing code to better select the curvature vs. length profile of the optic. Presently we are able to make parabolic or elliptically shaped capillaries that are very close to the design shape and to do so reproducilbly (figure 4). The capillary figure determines the minimum and maximum slopes needed and in turn influences the maximum furnace length and the ratios of the furnace velocity to the glass velocity. The key point in making specific profile shapes is to run an open loop control method that based on a conservation of mass principle. Shown schematically in figure 3, the furnace is held fixed and glass at a starting diameter d_0 is pushed into the furnace at a rate v_0 and drawn out with a velocity v_f at a reduced diameter d_f. If the mass of the glass in the furnace remains constant, then it must be that the rate of glass entering the furnace and that leaving the furnace are equal, ie $d_f^2 \times v_f = d_0^2 \times v_0$ or that $d_f = d_0 \sqrt{v_0/v_f}$. This is the controlling equation for our puller (though in the real instrument, our furnace moves and the part to be drawn remains fixed.)

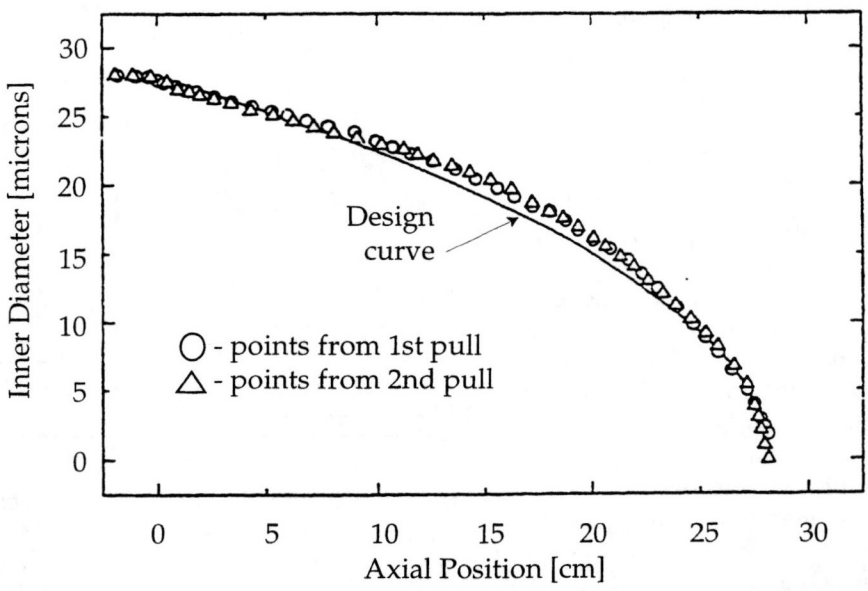

Figure 4: Profile of the inner diameter versus axial position showing measurements from two capillaries and the design curve[11].

The optimization of the drawing process is still underway. To better complete the optimization process, we need glass with a smoother finish, better control of furnace temperature and motion during drawing, grinding of tips to optimal shape, better methods of mounting capillaries, more accurate metrology to measure the shape of parts, and better programs for modeling both the drawing process itself and the way that x-rays reflect from real glass surfaces.

SINGLE-BOUNCE IMAGING CAPILLARIES

In early 1996, reports begin circulating that TGCs were fabricated that performed quite well as cylindrical mirrors. From the University of Melbourne (Australia), Balaic et al. demonstrated[12] that a capillary with a parabolic taper profile could be used as a single-bounce focussing element (figure 5a). In essence, the single-bouce capillary may have a similar figure as the glass concentrator, except that the tip has been cut off; in doing so a significantly larger working distance is created and the geometry only accommodates one bounce. The second publication[13] from the Australian group demonstrated a focussed spot of 40-50 microns in size, with monochromatic light, and a gain factor of 700. The gain was defined as the factor of increase in photon flux through a small 5 micron by 5 micron aperture, with versus without, the TGC. The performance of the TGC was shown to work over a wide energy range with photon energies from 8 to 20 keV. Another distinct feature of the strong focussing by the TGC was that the convergence angle of the x-rays onto the focal position was approximately 6 milliradians FWHM (~.3 degree).

It was readily apparent that gains of this sort could be immediately useful to a wide audience of users at CHESS. The same DB-1 glass pulling tower used to fabricate small-bore concentrators was quickly put to the task on large-bore stock. In due course, we were able to control the gross shape of the capillary and had moderate success at achieving an approximately parabolic taper. The first TGC had an 800 micron input diameter and was trimmed to have a 300 micron exit hole. As with any complex system of lenses, the size of the focussed spot produced by the optic depends quite sensitively on the source size and the arrangement of optical components between the source and TGC. To study these effects, test were carried out on a variety of unfocussed and focussed bend magnets and wiggler source beamlines, resulting in measured focussed spots ranging from 20 to 50 microns in diameter (FWHM). The smallest spot sizes were achieved with monochromatic light produced by a double-bounce Si(111) monochromator on the D-line bend magnet station (source size approximately 2 mm H by 0.7 mm V FWHM). Figure 5b shows a 3D contour plot of the intensity in the focal plane measured by a raster scan of a 10

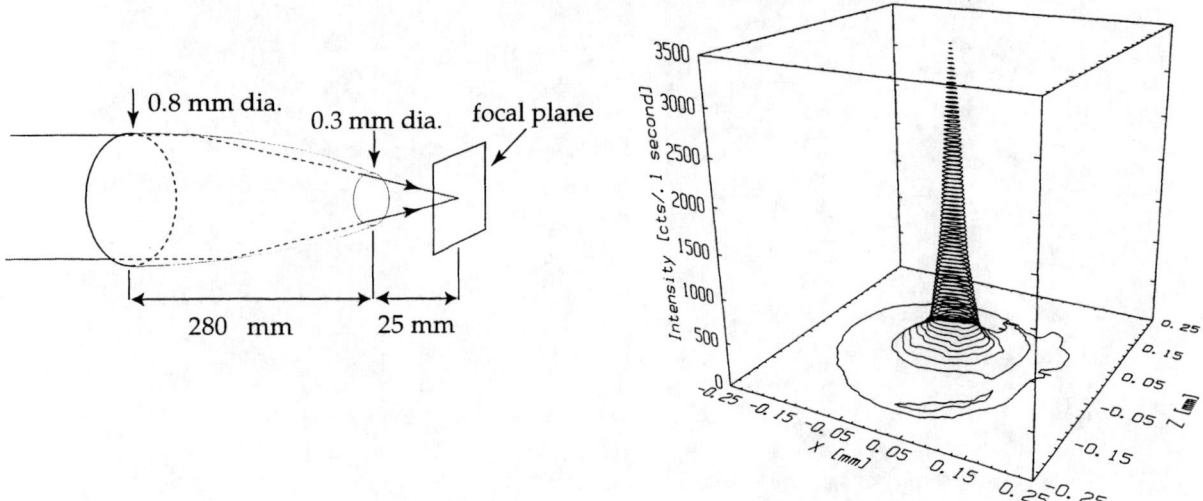

Figure 5: (a: left) Schematic of x-ray beams bouncing off the inside of a wide-bore TGC and converging onto a focal point. (b: right) A 3D contour plot of the intensity passing through a 10 micron aperture as the aperture is scanned in the focal plane of a TGC fabricated and tested at CHESS. The intensity scale is given in units of ion chamber current; the central peak corresponds to an x-ray intensity of 6×10^{11} photons/second.

micron diameter aperture. To convert the vertical intensity scale into gain we note that if the 10 micron round aperture is placed in the central beam, the measure intensity passing through the aperture decreases by a factor of 150 when the TGC is removed.

This measured gain is less than the calculated ratio of illuminated areas: 800 microns input diameter versus 50 microns focal spot size should provide an intensity gain of $(800/50)^2 = 256$, whereas we measure only 150. In addition, the calculated spot size for the D-line source size is the smallest of any CHESS beamline, 0.7 mm vertical by 2.1 mm horizontal, so we would calculate that a 100:1 focal geometry should produce an focal spot of 7 V x 21 H microns. The best we've measured is 20 V x 30 H microns. Several factors contributing to these discrepancies include a slightly imperfect TGC shape and surface roughness. The large bore tubes are sufficiently large that we were able to measure directly the x-ray reflectivity of the inner surface. Short lengths of a tube were illuminated by a 10 micron-sized monochromatic beam in a Fresnel reflectivity apparatus. Among the glass used in this study, an average reflectivity of 75% was measured for 12 keV x-rays. While this glass is sufficiently smooth for single-bounce applications, glass with less surface roughness would be needed for multibounce glass concentrators.

It has proven more useful, over the past year, to use the wide-bore TGC downstream of a 1.5% energy bandpass multilayer mirror. The multilayer passes nearly 100 times the total intensity of the Si(111) monochromator, but the slope errors inherent in this particular multilayer produce a beam with a higher angular spread. As a result, the smallest focal spots from the multilayer-TGC combination are 50 microns FWHM. The tremendous advantage of this combination is that the focussed beam total flux at 160 milliAmpere CESR current is 6×10^{11} photons/second. This intensity is nearly <u>equal</u> to the total beam from passed by 300 micron collimator at the CHESS double-focussed wiggler stations equipped with a silicon (111) monochromator! The result is that for those experiments that can tolerate the energy bandpass and the angular spread of the tightly focussed beams, the multilayer-TGC combination provides a bend-magnet alternative to the oversubscribed high-flux wiggler stations.

Given these high flux densities (and we expect another decade of improvement with straightforward technical improvements), the Microbeam Facility at CHESS can support and enhance the research programs of a number of CHESS users and potential users. In particular, measurements that can make use of the wide-bandpass and angular spread of the beam include small- and wide-angle scattering from partially ordered liquid systems such as phopholipids, proteins, synthetic and natural fibers (see figure 6). Another use for these highly concentrated beams includes those measurements on very

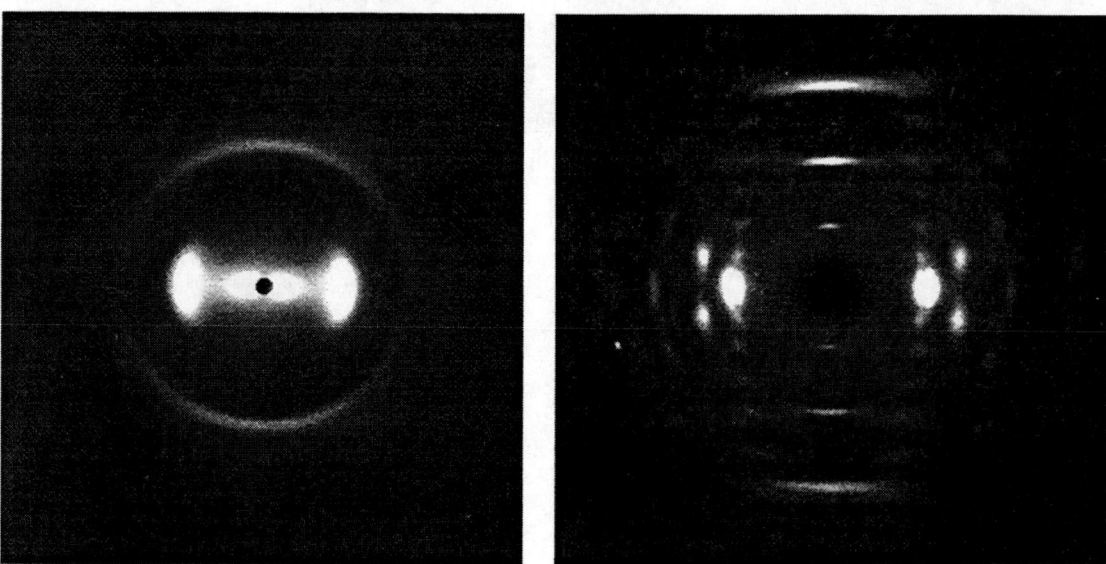

Figure 6: (left) X-ray diffraction pattern from a bundle of 5 micron diameter carbon fibers with 12 keV x-rays, an exposure time of 3 minutes, and a film-to-sample distance of 30 mm.) (right) Scattering from bundle of 12 micron Kevlar fibers. (Same conditions as left image except the exposure time was 10 minutes.)

small specimens, such as low-resolution powder diffraction from samples in diamond-anvil cells, small protein crystals, highly dispersed dyes (either powder or 10 micron-sized single crystals). It is also clear from early feasibility studies on these various systems that there will be no single TGC design that is best for every measurement. Being able to fabricate TGCs of almost arbitrary shape is important in that we can tailor TGC shape, and therefore the characteristics of the x-ray beam, to best match a number of different experimental conditions.

CONDENSING CAPILLARY OPTICS

Condensing capillaries are useful for making the smallest diameter x-ray beams. These devices can concentrate x-rays of all energies up to a cutoff-energy that is dependent on the capillary material and its shape, i.e., it is inherently a wide-bandwidth component. Condensing capillaries also make a well defined small beam, especially if leaded glass is used. One capillary, fabricated on a gravity puller, with a somewhat irregular shape, made a 50 nm (.05 µm) beam diameter[7] at 8 keV with an intensity gain of 960. The capillary shape can be adjusted to optimize the beamsize, divergence, and gain for a particular experiment on a specific x-ray tube or synchrotron radiation source. We believe that with further agressive work, it may be possible to reach 20 nm or smaller beamsizes.

One disadvantage of capillaries is that the sample needs to be positioned no more than 10 to 100 beam diameters away from the tip in order for the x-ray spot size to be similar to the exit size of the tip. For a 1 micron diameter beam, the sample would be positioned somewhere between 10 and 100 microns from the capillary tip.

An example application of a condensing TGC included an x-ray fluorescence study[14] of the patterned Er and Ti dopants in a lithium niobate wafer. 95 Å of Ti and 18.5 Å of Er were diffused into the wafer to a depth of a few microns to construct an optical beamsplitter. The buried dopant profile was determined with 10.1 keV x-rays at BWB1 beamline at Hasylab as shown in the schematic diagram of Figure 7. In a 5 second counting time per point scan (fig. 8), we achieved a sensitivity of detecting less than a one Angstrom thick layer of Erbium which corresponds to a sensitivity of better than 4×10^{-15} grams or 4 femtograms. This amounts to less than 1.5×10^7 Er atoms in the x-ray beam. We believe that this high sensitivity to elements that fluoresce above about 1 keV will make it possible to determine the thickness of buried metal conductors in integrated circuits, etc.

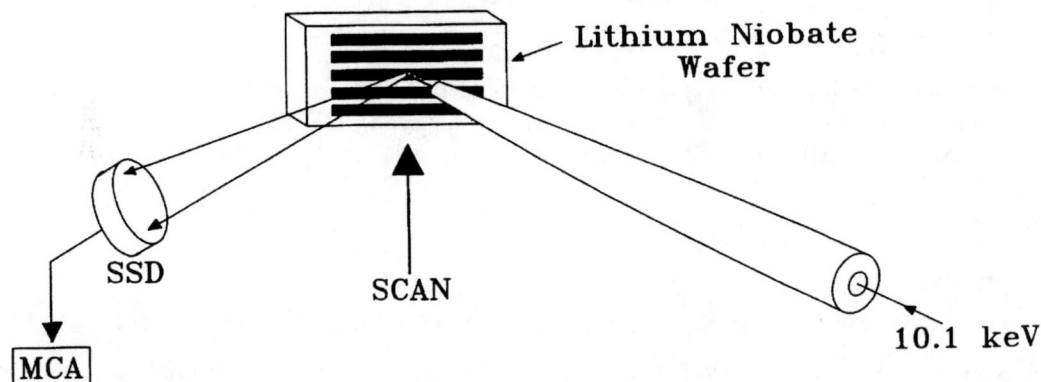

Figure 7. In the HASYLAB experiment, the Lithium Niobate sample was placed at a 45 degree angles to the capillary axis 175 microns away from the tip. The illuminated area on the sample was a 1.8 µm (vertically) by 2.5 µm (horizontally). Upstream optics included two mirrors and a 1% BW, 150 periods of W/C, d=30 Å multilayer.

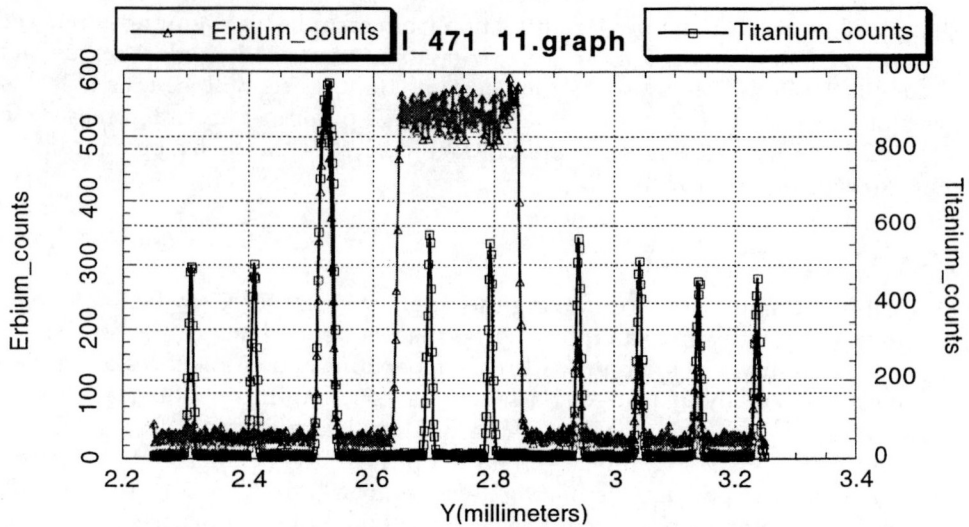

Figure 8. An experimental count rate of 100 cps was observed for the Er lines that were produced by an incident beam flux of 5.8×10^7 x-rays/sec at BW1 beamline (Hasylab, 4.44 GeV, 88 ma) behind a multi-layer monochromator. Data from capillary CHI015 made from a 28 cm long soda-lime glass with a 0.8 µm opening at the tip

CONCLUSIONS

The computerized glass capillary puller at CHESS can produce good quality optics for making hard x-ray beams on the scale from 0.1 to 50 microns. A number of technical improvement are yet to be made to optimize the optics. More than a dozen groups are making specialized pullers of various sorts and conducting x-ray experiments. We expect the capillary optics field will rapidly mature to reveal the fruitful methods of glass selection, pulling techniques, and even in the design and modeling phases. We believe that these glass capillary components will continue to have a strong role in the microbeam field where they can be used independently or in conjuction with other optical components.

ACKNOWLEDGEMENTS

We acknowledge helpful discussions with Jim Patel and Ed Franco. This work is based upon research conducted at the Cornell High Energy Synchrotron Source (CHESS), which is supported by the NSF under award number DMR-9311772, by NIH under Grant RR-10646, and by HASYLAB at DESY.

REFERENCES

[1] H. A. Padmore, M. R. Howells, S. Irick, T. Renner, R. Sandler, Y-M. Yoo, SPIE vol. **2854**, 145-156 (1989).

[2] P. J. Eng, M. Rivers, B. X. Yang and W. Schildkamp, Proceedings of the International Society for Optical Engineering, SPIE Vol. **2516**, 41 (1995).

[3] A. Thompson, private communication.

[4] A. A. Kraxnoperova, Z. Chen, F. Cerrina, E. Difabrizio, M. Gentili, W. Yun, B. Lai, and E. Gluskin, Proceedings of the International Society of Optical Engineering, *X-ray Microbeam Technology and Applications*, SPIE Vol. **2516**, 15 (1995).

[5] A. Snigirev, V. Kohn, I. Snigireva, and B. Lengeler, Nature **384**, 49-51 (1996).

[6] L. Wang, B. K. Rath, W.M. Gibson, J. C. Kimball, and C. A. MacDonald, J. Appl. Phys. **80** (7), 3628 (1996); M. A. Kumakhov and F. F. Komarov, Physics Reports **191**, No. 5, 289-350 (1990).

[7] D. H. Bilderback, S. A. Hoffman, and D. J. Thiel, Science Vol **263** (1994)201-203.

[8] D. J. Thiel, D. H. Bilderback, and A. Lewis, Rev. Sci. Instrum. Vol **64** (1993) 2872-2878; D. J. Thiel, D. H. Bilderback, and A. Lewis, proceedings of SPIE, proceedings of SPIE, Vol **1740**, (1992)248-251.

[9] See the proceedings of the First International Developers Workshop on Glass Capillary Optics for X-ray Microbeam Applications (1996), available from CHESS, Wilson Laboratory, Cornell University, Ithaca NY 14853

[10] D. H. Bilderback, R. Pahl, and R. Freeland, CHESS Newsletter, 1995, pages 41-43.

[11] R. Pahl and D. H. Bilderback, proceedings of SPIE, Vol **2805**, (1996) 202-211.

[12] D. X. Balaic, K. A. Nugent, Z. Barnea, R. F. Garrett and S. W. Wilkins, J. Synchrotron Rad. (1995) **2**, 296-299.

[13] D. X. Balaic, Z. Barnea, K. A. Nugent, R. F. Garrett, J. N. Varghese and S. W. Wilkins, J. Synchrotron Rad. (1996) **3**, 289-295.

[14] Microbeam Monocapillary Optics for Detecting Trace Amounts of Er and Ti Dopants Patterned into a Lithium Niobate Wafer used as an Optical Waveguide", D. Bilderback, E. Fontes, R. Pahl, D. Novikov, G. Materlik, and R. Frahm, Hasylab Annual Report **I** (1996) 952-953, Hamburg, Germany.

Beyond Sunshine: Hard X-rays for Precision Microfabrication

Erik D. Johnson, D. Peter Siddons, and J. Christopher Milne

National Synchrotron Light Source
Brookhaven National Laboratory
Upton, NY 11973
and

Henry Gückel and Jonathan L. Klein

University of Wisconsin Madison
Madison WI 53706

For several years we have explored the use of hard x-rays for a broad range of lithographic applications. The high energy available from the NSLS x-ray ring (E>15 keV) allows the exposure of resist up to several cm thick, while maintaining micron level precision. The high flux and close proximity to the source at this machine make it possible to achieve workable exposures on realistic time scales, enabling production work. In addition to the conventional two-dimensional exposure schemes, we have demonstrated methods for achieving fully figured three dimensional objects with internal re-entrant geometry. Users from outside BNL have been sufficiently successful with their work at our prototype beamline (X-27B) that we have initiated the construction of a dedicated exposure station (X-14B) for High Aspect Ratio Precision Manufacture. An overview of our previous work as well as the current status of the new beamline will be described.

X-ray lithography at CXrL - 3D nanostructures

Yuli Vladimirsky

Center for X-ray Lithography
College of Engineering
University of Wisconsin-Madison
Madison WI 53706

The Center for X-ray Lithography (CXrL), established in 1983, has worked specifically to develop the scientific foundation for new manufacturing applications based on the use of a technique called X-ray lithography (XRL). By using X-rays, from an electron storage ring for the transfer process, instead of the more conventional visible or ultraviolet light, it is possible to define ultra-small 3-D and high aspect ratio structures for semiconductor, communication, navigation, sensor and actuator applications. XRL allows each device in a system to contain more components, acquire and process information more quickly, and consume less power. Research in the center addresses all phases of the XRL process, including mask design, X-ray resist development, and computer modeling of the multiple imaging and processing steps, as well as fabrication of sophisticated micro- and nano-structures. The structures fabricated at the CXrL include IC devices, infrared filters, quantum well infrared photodiodes, quantum dots, thick Fresnel phase zone plates for 0.6 to 1.5 Å X-rays, gas micro-turbines, tactile actuators, 100 GHz wave-guides, antennae and filters, computer controlled atomizer, etc. Details of work on some of these structures will be presented.

X-ray fluorescence correlation spectroscopy for studying particle dynamics in condensed matter

Jin Wang [*], Ajay K. Sood [*,¶], Parlapalli Satyam [*], Yiping Feng [*], Xiao-zhong Wu [§], Zhonghou Cai [*], Wenbing Yun [*] and Sunil K. Sinha [*]

[*] Experimental Facilities Division
Argonne National Laboratory, Argonne, IL 60439, USA

[§] Department of Physics
Northern Illinois Unversity, Dekalb, IL 60115 and

Materials Science Division
Argonne National Laboratory, Argonne, IL 60439, USA

[¶] On leave from Department of Physics, Indian
Institute of Science, Bangalore 560012, India

Photon correlation spectroscopy probing fluctuations in scattered or fluorescent intensity to study particle dynamics in fluids is by now well established in the visible light regime. With the advent of high-brilliance synchrotron radiation sources, correlation spectroscopy utilizing scattered radiation has recently been extended to the X-ray wavelength regime by using spatially coherent X-rays to study the time fluctuations of the corresponding speckle patterns. In this presentation, we report the development of a new technique, X-ray fluorescence correlation spectroscopy (XFCS) for elucidating the dynamics of particles. This technique does not require coherent beams but relies on intense microfocused X-ray beams. Further, it is element specific. As a demonstration of this method, the dynamics of gold colloidal particles and aggregates undergoing diffusion and sedimentation in water was studied by measuring the time autocorrelation of the gold fluorescence intensity from a small illuminated volume. The values of the translational diffusion constants and sedimentation velocities obtained are in excellent agreement with theoretical estimates and other measurements. Further potential applications of the technique are discussed.

A manuscript about this work will be published. This work is supported by the U. S. Department of Energy, BES-Materials Science, under contract number W-31-109-ENG-38 and is supported partially by the State of Illinois under HECA.

A Beamline for Micromachining and Microcharacterization at the APS

B. Lai, W. Yun, D. C. Mancini, F. DeCarlo, D. Shu, J. Chang

*Advanced Photon Source, Argonne
National Laboratory, Argonne, IL 60439*

Beamline 2-BM at the Advanced Photon Source had been designed for developing micromachining techniques based on deep x-ray lithography and also for micro-characterization of optics and samples. With a critical energy of 19.5 keV and a highly collimated beam, the APS bending-magnet source is well suited for fabricating thick photoresist structures (> 1 mm) with high precision. The 2-BM beamline was designed to exploit these source characteristics and to provide flexible spectral tuning in order to accommodate different mask/resist thicknesses and to study the effects of the x-ray energy on the lithography process. The beamline will also be used for developing micro-characterization techniques. This includes characterization of microfocusing optics such as zone plates and developing instrumentation for techniques such as x-ray microprobe and microtomography. For this purpose, two monochromators, one using crystals and one using multilayers, will be used to cover the 1-35 keV regime with different energy bandwidths. Beamline design, end-station layout, and recent results will be presented.

Acknowledgment. This work is supported by U.S. Department of Energy, BES-Materials Sciences, under contract no. W-31-109-ENG-38.

X–ray Microdiffraction Studies of an Integrated Laser–Modulator System

W. Rodrigues[*,†], Z. Cai[*], W. Yun[*], H.–R. Lee[*], P. Ilinski[*], E. Isaacs[**], and J. Grenko[††]

[*]*Experimental Facilities Division, Argonne National Laboratory, Argonne, IL 60439*
[†]*Department of Physics and Astronomy, Northwestern University, Evanston, IL 60208*
[**]*Bell Laboratories, Lucent Technologies, 700 Mountain Avenue, Murray Hill, NJ 07974*
[††]*Microelectronics Group, Lucent Technologies, 9999 Hamilton Blvd., Brienigsville, PA 18031*

Abstract. We report the use of a spatially resolved x–ray microdiffraction technique for the structural study of an integrated laser–modulator system. The monochromatic (11 keV) x–ray beam microfocused to less than 1 μm in the vertical direction was obtained using a phase zone plate. The photon flux at the focal spot exceeded 3×10^{10} photons/s/0.01%bw/μm^2. The intense flux density and high spatial resolution of the focused beam was used to study the structure of a laser–modulator system, which is a 1–μm–wide and 1–mm–long multi-quantum well structure on an InP substrate. The superlattice d-spacing and the strain field in the direction normal to the diffracting planes were mapped as a function of position along the length of the device.

INTRODUCTION

With the availability of coherent, high–brilliance synchrotron radiation and advanced microfocusing optics (1), it is now possible to characterize materials at submicron length scales by x–ray diffraction. The x–ray diffraction technique is a nondestructive method and can be used to study crystallographic strain with high precision. The ability of x–rays to penetrate matter also makes it suitable for studying buried structures. Modern electronic and opto–electronic devices are smaller and more complex. Small changes in the structure within the device could change their performance considerably. Using a hard x–ray microprobe, one can directly map out the structure of an individual device over its entire volume nondestructively. In this article we report the use of a x–ray microbeam obtained by zone plate focusing to map the crystallographic strain and multilayer period in an opto–electronic device.

THE SAMPLE

The sample is an electro–absorption modulator/laser (EML), opto–electronic device supplied by Lucent Technologies. The EML is a monolithically integrated $In_{1-x}Ga_xAs_yP_{1-y}$ multiquantum well (MQW) structure grown on an InP substrate, as shown in figure 1. The width of the active region in the laser and the electro–absorption modulator regions is 1 μm. The device is grown by metal–organic vapor phase epitaxy (MOVPE) (2). The presence of SiO_2 pads patterned on the InP substrate causes the MQW to grow thicker in the laser region compared to that in the modulator region of the device.

EXPERIMENTAL

The experiment was performed at the 2–ID–D beamline at the Advanced Photon Source (APS) using monochromatized radiation from an undulator. The beamline has been designed to take advantage of the high coherence and low emittance of the APS to focus monochromatic x–rays to submicron size. The beamline has been described in detail previously (3). A schematic of the x–ray microscope is shown in figure 2. The microfocusing optic is a phase zone plate (PZP) with a focusing efficiency of 33% at 8 keV x–rays (4). The energy of the monochromatic beam used was 11 keV. The focal length of the PZP is 6.88 cm at that energy, and its diameter is 50 μm. The function of the order–sorting aperture is to allow only the first–order focused beam to reach the sample. The beam was focused at the center of a two–circle diffractometer arranged in horizontal geometry. The sample, mounted on an XYZ stage,

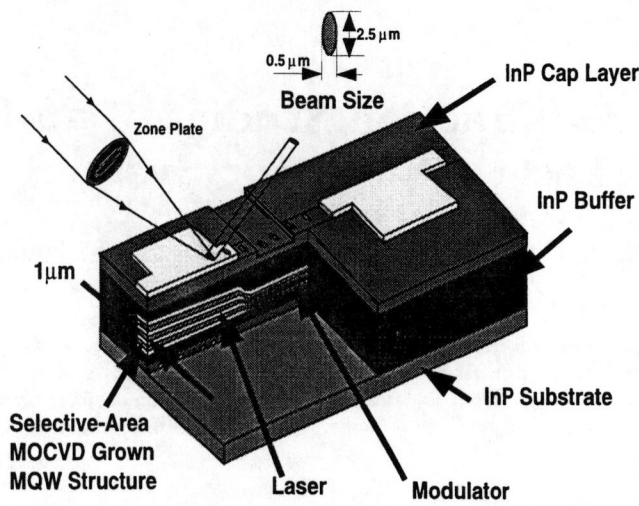

FIGURE 1. Schematic of the laser/modulator device

was accurately positioned in the center of the diffractometer with the help of two optical microscopes, one with its optical axis aligned with the azimuthal axis of the diffractometer and another lying in the diffraction plane.

To avoid any experimental artifacts in the diffraction data, the sample has to be very stable and stay in the center of focused beam at all times during the scan. In an experiment involving a θ–2θ scan, the sample may not be exactly in the center of the diffractometer, and the diffractometer usually has some wobble about its azimuthal axis. The net effect of all this is that the beam walks over the sample during the scan. This beam walking could be of several microns per degree of θ motion at some angles. In normal experiments, this is not very critical as the beam size is relatively large. To avoid this problem, one may scan the incident energy instead of the incident angle. In our experiment we could not do that because the zone plate focal length and focusing efficiency are not constants but functions of the incident energy of the x-rays. To solve the problem we used the high-resolution z-stage of the sample holder to compensate for any movement of the sample with respect to the axis of the diffractometer, by monitoring the Ga K fluorescence coming out of the sample. Figure 3 shows the Ga K fluorescence during a typical scan with z-compensation compared to without z-compensation.

The intrinsic zone plate resolution, which is 1.22 times the outermost zone width (5), is the diffraction-limited spot size that can be obtained. In this experiment the beam spot size was 0.5 μm in the vertical direction and about 1 μm in the horizontal direction. The footprint of the beam on the sample in the vertical direction was 0.5 μm and was 2.5 μm in the horizontal direction. It was larger in the horizontal direction because of the incident angle (Bragg). The divergence of the beam was 0.7 mrad. The incident photon flux was estimated to be about 3×10^{10} photons/s/0.01%bw/μm^2. Bragg scans (θ–2θ) were obtained at different points along the length of the sample. The position of the beam on the sample was continuously monitored during the Bragg scans by recording the Ga K fluorescence.

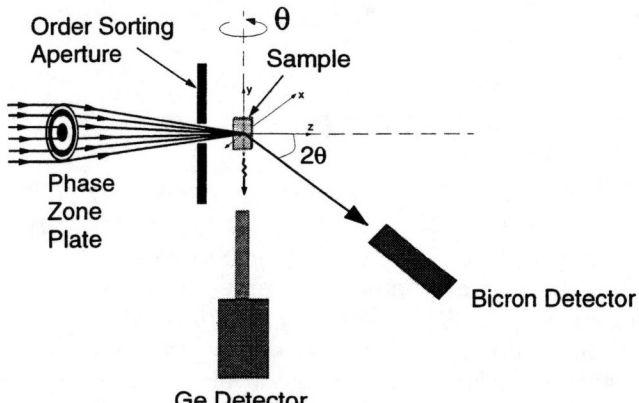

FIGURE 2. Schematic of the hard x-ray microscope

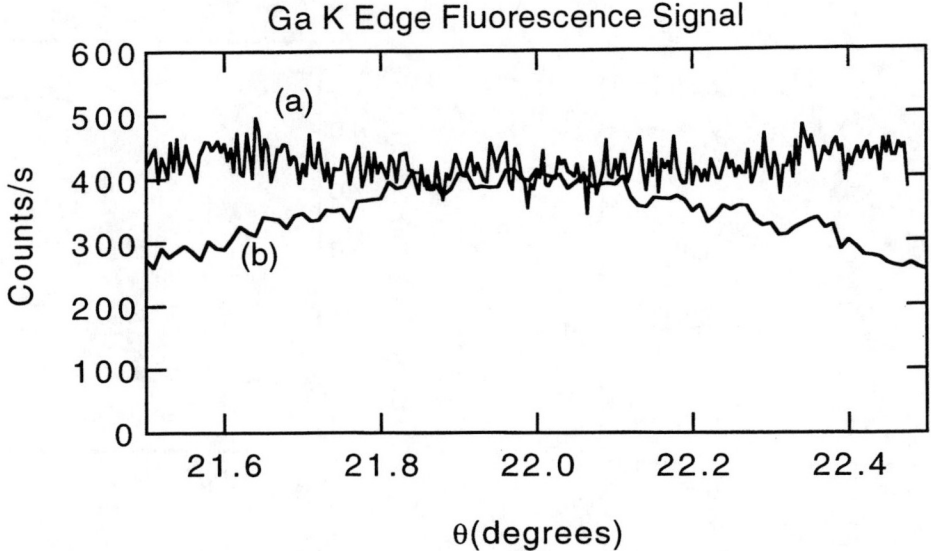

FIGURE 3. Ga K edge fluorescence signal during a Bragg scan (a) with z-compensation and (b) without z-compensation

RESULTS AND DISCUSSIONS

Figure 4 shows the θ–2θ scans of the diffractometer starting from the laser region and going into the modulator region. In addition to the InP (004) Bragg peak, satellite peaks from the MQW structure are also seen. Figure 5a shows the variation in the MQW period across the boundary between the laser and modulator regions. The period d_s was calculated by measuring the separation ϕ between the satellite peaks and using Bragg's law

$$\lambda = 2d_s \sin\phi. \tag{1}$$

The thickness of the MQW is about 20% smaller in the modulator region compared to that in the laser region. Because of the lattice mismatch between the epilayer and the substrate, the zeroth order peak of the MQW is separated from the InP Bragg peak. The separation of this peak gives the degree of mismatch or strain relaxation in the direction normal to the surface. Figure 5b shows the degree of mismatch as a function of the device position. The laser region being thicker is more strain relaxed or has larger mismatch than the modulator region, which is relatively thinner.

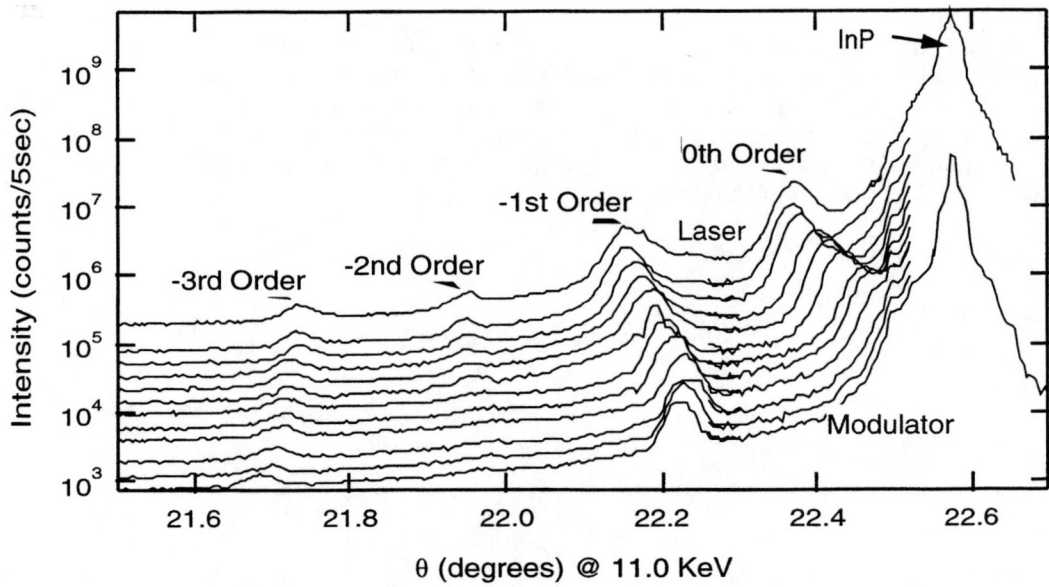

FIGURE 4. Bragg scans along the length of the device

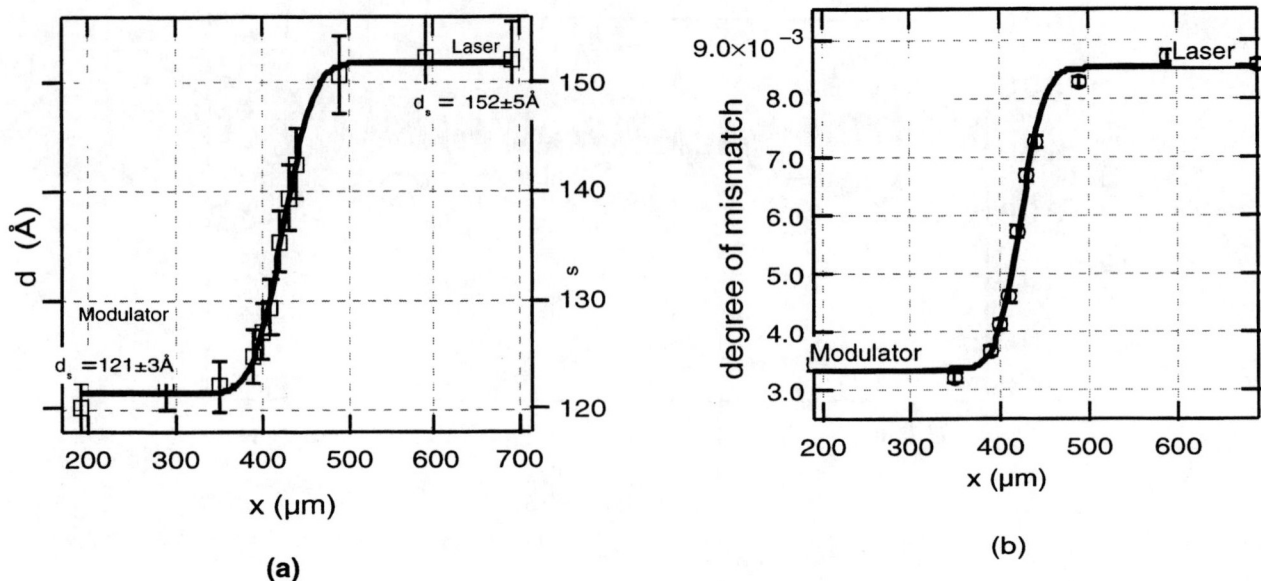

FIGURE 5. Variation of (a) MQW period and (b) lattice strain along the length of the device

This experiment demonstrates the high-resolution imaging capability of the x-ray microscope using phase zone plate as the focusing optic. The submicron spatial resolution of zone plates and the intense flux available from third-generation synchrotron sources like the APS can be used to nondestructively characterize miniature electronic and opto-electronic devices.

ACKNOWLEDGMENTS

We would like to thank Bell Labs, Lucent Technologies for providing us with the sample and D. Legnini for his help with the scanning software. This work supported by U. S. Department of Energy, BES. Material Sciences, under contract no. W-31-109-Eng-38

REFERENCES

1. Yun, W., Viccaro, P. J., Lai, B., and Chrzas, J., *Rev. Sci. Instrum.* **63**, 582 (1992).
2. Galeuchet, Y. D., and Roentgen, P., *J. Crystal Growth* **107**, 147 (1991).
3. Yun, W., Lai, B., Shu, D., Khounsary, A., Cai, Z., Barraza, J., and Legnini, D., *Rev. Sci. Instrum.* **67**, (9) (1996) CD-ROM.
4. Lai, B., Yun, W., Legnini, D., Xiao, Y., Chrzas, J., Viccaro, P. J., White, V., Bajikar, S., Denton, D., Cerrina, F., Di Fabrizio, E., Gentilli, M., Grella, L., and Baciocchi, M., *Appl. Phys. Lett.* **61**, 1877 (1992).
5. Myers Jr, E. O., *Am. J. Phys.*, **19**, 359 (1951).

Exposure station with precision scanning stage for deep x-ray lithography

Derrick C. Mancini, Francesco DeCarlo, Barry Lai

Advanced Photon Source
Argonne National Laboratory
9700 S. Cass Ave.
Argonne, IL 60439

An exposure station with a precision scanning stage has been designed and constructed for use at the Advanced Photon Source for deep x-ray lithography. The precision scanning stage consists of four motion stages—two translations and two rotations. There is a theta rotation in the horizontal plane at the base, which allows precise setting of the angle of inclination of the x-rays to the substrate for inclined, trapezoidal, and conical exposures. The horizontal X travel is mounted on the theta rotation and provides accurate positioning in the horizontal direction to allow field stitching and general alignment. The vertical Z travel is mounted on the X travel and is used to scan the mask and substrate through the x-rays repeatedly during exposure. The phi rotation is mounted on the Z travel and can be used for generating pyramidal and conical structures. Total absolute worst case angular error due to the sum of the stage motions including theta axis wobble, X axis pitch, yaw, and roll, and Z axis pitch, yaw, and roll should be 100 microradians. The entire scanning stage is mounted on a precision optical table that can be aligned to within 5 microns in position and 5 microradians in angle with full 6-degrees of freedom.

The motion stages are driven by stepping motors for positioning and DC servo motors for the vertical Z scan and include encoder feedback. The motors are controlled using the EPICS distributed control system. All of the other beamline components and the optical table are also controlled by EPICS, and this allows complete integration of the operation of the scanning stage with the optical table, filters, slits, mirrors, and shutters to provide the user with full control of the exposure from a program running on a computer workstation.

A temperature-controlled fixture is mounted on the f rotation, and the mask and substrate to be exposed are held by this fixture. A small housing with a Kapton window mounts onto the fixture to enclose the mask and substrate in a controlled helium environment during exposure.

Acknowledgement. This work is supported in part by U.S. Department of Energy, BES-Materials Sciences, under contract no. W-31-109-ENG-38.

X-ray Microdiffraction Studies to Measure Strain Fields in a Metal Matrix Composite

Heung-Rae Lee*, Wenbing Yun*, Zhonghou Cai*, William Rodrigues*, and Davis S. Kupperman†

*Experimental Facility Division at Advanced Photon Source and †Energy Technology Division,
Argonne National Laboratory,
Argonne, Illinois 60439

An x-ray diffraction microscope has been used to map the strain field in a fiber-reinforced composite material. The monochromatic x-ray (11 keV) beam was focused by a phase zone plate to produce a focal spot of 1 x 4 μm^2 on the specimen. The change in the peak position of diffraction patterns due to interatomic spacing change, caused by stress in the sample, was measured by using a two-dimensional CCD detector. The radial residual strain field in the fiber-reinforced composite (SCS-6/Ti-14Al-21Nb) was measured from diffraction patterns with a sensitivity of ~10^{-4} and an average standard deviation of 9.4 x 10^{-5}.

INTRODUCTION

X-ray diffraction studies with the use of synchrotron radiation source have become a very useful tool to understand the interatomic structures of materials. It was recently demonstrated that the high brilliance properties of third-generation synchrotron radiation sources like the Advanced Photon Source (APS) allow one to use a microfocusing technique with a zone plate and thus generate a high monochromatic flux with diameters down to a ~1 μm (1,2). The use of a microfocused x-ray beam for conducting x-ray diffraction studies makes it possible to measure residual strain/stress distributions at microscale levels in composites.

The measurement of strains can be obtained by investigating the relative shift of a peak position of a diffraction pattern. The relative shift is measured from sequential data for different positions of a specimen; this shift is due to interatomic spacing change caused by stress in the specimen. Such a relatively simple measurement provides us with quantitative analysis of the strain distribution in the specimen.

The proposed x-ray diffraction microscope (XDM) used in this work, is a straightforward experimental setup, in which the diffraction patterns are recorded by a CCD camera. We will demonstrate that the importance of XDM at the APS is in allowing a fast measurement and analysis of residual strain distribution through a specimen. The technique permits a precise measurement with a sensitivity of ~10^{-2} degree in diffraction angle (2θ) corresponding to that of ~10^{-4} in strain. The technique is then used to measure the strains near silicon carbide fibers (SCS-6) in a Ti-14Al-21Nb matrix.

EXPERIMENTAL SETUP

The experiment was performed at the 2-ID-D undulator beamline at the Advanced Photon Source (3). Figure 1 shows a schematic view of the experimental setup. A double-crystal monochromator, which is composed of two (111)-Si crystals, was used to select a 11 keV (λ=1.127 Å). A zone plate was used to focus the monochromatic 11 keV x-ray beam ($\Delta\lambda/\lambda = 10^{-4}$) and then produce a focal spot size of ~1 x 4 μm^2 with a flux of 1.3x10^{11} photons/sec. The microfocused, monochromatic 11 keV x-ray beam was used to observe Debye diffraction patterns of the fiber-reinforced composite. The sample was fixed on a two-dimensional (x-y) stage that is attached to a rotational (θ) stage, and a CCD detector is mounted on the 2θ stage. Both θ- and 2θ-stages are concentric, and the center of rotation of the stages is well defined by using the vertical microscope for accurate sample alignment. With the use of horizontal and vertical microscopes, the region of interesting of the sample was positioned at the center of beam and simultaneously at the center of rotation. Further, a Ge detector, which produces a fluorescence spectrum, was used for the more precise determination of the center of a fiber/matrix interface (within a ~5 μm range).

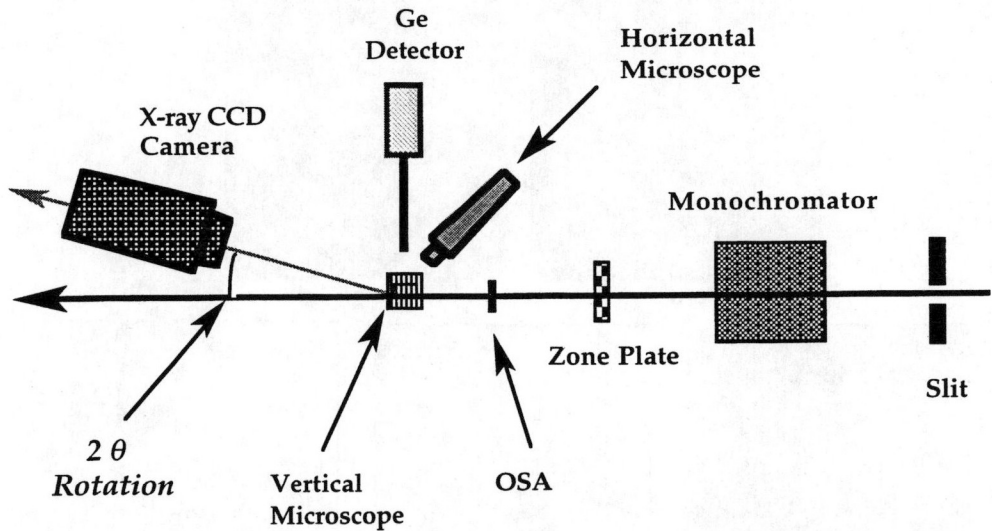

FIGURE 1. Schematic of experimental setup: The x-ray (11 keV) beam is selected by a double-crystal monochromator and focused by a zone plate that has a 54 cm focal length at 11 keV. An order sorting aperture (OSA) selects only the first-order focused beam and then produces a focal spot size of ~1 μm.

The sample for residual strain measurement was a fiber-reinforced composite (SCS-6/Ti-14Al-21Nb) (4). The composite was fabricated by hot isostatic pressing consolidation of an intermetallic Ti-14Al-21Nb (wt%) alloy with silicon carbide fibers of 140 μm diameter. Controlled cooling was done at less than 2°C/minute from the hold temperature of 950°C, and then the composite was cooled to ambient at ~20°C per minute under a pressure of 205 MPa. The sample, as shown in figure 2, was positioned on the sample stage in order to make the orientation of silicon carbide fibers parallel to the beam direction.

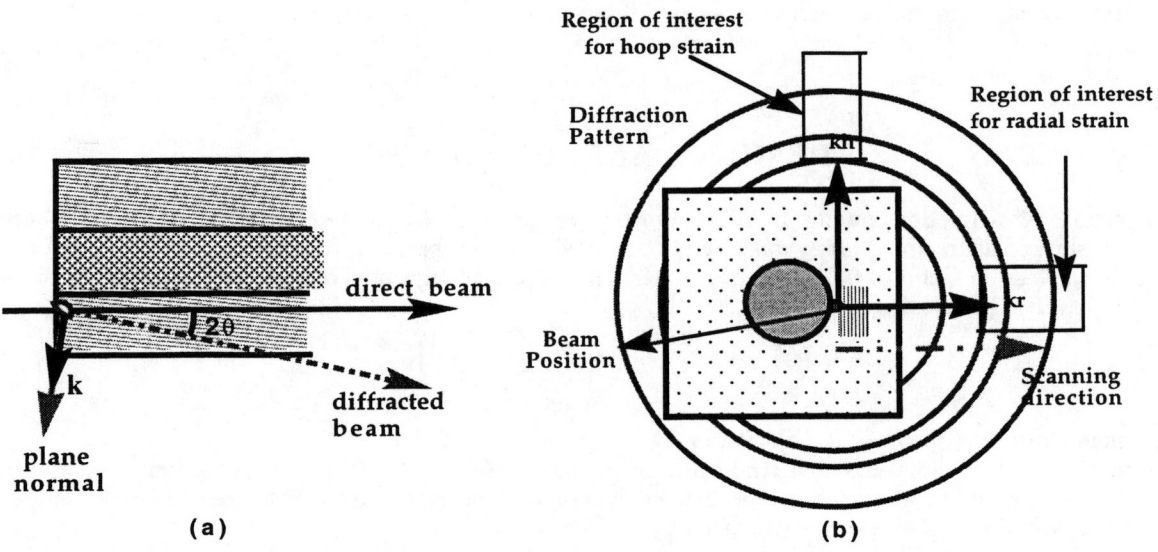

FIGURE 2. Orientation of sample and scanning direction: The sample is oriented parallel to the direction of the beam. For measurement of the radial strain gradient, the sample was scanned in the horizontal direction with a 10 μm step size from the center of the fiber/matrix interface. (a) is the side view of the sample orientation. (b) is the front view of the sample orientation. The experiment was perform by horizontal scanning of the sample to measure the gradient of radial strain.

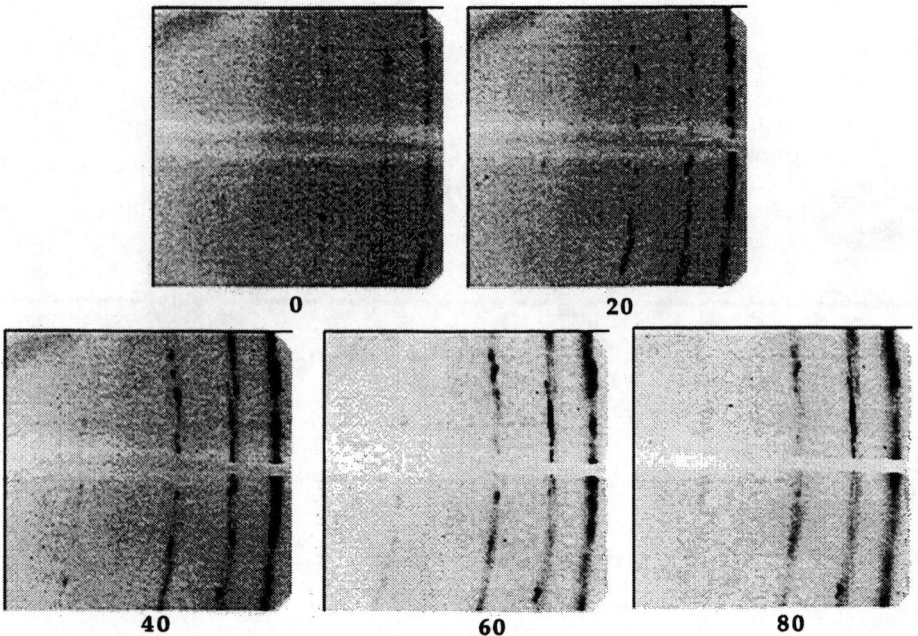

FIGURE 3. Examples of diffraction patterns: Diffraction patterns were taken at every 10 μm movement of the sample from the center of the interface by the CCD camera, which was located at 25 degrees in 2θ and at 282 mm distance from the sample.

For the measurement of strain gradient, the sample was scanned in the horizontal direction with a 10 μm step size from the center of a fiber/matrix interface. Diffraction patterns of the composite were recorded on the cryogenically cooled CCD camera. The CCD camera, which has a 1024 x 1024 pixel format with pixel dimensions of 19 x 19 μm², is coupled with a tapered-fiber glass that has a demagnification of 3. The CCD camera was located at 25 degrees in 2θ and at 282 mm distance from the center of rotation of the 2θ-stage (i.e., the sample), and hereby the sensitivity of strain measurement is ~3.85x10^{-4}/pixel. Acquisition time for each picture was 20 seconds.

RESULTS AND DISCUSSION

The pictures of diffraction patterns at different matrix positions are shown as examples in figure 3. For each picture, the relative pixel shift of diffraction was averaged and converted into the change in Bragg reflection (2θ=φ) by multiplying with a geometric conversion factor of 0.0105 degree/pixel. Thereafter, the sequential relative strain at a certain distance from the center of interface can be calculated by

$$\Delta d / d_o = -\Delta \phi / 2 \cot(\phi / 2), \quad (1)$$
$$\Delta \phi = 0.0105 \times \Delta p$$

where Δp is a number of pixels and $\Delta \phi$ is a change in 2θ.

Figure 4 presents a radial residual strain profile in the Ti-14Al-21Nb matrix as a function of distance from the fiber/matrix interface. A solid line represents the strain calculated by the general form of the residual radial strain from the prediction of a simple cylinder elastic model (5), given by

$$\sigma_{radial} = A \left[1 - b^2 / r^2 \right], \quad (2)$$

where b is the radius of the matrix sleeve, r is a distance in the radial direction, and A is a constant determined by the characterization of a sample. The radial residual strain developed in the matrix was measured from diffraction patterns with a sensitivity of ~10^{-4} and an average standard deviation of 9.4 x 10^{-5}. Reasonable agreement between measured and calculated strains is shown, but the slight conflict, close to the interface, might be caused by relaxation, by plasticity, and (or) by radial cracking and spalling upon sample cooling (4,6).

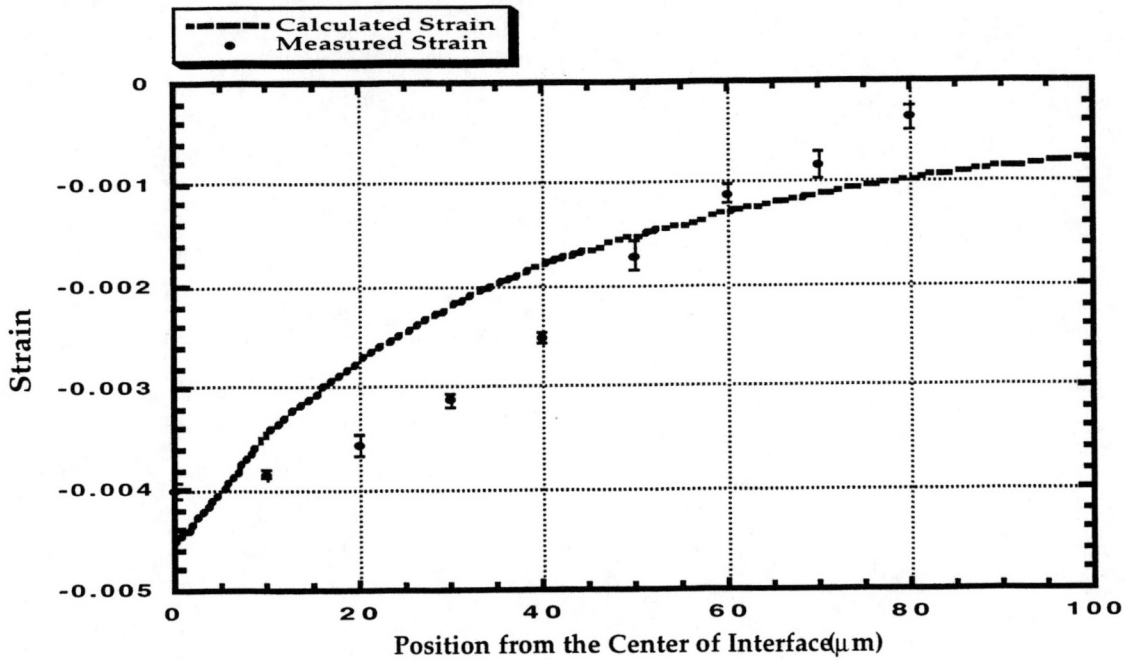

FIGURE 4. Profile of radial residual strain as a function of distance from the fiber/matrix interface: The measured radial residual strains (i.e., solid dots) in Ti-14Al-21Nb matrix have reasonable agreement with the calculated values (i.e., dotted line)

CONCLUSION

The XDM at the 2-ID-D beamline is a straightforward experimental setup, which has the capability to acquire diffraction patterns quickly with a highly brilliant monochromatic x-ray beam. In addition, the microfocusing method using a zone plate enables the XDM to produce a high spatial resolution of ~1 μm in measurements. The XDM allows one to quickly measure and analyze diffraction patterns to study lattice structures and understand physical phenomena inside materials.

ACKNOWLEDGMENT

This work was supported by the U.S. Department of Energy, BES-Material Sciences, under contract no. W-31-109-Eng-38.

REFERENCES

1. Lai, B., Yun, W., Xiao, Y., Yang, L., Legnini, D., Cai, Z., Krasnoperova, A., Cerrina, F., DiFabrizio, E., Grella, L., and Gentili, M., Rev. Sci. Instrum., **66**, 2287-2289 (1995).
2. Wang, J.-D., Kagoshima, Y., Miyahara, T., Ando, M., Aoki, S., Anderson, E., Attwood, D., and Kern, D., Rev. Sci. Instrum., **66**, 1401-1403 (1995).
3. Yun, W., Lai, B., Shu, D., Khounsary, A., Cai, Z., Barraza, J., and Legnini, D., Rev. Sci. Instrum., **67**, CD ROM (1996).
4. Wright, P.K., Sensmeier, M.D., Kupperman, D.S., and Wadley, H.N.G., "Thermal stress effects in intermetallic matrix composites," in NASA Contractor Report 191191, 1993.
5. Chawla, K.K., *Composite Material, Science and Engineering*, New York: Springer-Verlag, 1987, pp. 189-196.
6. Flinn, P.A., American Institute of Physics, 73-88, (1992).

HIGH-TECH: BEAM STABILITY, BEAMLINE HARDWARE, AND SOFTWARE CONTROL

Progress of the APS High Heat Load X-ray Beam Position Monitor Development

Deming Shu, Juan Barraza, Hai Ding, Tuncer M. Kuzay, and Mohan Ramanathan

Experimental Facilities Division
Advanced Photon Source
Argonne National Laboratory
Argonne, Illinois 60439, U. S. A.

Abstract: Several novel design developments have been established for the Advanced Photon Source (APS) insertion device (ID) X-ray beam position monitor (XBPM) to improve its performance:
--- optimized geometric configuration of the monitor's sensory blades;
--- smart XBPM system with an intelligent digital signal processor, which provides a self-learning and calibration function; and
--- Transmitting XBPM with prefiltering in the commissioning windows for the front end.
In this write-up, we summarize the recent progress on the XBPM development for the APS ID front ends.

1. INTRODUCTION

A third-generation synchrotron radiation source, such as the 7-GeV Advanced Photon Source (APS), generates high brilliance and intense synchrotron radiation from its insertion devices (IDs). There are many challenging tasks in the design of the ID beamline instrumentation that relate to high heat load and high heat flux problems. One of such component is the X-ray beam position monitor (XBPM) for the ID front ends and beamlines. The design requirements for APS front-end X-ray beam position monitors (XBPM) are such that they must withstand the high thermal load (up to 600 Watts / mm^2) and be able to achieve submicron spatial resolution while maintaining their stability.

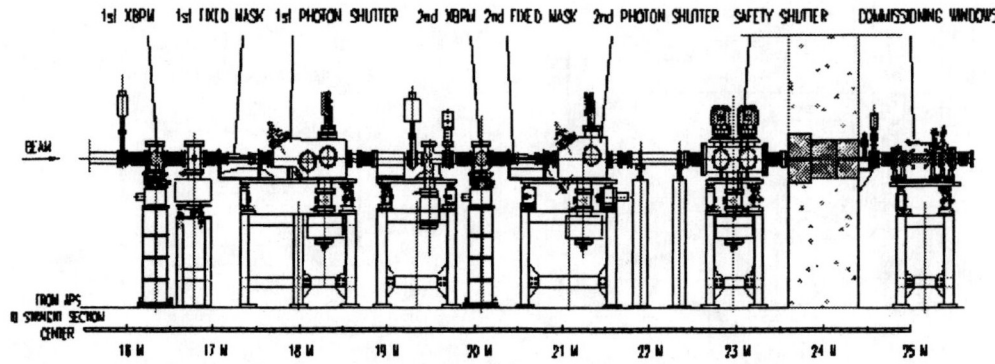

FIGURE 1. Schematic of the APS undulator beamline front end.

At the APS, each beamline front end has two XBPMs to monitor the X-ray beam position for both that vertical and horizontal directions. The XBPMs measure photoelectrons generated by the sensory blades and deduce the beam position by comparison of the relative signals from the blades. As shown in Fig. 1, both the first and second XBPM are located upstream of the user photon shutter (PS2) so that they are functional whether the user shutter is open or closed [1]. The major advantage of the XBPM is its high positioning sensitivity. Besides that, compared to the particle beam position monitors in the storage ring, the front end XBPMs have much higher sensitivity to the X-ray beam angular motion simply because they are located far away from the source.

Additional design challenges for a conventional photoemmision type XBPM are the bending magnet contamination of the signal and its sensitivity to the ID gap variations. Work at other synchrotron radiation laboratories has shown that contamination signals caused by the bending magnet (BM) emitted radiation become a major problem [2]. Problems are exacerbated for the XBPM when the insertion devices (IDs) operate with different magnet gaps, because the percentage level of the contamination will be a variable.

There are several novel design developments established for the APS ID XBPM to improve its performance:

--- Optimized geometric configuration of the monitor's sensory blades.

--- Smart XBPM system with an intelligent digital signal processor, which provides a self-learning and calibration function.

--- Transmitting XBPM with prefiltering in the commissioning windows for the front end.

In this write-up, we summarize the recent progress on the XBPM development for the APS ID front ends in the mitigation of the problems explained in the foregoing.

2. THE FRONT END XBPM STRUCTURAL DESIGN

Since 1991, a number of the APS high heat load XBPM prototypes using CVD diamond as the blade material were tested successfully at CHESS and NSLS. Both analytical and experimental results proved that CVD diamond is a good choice for the APS high heat load XBPM blade material because of its superior thermophysical properties, such as: high thermal conductivity, low thermal expansion coefficient, good mechanical strength and stiffness under heat. Submicron position sensitivity was also demonstrated by the APS XBPM prototype using CVD diamond blades during CHESS and NSLS tests [3].

Fig. 2 shows the structure of the first XBPM (upstream) main assembly on the APS undulator beamline front end. In this

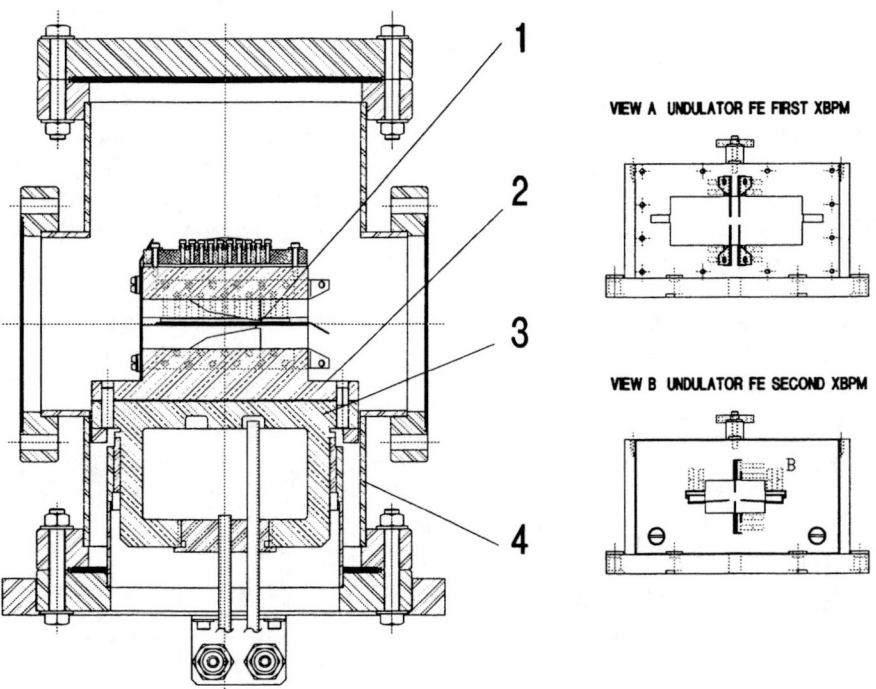

FIGURE 2. Structure of the XBPM main assembly for the APS undulator beamline front end, (1) CVD diamond blades, (2) monitor body, (3) water cooling base, (4) vacuum chamber. (view a for first XBPM monitor body, view b for second XBPM monitor body)

design, four 150-μm-thick CVD diamond blades (1) were coated with 1 μm of gold. The blades are mounted vertically in pairs on the monitor body (2), which is made of oxygen-free copper (OFHC) and is cooled by a water cooling base (3) from the bottom. The vacuum chamber (4) and the cooling base are designed for ultrahigh-vacuum (UHV) condition.

To eliminate the blade shadowing problems, the second XBPM (downstream) has a different blade placement configuration. As shown in Fig. 2b, the second XBPM has one pair of vertical blades, and one pair of "tilted" horizontal blades. This configuration reduces the signal contamination level from the BM-emitted radiation.

The XBPM monitor has the capability to apply a bias voltage. However, the test results show that a zero bias is acceptable and has the advantage of reducing the thermal resistance caused by the bias insulator. The geometrical configuration of the APS XBPM provides a low noise environment for photoelectron current output. The XBPM was sensitive enough to read out the photoemmission signal (about 0.6 nA) from a BM source while the APS storage ring had only a 24 µA electron beam stored at the first APS X-ray test on March 26, 1995.

3. STABILITY OF THE XBPM SUPPORTING STAGES

As shown in Fig. 3, the XBPM main assembly (1) is supported by a precision supporting stage (2), which is mounted on top of a mounting post (3). The post is made of steel, filled on the inside with sand, and thermally insulated on the outside by ceramic cloth. This post design is very resilient to short-term temperature fluctuations.

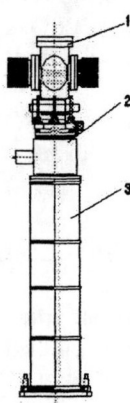

FIGURE 3. Front end XBPM with supporting stages, (1) XBPM main assembly, (2) precision supporting stages (3) mounting post.

The XBPM stage assembly consists of stepping-motor-driven vertical, horizontal, and rotational stages. Test measurements using a Laser Doppler Displacement Meter (LDDM) prove that the vertical stage attained a resolution of <0.2 µm with 1 µm repeatability under a 200 lb load [4]. Preliminary on-situ vibration tests show that the XBPM main assembly maintains less than 0.1 µm rms vibration displacement level with the cooling water on.

4. SMART XBPM SYSTEM (SBPM)

The optimized geometric design for the blades helped reduce the BM contamination. For instance, on the first XBPM on the APS 1-ID front end, the BM contamination has been determined to be about 10% of the signal from the 2.4 m undulator A with a 15.8 mm magnet gap. However, the contamination level will be much higher when the undulator gap is opened more. The regular XBPM calibration process can only provide signal correction for one set of conditions. During normal operations, the insertion devices function at varying storage ring current, particle orbit and a variety of ID gaps. In addition, because of the expected imperfections in the ID magnetic field distribution, each ID and its location on the storage ring has its own "personality".

To offset the XBPM sensitivity to such operational variables, a newer XBPM system has been designed and a prototype built and tested for the APS. This new XBPM system has an intelligent signal processor, which provides a self-calibration function to serve as a noise and contamination signal rejecter to improve the system sensitivity and reliability [5].

The new APS XBPM system configuration is depicted in Fig. 4. It includes:

A. a pair of photo-electron emission-style beam position monitors using CVD diamond blades for undulator beamline front ends

B. a set of photo-electron current preamplifiers

C. a preamplifier auto-ranging controller and digitizer [6]

D. a digital signal processor (DSP) with EEPROM data base and ID source input interface for normalization [7]

E. a system controller with motor driver and encoder interface for XBPM calibration processes

The new system, the so-called smart photon beam position monitor system (SBPM), has a built in EEPROM memory that is large enough to "remember" a complete calibration database covering all of the possible operating conditions. During the calibration mode, the monitor system controller initializes a series of automatic scan motions for the XBPM with different ID set-up information, and record them into the EEPROM database array. With the XBPM operating, the system corrects the normalized output according to the ID setup information and the calibration database. So that, with this novel system, the XBPM is always calibrated.

The heart of the smart system is a digital signal processor TMS320c40 from Texas Instruments Inc. [8], which is a floating-point processor designed specifically for digital parallel processing and real-time embedded applications. The key features of the TMS320c40 device, especially those to be used in the SBPM system, are the following: a high-performance DSP CPU with 40-ns instruction cycle times; a 40/32bit single-cycle floating-point/integer multiplier for high performance in computationally intensive algorithms; a six-channel DMA co-processor for concurrent I/O and CPU operation; six communication ports for high-speed inter communication; 512byte on-chip program cache and 8 kbytes on-chip dual-access/single-cycle RAM; two identical external data and address buses supporting a shared memory system with high data rate, etc..

In the operating mode, the DSP gets the XBPM signal data from the pre-amplifier/digitizer through one of the communication ports and groups them into an input buffer array. Then the DSP calculates the data under the control of a signal normalization program, which is using the external EEPROM database for reference. After a step-by-step approaching process, the final beam position data (a pair for the beam positions at the first XBPM location and a pair for the beam angular displacement) is transmitted to a signal output buffer. There are two types of output data format available for users: 24-bit digital parallel output and 4-20 mA current loop for analog output. Both digital and analog output will keep the final beam position signal with a DC-50-Hz or a DC-300-Hz bandwidth (depending on the type of microprocessor to be used in the preamplifier controller).

5. TRANSMITTING XBPM FOR THE FRONT END COMMISSIONING WINDOWS (TBPM)

During the beamline and front end commissioning activities, the final fine tuning of the storage ring and/or final adjustment of the front-end components is attempted. Based on the measurement data for the beam position in two locations

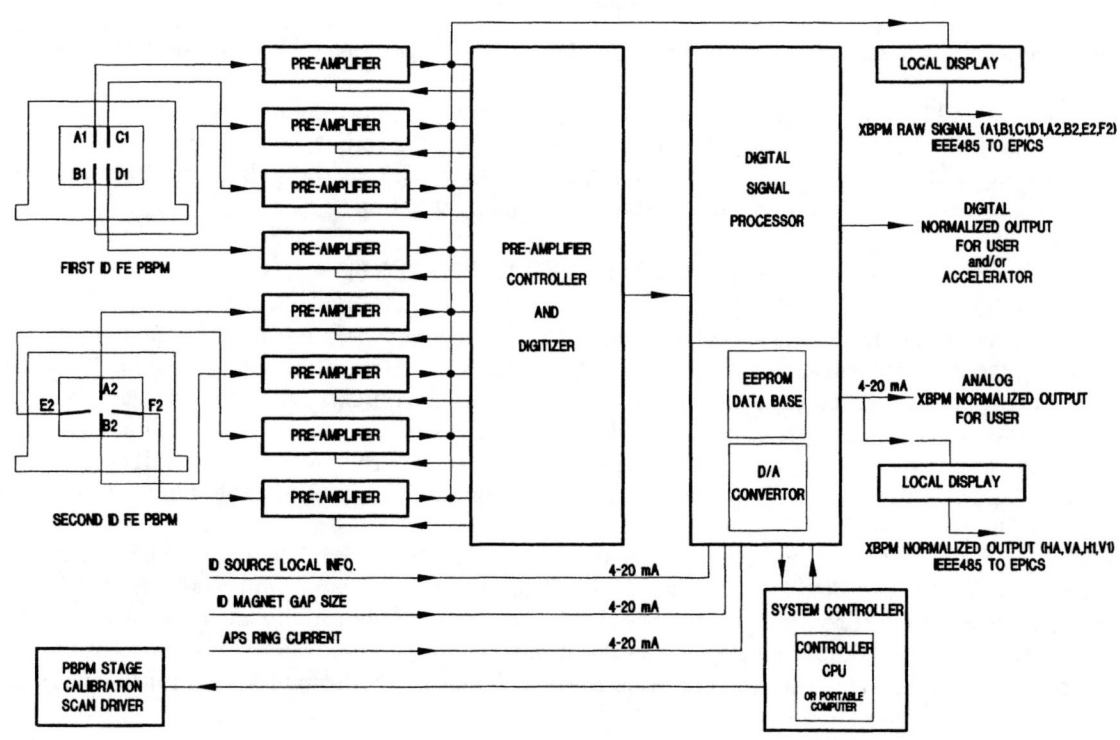

FIGURE 4. Schematic of the APS ID front end smart XBPM system.

in an experimental station and in comparison with the calibration scan data from the front end X-ray BPMs, a new zero position is set after the synchrotron radiation beam commissioning.

A CVD diamond filter, which is a 25.4-mm diameter disk mounted on the downstream side of the fixed mask, is also designed as a transmitting x-ray beam position monitor (TBPM) for the APS commissioning window system [9]. The basic concept of the TBPM is to mount the monitor blade perpendicular to the synchrotron radiation beam and design the blade and its low-Z metal coating thickness in such a way that most of the X-ray beam is transmitted through the blade (just like filter or window). In this design, the 160-μm-thick CVD-diamond disk is coated with four electronically isolated aluminum quadrant patterns. The thickness of the aluminum coating is about 0.2 μm. The photoelectron emission signal is collected by a terminal interface disk, which is made from thin alumina and is coated with silver. This design concept provides the possibility of integrating the filter with TBPM functions. The beam position information from the TBPM in the commissioning window is very valuable to the front end commissioning and smart XBPM system initial calibration.

6. DISCUSSION

To date three smart XBPM systems have been installed on the APS ID front ends and they are operational. On-line preliminary tests began in August 1996. Rest of the 20 ID front ends is expected to be furnished with SBPM system with DSP within a year. Based on the experience from the prototype operation, we will determine the time duration of the calibration period and optimize the database structure. Automatic calibration is necessary if the particle beam orbit changes frequently. If needed, the beam position at the neighboring bending magnet front end may also be used as another database reference input.

ACKNOWLEDGEMENTS

We acknowledge the help in the XBPM tests by Dr. Dean Haeffner, and Messrs. Mark Keeffe, Michel Lehmuller, Tim Cundiff and Martin Smith. This work was supported by the U.S. Department of Energy, BES Materials Sciences, under Contract No. W-31-109-Eng-38.

REFERENCES

1. D. Shu and T. M. Kuzay, Nucl. Instru. and Meth. A 347 (1994) 584.
2. T. Warwick et al. Rev. Sci. Instru. 66(2) Feb. 1995.
3. D. Shu, B. Rodricks, J. Barraza, T. Sanchez and T. M. Kuzay, Nucl. Instru. and Meth. A 319 (1992) 56.
4. T. Kuzay, ANL/APS TB-5, 1993.
5. D. Shu and T. M. Kuzay, Smart X-ray Beam Position Monitor System for the Advanced Photon Source, SRI95.
6. F. Meng, Unpublished M. Sc Thesis, IIT, May 1995.
7. X. Wu, Unpublished M. Sc Report, IIT, May 1996.
8. TMS320c4x User's Guide, Texas Instruments Inc. 1993.
9. D. Shu and T. M. Kuzay, Design of the Commissioning filter/Mask/Window Assembly for Undulator Beamline Front Ends at the Advanced Photon Source, SRI95.

High heat load fixed primary aperture for an undulator beamline with integral beam position monitors

G. Rosenbaum and T. Fornek

Structural Biology Center
Argonne National Laboratory
Advanced Photon Source
9700 S. Cass Ave., Argonne, IL 60439

A fixed primary aperture has been designed and built to limit the angular extent of radiation from an APS undulator onto a liquid nitrogen-cooled monochromator crystal. The design combines high heat load capability, simplicity and ease of manufacture, and integral thermal beam position monitors.

The aperture is designed to absorb 9 kW, the amount produced by the APS undulator A at 300 mA beam current after passing through apertures in the front end.

The device consists of a Cu-cylinder of 150 mm length and about 90 mm diameter with a tapered rectangular opening converging to 4 mm x 2 mm (horizontal x vertical) at the narrowest point. Radial saw cuts along the cylinder axis from the outside in create fins which provide a large Cu-to-water interface. A cylindrical shell with water inlet and outlet manifolds is brazed to the outside of the cylinder. Conflat flanges are brazed to both ends for vacuum connection.

Thermocouples inserted into holes drilled through the fins from top, bottom, left and right at the location of the narrrowest opening provide differential temperature readings which are calibrated and translated into beam position. Sensitivity is heat load dependent and better than 0.1 mm at small and medium undulator gaps and better than 0.2 mm even at large gaps.

This work was supported by the US Department of Energy, Office of Health and Environmental Research, under Contract W31-109-ENG-38.

Mirror Mounts Designed for the Advanced Photon Source SRI-CAT

D. Shu, C. Benson, J. Chang, J. Barraza, T. M. Kuzay, E. E. Alp, W. Sturhahn,
B. Lai, I. McNulty, K. Randall, G. Srajer, Z. Xu, and W. Yun

Experimental Facilities Division
Advanced Photon Source
Argonne National Laboratory
Argonne, Illinois 60439, U. S. A.

Abstract: Use of a mirror for beamlines at third-generation synchrotron radiation facilities, such as the Advanced Photon Source (APS) at Argonne National Laboratory, has many advantages. [Yun et al., Rev. Sci. Instrum. 67(9)(1996)CD-ROM] A mirror as a first optical component provides significant reduction in the beam peak heat flux and total power on the downstream monochromator and simplifies the bremsstrahlung shielding design for the beamline transport. It also allows us to have a system for multibeamline branching and switching. More generally, a mirror is used for beam focusing and/or low-pass filtering.
Six different mirror mounts have been designed for the SRI-CAT beamlines. Four of them are designed as water-cooled mirrors for white or pink beam use, and the other two are for monochromatic beam use. Mirror mount designs, including vacuum vessel structure and precision supporting stages, are presented in this paper.

1. INTRODUCTION

In the beamline design for the Synchrotron Radiation Instrumentation Collaborative Access Team (SRI-CAT) at the Advanced Photon Source (APS) Argonne National Laboratory, x-ray mirrors are widely used for beam focusing and/or low-pass filtering. As shown in Table 1, six different mirror mounts have been designed for the SRI-CAT beamlines.

TABLE 1. Mirror mounts designed for SRI-CAT beamlines

Mirror Mounts	Location	Beam Type	Reflection	Mirror Type
Y2-20	2-ID-A	Undulator White Beam	Horizontal	Water-cooled silicon flat with multiple coating strips
Y4-20	2-ID-A	Undulator Pink Beam	Horizontal	Water-cooled silicon spherical with three mirrors switching
Y5-20	1-BM-A	BM White Beam	Vertical	Water-cooled silicon flat with commercial bender
Y3-20	2-BM-A	BM White Beam	Vertical	Water-cooled silicon flat multiple coating strips
Y3-30	2-BM-B	BM Pink or Mono Beam	Vertical	Water-cooled silicon flat
Y7-30	3-ID-B	Undulator Mono Beam	Vertical	Zerodur cylindrical with commercial bender

Use of mirrors for beamlines at synchrotron radiation facilities is a well-known technique,[1,2,3] however, there are still many new challenging tasks in the design of the mirror mounts at third-generation synchrotron radiation facilities, such as the APS.

First, as the first optics on the beamline at the APS, the power density from the beam on the mirror is enormous. Six kW of the total emitted power with 160 W/mm^2 peak heat flux is generated at 30 m from the source when the undulator gap is closed.[4] Thermal distortion control for the mirror optical surface therefore becomes a major task, and the mirror cooling system makes the mirror mount design very complicated.

Second, the positioning precision and stability for the mirror mount design at the APS are critical, not only because the mirror usually is located at about 27 m or further from the source, but also because the high-brilliance synchrotron radiation source is in the 100 μm size. Microradian or even sub-microradian resolution is needed for the mirror mount primary angular motion control. Meanwhile, the dynamic characteristics of the mirror mount structure have to be analyzed to insure that the system is stable under the normal disturbance coming from the environment and cooling system.

Third, it is important to design a beam mask integrated with the mirror mount for missteered beam control to prevent component damage. For a multi-mirror system, especially for undulator beamline branching and switching, the design of such a masking system becomes a challenging task.

In this article, the modular mirror mounts designed for the SRI-CAT are presented. The mirror vacuum chamber, as well as its precise supporting and positioning system, is described. A multi-mirror system designed for the undulator beamlines branching and switching, including its beam masking system is also addressed.

2. MIRROR MOUNTS FOR A SINGLE MIRROR

A typical single mirror mount assembly is shown in Fig. 1. It consists of two main components: a vacuum chamber with in-vacuum kinematic mirror mounting structure and a modular support table, which performs the mirror primary positioning and angular adjustment functions by moving and/or rotating the whole mirror chamber.

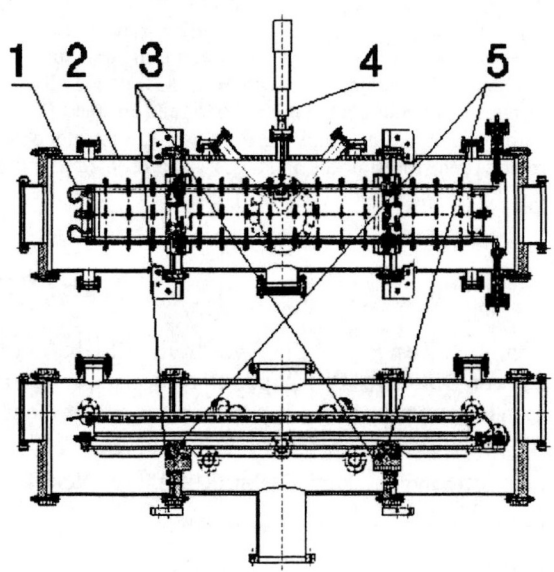

FIGURE 1 Single mirror mount assembly Y3-20.   **FIGURE 2** Top and side views of the Y3-30 chamber.

Mirror Chamber for SRI-CAT 2-BM-A (Y3-30)

The first mirror of SRI-CAT 2-BM beamline is located upstream of the double multilayer monochromator in the first optics enclosure. The main purpose of this mirror is to separate the synchrotron radiation from the bremsstrahlung. Thus, it greatly facilitates the transport of the pink beam into the end station. Deflecting the x-rays by 0.3° will create a separation of 3.3 cm between the x-rays and the primary Bremsstrahlung beam at the location of the integral shutters and stop. The separation provides sufficient clearance for placing a bremsstrahlung shield against the direct beam.

The mirror is a polished Si substrate with two coating stripes: a Pt coating for 33 keV high-energy cutoff and a Cr coating for 20 keV high-energy cutoff. The level of the high-energy cutoff can be selected by translating the mirror horizontally. The vertical acceptance is limited by the mirror size to 0.12 mrad for a 1.2-m long mirror. A simple side-cooling scheme is used to remove the heat load, which is possible because the peak heat flux is 0.0032 W/mm^2 on the mirror surface with a total power 125 Watts.

The Y3-30 mirror chamber has been designed to hold the first mirror with kinematic mounts inside vacuum chamber as shown in Fig. 2. To maximize the system stability, the vacuum chamber is designed such a way that both the supporting points (inside and outside of the vacuum chamber) are located on the same base and directly on the vacuum flanges. The kinematic mounts decouple the mirror from the vacuum chamber to avoid thermal and vacuum distortion, stresses resulting from changing of the environment. Three balls of the kinematic mount are located separately on top of two translation stages with cross roller bearings. These two translation stages provide 60 mm horizontal movement. They are connected to a linear motorized actuator through a long plate. "S" shaped spring plates as shown in Fig. 3 clamp the bottom groove of the mirror to the kinematic mounts. Disk-style spring washers are used to minimize the clamping forces applied to the mirror.

The water cooling tube is silverbrazed to two pieces of an OFHC copper bar, which are clamped to two sides of the mirror. The entrance and exit ends of the copper tube are supported to minimize vibrations caused by the cooling water. Also two spring plungers hold the mirror at both ends to reduce the vibration of the mirror. Indium foils are placed between mirror and the copper cooling bar to increase the thermal conductivity of the interface.

Mirror Chamber for SRI-CAT 2-BM-B (Y3-20)

The Y3-20 mirror chamber for SRI-CAT 2-BM-B experimental station is similar to the Y3-30 mirror chamber in vacuum chamber design as well as in kinematic mounting and side cooling structures. However, the Y3-20 mirror has only one coating, and it does not move horizontally. In addition, the Y3-20 mirror chamber can be rotated in pitch from 0 – 1.5 degrees to select the appropriate spectral region for experiment needs.

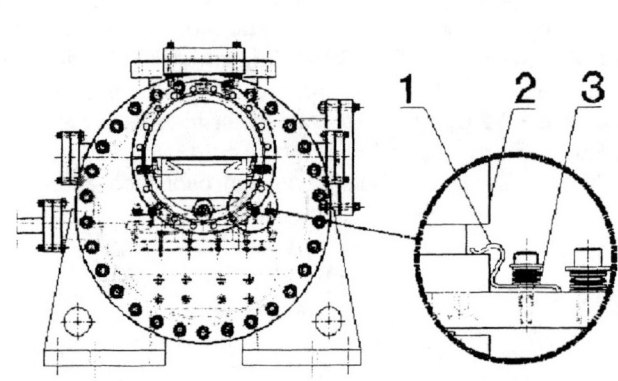

FIGURE 3 Y3-30 mirror clamping details: (1) "s" shaped spring plate, (2) mirror, (3) disk spring washer.

FIGURE 4 Y5-20 mirror mounts at 1-BM-A

Mirror Chamber for SRI-CAT 1-BM-A (Y5-20)

As shown in Fig. 4, the first mirror for the SRICAT 1-BM beamline (Y5-20) is installed approximately 28 meter from the source upstream of a double-crystal monochromator in the first optics enclosure. This mirror enables the user to collimate the synchrotron beam vertically with the beam reflecting in an upward direction. Beam collimation is achieved though a pneumatic bending mechanism with a variable tangential radius of 15 km to infinite.

The material of the main mirror body is silicon with a 30 nm coating of palladium. This makes it possible for the mirror to act as a filter with a high-energy cutoff at 24 keV. With a total power of 320 W deposited on the mirror surface, moderate water cooling is required. A simple side-cooling scheme was chosen. The horizontal beam acceptance is limited by the width of the mirror to 3.7 mrad. Vertically the full beam can be accepted.

Other than the purpose and location of the access ports and feedthroughs, the design of the vacuum chamber, kinematic mount, and external kinematic support for this mirror is similar to the designs of the Y3-20 mirror mount system. While the chamber, mount, and support are designed at APS, the bending mechanism and the mirror body were developed and fabricated by the ZEISS company. This assembly is ultrahigh vacuum (UHV) compatible.

Mirror Chamber for SRI-CAT 3-ID-B (Y7-30)

The SRI-CAT 3-ID-B mirror allows the user to collimate or focus the synchrotron beam in both the vertical and horizontal directions, by providing two separate cylindrical grooves. This mirror is installed in the 3-ID-B hutch at approximately 34 meters from the source. It is designed for use with monochromatic beam, which eliminates the need for water-cooling.

To achieve focal capabilities, a pneumatic bending mechanism with a variable tangential radius between 11 km and 33 km was chosen. The mirror reflects in the upward direction, which allows for the bending mechanism to operate in the vertical direction. The horizontal collimation and focusing is accomplished by the above mentioned two grooves in the mirror. Each groove can be moved into the beam as needed, utilizing three externally located horizontal stages that also kinematically support and align the entire mirror / housing assembly.

The material of the main mirror body is Zerodur. Both grooves are coated with > 30 nm palladium. Each groove is 800 mm long, which therefore allows for a maximum vertical acceptance of 2 mrad.

Except for additional access ports and feedthroughs, the design of the vacuum housing, kinematic mount, and external kinamatic support for this mirror is similar to that for the Y3-20 mirror mount system. While the housing, mount, and support were designed at the APS, the bending mechanism and the mirror body were developed and fabricated by the ZEISS company. This assembly is UHV compatible.

Mirror Chamber for SRI-CAT 2-ID-A (Y2-20)

The first mirror mount for SRI-CAT 2-ID-A was designed for a 1200-mm-long water-cooled silicon mirror (Y2-20). On the horizontally deflecting mirror surface, three parallel stripes of different coating materials (Si, Pt, and Rh) provide various high-energy filtering functions. The UHV mirror chamber is set on a stepping-motor-driven kinematic mounting table, which has an 50 mm vertical motion range to translate the tank and, therefore move the different reflecting surfaces into the beam. The mirror deflects the incident undulator radiation by 0.3°. The deflected beam has a maximum cut-off energy of 33 keV, using the Pt coating for the 2-ID-D/E branch. The original design of the Y2-20 mirror contained an internal direct water cooling system. Due to a manufacturing difficulties with the internal cooling channel structure, an alternate side-cooling silicon mirror has been installed and is now operational shown in Fig. 5. Several papers have been published about both direct-cooling and side-cooling structure for this mirror.[5,6]

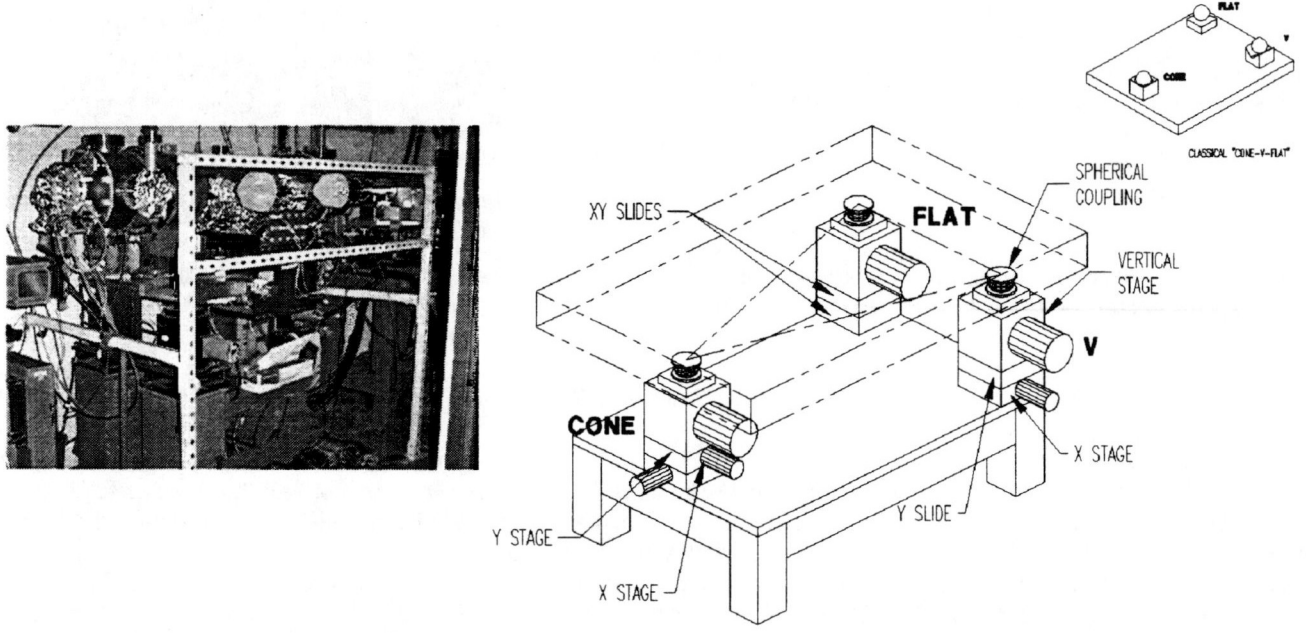

FIGURE 5 Y2-20 mirror mounts at 2-ID-A.

FIGURE 6 Schematic of the "cone-v-flat" equivalent rolling kinematic mounting structure.

Modular Designed Support and Positioning Table for Single Mirror

A common feature of the SRI-CAT single mirror support is the use of modular and standardized components, which resulted in economic use of engineering effort and design. In addition, each support incorporates a "cone-v-flat" equivalent rolling kinematic mounting structure, which provides high positioning precision and high load capacity for stable and minimally constrained support.[7]

Fig. 6 shows a schematic of the "cone-v-flat" equivalent rolling kinematic mounting structure. The top frame is the mounting base for the mirror chamber, which is attached via three self-aligned ball bearings to three precision vertical positioning stages.[8] Each vertical stage is mounted on top of a pair of orthogonal stacked horizontal stages/slides. Table 2 shows the support table specifications for the mirror mounts Y3-20, Y3-30, Y5-20, and Y7-30.

Similar to the support table described above, the Y2-20 support consists of a set of precision stages mounted on three steel columns that are filled with sand for enhanced thermal and vibration stability. To improve the angular positioning resolution for the horizontally deflecting mirror, the mirror chamber is attached to a rotation platform assembly. Horizontal rotational movement is performed with a differential stage actuator by pushing/pulling through a spring-preloaded spherical point contact. A 0.025 µrad resolution has been achieved in a 50 µrad horizontal rotation range with a 1000-kg load.[9] The dynamic properties of this support system have been studied via theoretical and experimental techniques by Basdogan et al.[10]

Table 2. Y5-20, Y3-20, Y3-30, Y7-30 Mirror Support Specifications

Load capacity	1000 kg
Degrees of freedom	4
Vertical travel range	25 mm
Horizontal travel range	25 mm
Vertical motion resolution	0.64 µm
Vertical motion repeatability	5 µm
Horizontal motion resolution	3.2 µm
Horizontal motion repeatability	10 µm
Straightness of trajectory	200 µrad/25mm
Pitch angle resolution	0.6 µrad
Yaw angle resolution	3.0 µrad

3. MIRROR MOUNTS FOR MULTI-MIRROR SYSTEM (Y4-20)

Multi-Mirror System for SRI-CAT 2-ID Beamlines

The second mirror chamber at SRI-CAT 2-ID-A is a vacuum vessel for three mirrors that switch the beam into the 2-ID-B and 2-ID-C beamline branches horizontally as shown in Fig. 7.

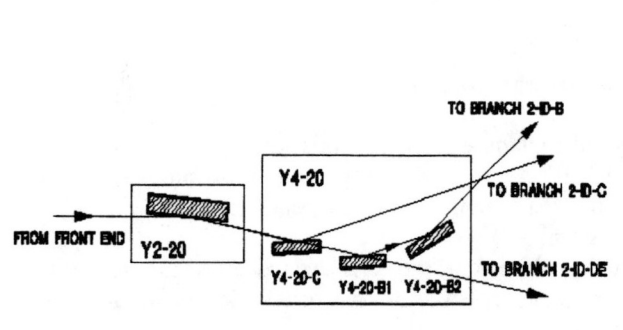

FIGURE 7 Schematic of 2-ID beamline branches.   **FIGURE 8** Y4-20 multi-mirror system at 2-ID-A.

The Y4-20-C mirror receives an incident beam that is reflected by Y2-20 and deflects it to 2-ID-C branch by an angle of 2.5° from the incident beam. The cutoff energy of Y4-20-C is 3 keV. To switch the beam into the 2-ID-B branch, the Y4-20-C mirror can be moved out of the beam. The Y4-20-B1 mirror then receives the beam reflected by Y2-20 and deflects it by 2.5° to the Y4-20-B2 mirror, which deflects the beam by another 2.5°. The 2-ID-B branch accepts the beam reflected by Y4-

20-B2. When all of the mirrors in the second mirror chamber (Y4-20) are moved out of the beam, the beam reflected by Y2-20 is delivered into the 2-ID-D/E branch.

Modularly Designed Mirror Support and Manipulator for 2-ID Multi-mirror System

A modular, UHV-compatible mirror support and manipulator system has been designed for the three mirrors in 2-ID Y4-20-B1, Y4-20-B2, and Y4-20-C, as shown in Fig. 8. There are three mirror-mount platforms inside the vacuum chamber. Each platform is attached via three self-aligned ball bearings to three vertical posts, which also function as vacuum feed-through components with welded bellows. Each vertical post is mounted on the top of a pair of orthogonal stacked horizontal stages/slides, then assembled on the precision vertical stage. Similar to the single mirror support table, this mirror manipulator is a "cone-v-flat" equivalent rolling kinematic mounting structure. To improve the system stiffness, an extra stiffener platform is attached via three self-aligned ball bearings to the base of the three vertical posts.

The water supply pipes for the mirror side-cooling structure are through the center hole of the manipulator post with a UHV mini-flange. No water-to-vacuum joints exist in this design. The specifications for this UHV mirror-mount actuator/stage system are shown in Table 2. The large vertical travel range of the manipulator allows users to move the mirror out of the beam or to choose a different coating strip on the mirror surface. The analysis of this high-precision UHV mirror manipulator for vibration stability has been presented previously.[11]

Table 3. Y4-20 Mirror support specifications

Load capacity	227 kg
Degrees of freedom	5
Vertical travel range	60 mm
Horizontal travel range	12 mm
Vertical motion resolution	0.3 μm
Vertical motion repeatability	3 μm
Horizontal motion resolution	0.1 μm
Horizontal motion repeatability	1 μm
Straightness of Trajectory	10 μrad/5mm
Pitch angle resolution	0.33 μrad
Roll angle resolution	2.4 μrad
Yaw angle resolution	1.1 μrad

Beam Masking for 2-ID Multi-Mirror System

To perform as a "fail safe" system, the beamlines must be designed to handle accidental missteering of the beam by the mirror, as well as the source, which can result in component damage. This can be achieved by limiting the angular motion of the mirror in conjunction with a carefully designed masking system. Maximum missteered beam ray-tracing analysis is needed to determine the mirror and mask system operational safety margin.

A total of eight masks are used for the 2-ID-A area, including the white-beam slit and the front-end masks for beam missteering control. Three of them are white-beam masks, which are water cooled and constructed with an enhanced heat-transfer mechanism developed for the APS front-end components.[12]

To prevent damage to the downstream beam transport components from the beam missteered by the mirrors in the Y4-20 mirror chamber, a mask is placed in each of the beamline branches in the photon shutter tank. In the 2-ID-D/E branch, the water-cooled mask is made of copper. For branches 2-ID-B and 2-ID-C, the mask is made of a tungsten alloy with indirect water cooling, so that the mask also functions as a bremsstrahlung collimator. In addition, two water-cooled masks are mounted in the beamline split-tank to protect the wall of the tank between the branches.

4. CONCLUSION

This paper discusses the modular mirror mounts designed for the SRI-CAT beamlines at the APS. Five different single mirror mounts and a multi-mirror system are presented. The use of modular and standardized components resulted in economic use of engineering effort and design.

Presently, the mirror mounts Y2-20, Y3-20, Y5-20, and Y7-30 have been installed on the beamline and are operational. The initial laboratory tests show that the mirror mounts achieved the expected design specifications. The mirror mount Y3-30 and the multi-mirror system Y4-20 are fabricated and in final alignment and installation stage. The mirror performance test data will be published in separate papers later.

ACKNOWLEDGEMENTS

We acknowledge the help to the SRI-CAT mirror mounts design and construction from Drs. Denny Mills, Efim Gluskin, Pat Den Hartog, Emil Trakhtenberg, Ali Khounsary, and Messrs. Bill McHargue, Greg Wienerslage, Joe Arko, Russ Otto, Kevin Knoerzer, Roger Ranay, Michel Lehmuller, Gerald Czop, and. Try Leng Kruy. This work was supported by the U.S. Department of Energy, BES Materials Sciences, under Contract No. W-31-109-Eng-38.

REFERENCES

1. W. Yun, A. Khounsary, B. Lai, K. J. Randall, I. McNulty, E. Gluskin, and D. Shu, "Advantages of using a mirror as the first optical component for APS undulator beamlines," Rev. Sci. Instrum., 67(9), CD-ROM (1996).
2. R. Z. Bachrach, R. D. Bringans, N. Hower, I. Lindau, B. B. Pate, P. Pianetta, L. E. Swartz, and R. Tatchyn, Nucl. Instrum. and Meth. Vol. 222, p70-79, (1984).
3. S. L. Hulbert, E. Rotela, and M. Shleifer, Nucl. Instrum. and Meth. A291, p337-339, (1990).
4. R. J. Dejus, B.Lai, E. R. Moog, and E. Gluskin, "Undulator A characteristics and specifications: enhanced capabilities," Argonne National Laboratory Report ANL/APS/TB-17 (1994)
5. A. Khounsary and W. Yun, "On optimal contact cooling of high-heat-load x-ray mirrors," Review of Scientific Instruments, 67(9), CD-ROM (1996).
6. W. Yun, B. Lai, Z. Cai, D. Shu, and J. Barraza, "Design of a coherence-based beamline for x-ray microfocusing at the APS," Rev. Sci. Instrum. 67 (9), CD-ROM (1996).
7. J. Barraza, D. Shu, and T. M. Kuzay, Rev. Sci. Instrum. 66(2), p1630 (1995)
8. D. Shu, J. Barraza, U.S. Patent # 5,526,903, "Low profile, high load vertical rolling positioning stage," June 18, 1996
9. J. Barraza, D. Shu, W. Yun, and T.M. Kuzay, "Design of a high precision mirror-rotation system at the Advanced Photon Source," Rev. Sci. Instrum. 67(9), CD-ROM (1996).
10. I. Basdogan, T.J. Royston, A.A. Sabana, J. Barraza, D. Shu, and T.M. Kuzay, "Analysis of high-precision optical positioning systems for vibration stability at the APS," to be published at SPIE, San Diego, CA, July 1997.
11. I. Basdogan, T.J. Royston, A.A. Sabana, D. Shu, and T.M. Kuzay, "Vibratory Response of a Mirror Support/Positioning System for the APS at ANL," presented at the SPIE conference, Denver, CO, Aug. 1996.
12. T. M. Kuzay, J. T. Collins, A. M. Khounsary, and G. Morales, Proc. ASME/JSME Joint Conf., Book No. I0309E-1991, pp 451-459.

Miniaturized Kappa goniometer for macromolecular crystallography

G. Rosenbaum and E. M. Westbrook

Structural Biology Center
Argonne National Laboratory
Advanced Photon Source
9700 S. Cass Ave., Argonne, IL 60439

A goniometer with kappa geometry has been designed and built specifically for macromolecular crystallography. The main feature is a miniaturized kappa stage made possible by the small weight of specimen and specimen holder.

The design goal was to: 1) eliminate interference between stage and area detector for specimen-to-detector distances of 100 mm and more; 2) minimize the sphere of confusion on expectation of dealing with very small crystals at third generation sources; 3) minimize the solid angle of shadow and inaccessible positioning of the sample due to interference of the stage with other objects in the sample area; 4) achieve a rotation speed of 10 degree/s at 0.5% constancy and 0.4 s acceleration time for 0.05 s exposures of 0.2 degree fine slice frames every 2 seconds, and 5) to achieve precise synchronization between rotation angle and shutter opening and closing

The kappa stage is mounted on a commercial high precision rotary table, designed for use in both horizontal and vertical orientation. This table provides the high precision rotation for data acquisition. The required crisp response and constant speed is delivered by a high output direct drive DC-motor, controlled by a closed-loop controller using feedback from a precision angular encoder. The kappa- and phi-motions are used for sample positioning only and are driven by miniature DC-motors equipped with integral encoders.

This work was supported by the US Department of Energy, Office of Health and Environmental Research, under Contract W31-109-ENG-38.

User-friendly Interfaces for Control of Crystallographic Experiments at CHESS

D. M. E. Szebenyi, A. Deacon, S. E. Ealick, J. M. LaIuppa, and D. J. Thiel

MacCHESS, Cornell University, Ithaca, New York 14853

Abstract. In designing a system to collect high quality diffraction data in an efficient manner, both hardware and software must be considered. This work focuses on the data collection software used at CHESS, the Cornell High Energy Synchrotron source, with emphasis on the interface between the user and the experimental components. For each type of detector used at CHESS, there is a graphical user interface (GUI) enabling the user to easily set up and run an experiment. For the CCD detector from Area Detector Systems Corp., this is a commercial product from ADSC, customized for CHESS. For the Princeton CCD detectors, a GUI has recently been developed to streamline communication between the user and the **TV6** program which controls the detector. For Fuji imaging plates, a new GUI controls operation of the oscillation camera, including the imaging plate carousel; scanning of plates is done using the software provided by Fuji. Although these GUI's are not identical, they have numerous similarities, making it easier for users to learn operation of a new detector. They also incorporate error-checking to avoid problems such as overwriting data files or collecting data with no x-rays. Common to experiments with all detectors is a GUI used for operations such as alignment of the optical table on which the oscillation camera is mounted. Integral to a good data collection system is the capability to process diffraction images, for evaluation of crystal quality, determination of data collection strategy, screening of potential derivatives, and so forth. The **mccview** graphical front-end has been developed to conveniently initiate processing programs, including preliminary routines (**correct, getbeam**), main analysis routines (**xdisp, denzo, scalepack**), and the strategy routine **m.simulate**.

INTRODUCTION

Crystallographers go to synchrotron sources to collect diffraction data on difficult projects. Their time at the beamline is usually brief, and the supply of good crystals may be limited. Hence, it is very important to such experimenters to be able to collect the best data possible in an efficient manner. Operators of beamlines at synchrotron sources have developed systems of hardware and software to satisfy this need (1, 2, 3). These systems are continually being improved, and we report here on the latest developments in the system at the Cornell High Energy Synchrotron Source (CHESS), with emphasis on the user interface.

A GENERAL DESIGN FOR A CRYSTALLOGRAPHIC USER INTERFACE

The goal is to design a user interface for collection of crystallographic data which is:

- easy to learn and use
- capable enough to handle all necessary operations
- tolerant of user errors
- able to communicate with the experimental hardware, i. e. able to send commands and to receive feedback on the current state of the equipment
- able to handle hardware error conditions appropriately
- modular, hence easily adapted for different experimental configurations
- compatible with an integrated approach combining data collection and analysis

The user interface is part of the general scheme of data collection and processing shown in Fig. 1.

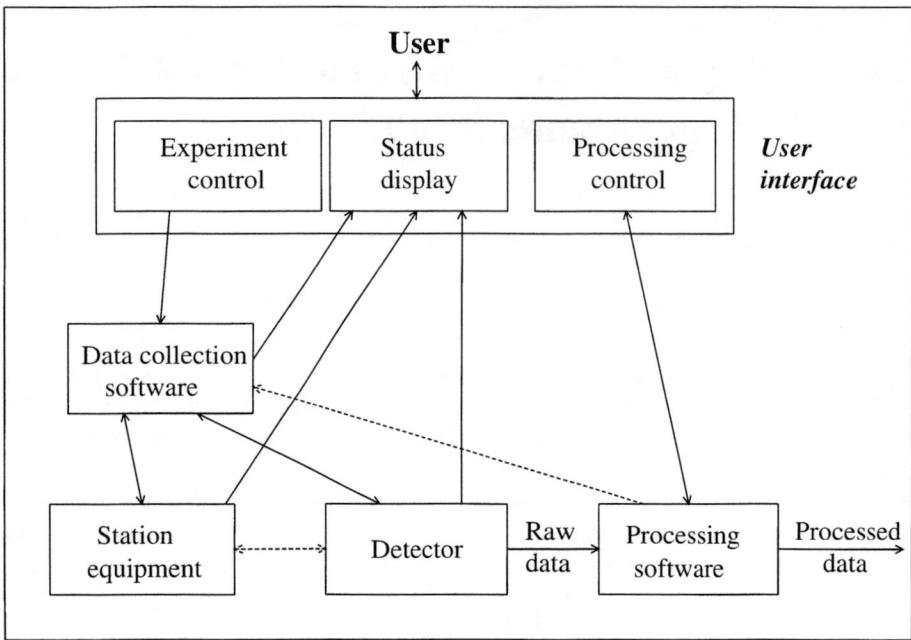

FIGURE 1. A general scheme for control of data collection and processing. The boxes represent logical entities, each of which corresponds to some combination of hardware and software, in some cases extending across multiple processors and associated equipment. For example, "station equipment" and "detector" include both actual equipment for taking data and associated low-level control programs. "Processing software" includes pre-processing functions such as calculation of the optimal strategy for collecting data. The arrows represent flow of information, including commands, responses, status reporting, and crystallographic data. Dashed connections may be indirect; for example, "detector" need not talk directly with "station equipment"; any necessary communication can be done through the "data collection software". Generally, feedback from data processing to data collection is done through the user, but a direct link from a strategy calculation program to the data collection engine is certainly possible and is indicated by a dashed line.

The user is an important part of the crystallographic experiment, even after a crystal has been mounted on the camera. Before starting data collection on a crystal, one must evaluate the quality of the crystal and determine whether it is worthwhile to collect these data at all. There is presently no good way to do this automatically; hence the interface shown in the top three boxes of Fig. 1 should be optimized to allow the user to:

- Easily take one or more initial diffraction "snapshot(s)".
- Display and manipulate the image(s), to evaluate the crystal quality.
- Decide on a strategy for data collection.
- Set up and initiate actual data collection, with minimum opportunity for mistakes.
- Process the data as fast as possible, for confirmation of its quality and completeness.

SPECIFICS OF THE CHESS SYSTEM

The CHESS Environment

Specific constraints for a user interface to be implemented at CHESS include:

- It must work through **spec**, a package from Certified Scientific Software, to control station equipment such as shutter, spindle motor, and optical table.
- Detectors at CHESS include two different makes of CCD and Fuji image plates. It is desirable that interfaces for the different detectors look and feel similar, although they need not be identical.
- It must work with the existing low-level control software for the various detectors, and take advantage of existing display programs. It should be able to work with local processing programs such as **m.simulate** and **mccview**.

At CHESS, "station equipment" at the A1, F1, and F2 stations includes a single-axis oscillation camera mounted on an optical table, controlled by **spec**. A small GUI called **CCDspec**, by J. LaIuppa, running on an H-P workstation, is used to control some **spec** functions which are independent of detector type, such as adjustment of the optical table position to maximize beam through the collimator. A central part of the "data collection software" is always the **rspec** communications hub, by J. LaIuppa. This routine accepts input from the user, either directly as commands typed in the "OpsTerm" window or indirectly through another program, and transmits it to **spec** for operation of the hardware. Replies are sent to the OpsTerm window and to the originating program, if any. **Rspec** runs on the H-P but can communicate with programs running elsewhere; several types of communication link are available. "Processing software" includes, among others, the strategy program **m.simulate** and the HKL processing package (**xdisp, denzo, scalepack**, by W. Minor and Z. Otwinowski). These run on DEC Alpha workstations. "Processing control" includes the **mccview** graphical front end as well as direct user interaction with the processing programs. Input to **m.simulate** and **xdisp** is through graphical interfaces, while **denzo** and **scalepack** accept typed commands. The "experiment control" and "status display" functions of the user interface are customized for the various detectors used at CHESS and are described in more detail below.

USER INTERFACES FOR DETECTORS AT CHESS

ADSC CCD: the ADX Suite

The CCD made by Area Detector Systems is operated by a suite of programs written by A. Arvai and C. Nielson of ADSC. **Ccd_dc_api** is the central coordinating program, communicating with the detector through **DET_API**, with the station hardware through **rspec** via **ccd_bl_a1**, and with the user through the **adx_ccd_control** GUI. Status display is provided by **adxv** (for images) and **ccd_status** (for experiment status), and **ccd_xform_api** handles transformation of raw data files to corrected images. The ADX system has been described in detail elsewhere (3).

Princeton CCD's: TV6 and mccdata

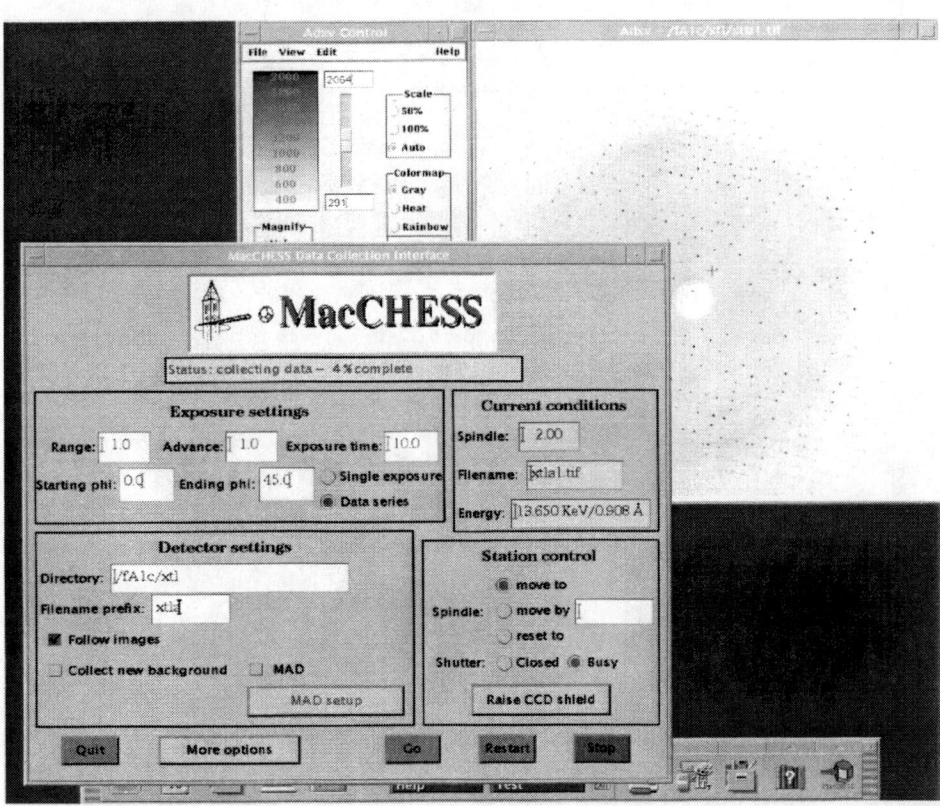

FIGURE 2. Computer screen during a data collection run; **mccdata** main window and **adxv** display of the current image are shown.

The 1K and 2K CCD's made by S. Gruner and coworkers are operated by a program called **TV6**, written by E. Eikenberry, which runs on a PC and encompasses both low-level detector control and "data collection software" functions. It communicates with **rspec** to control station equipment, and has a comand-line interface for "experiment control". Image display is provided by the **adxv** module from the ADX suite. A new program, **mccdata**, by D. Szebenyi, functions for "experiment control" and "status display". Features of the program include:

- The well-organized graphical interface is easy to understand and use.
- Setting up and initiating a single exposure or a data series is very straightforward. Multiwavelength runs, for MAD data collection, may also be created. Data collection can be interrupted and restarted at the same point, or an earlier point, in the series.
- A "current status" section of the main window keeps track of the current spindle angle, image file name, and x-ray wavelength.
- Error messages appear in pop-up windows, making them hard to miss.
- **Mccdata** communicates with **TV6** by means of command files written to a standard location on disk, and with **rspec** by means of a socket connection, for direct control of the shutter and spindle rotation when a data series is not in progress.

Preliminary tests indicate that users find this GUI convenient and helpful, and it is now the standard method of running the Princeton CCD's.

Fuji Image Plates: IMPspec

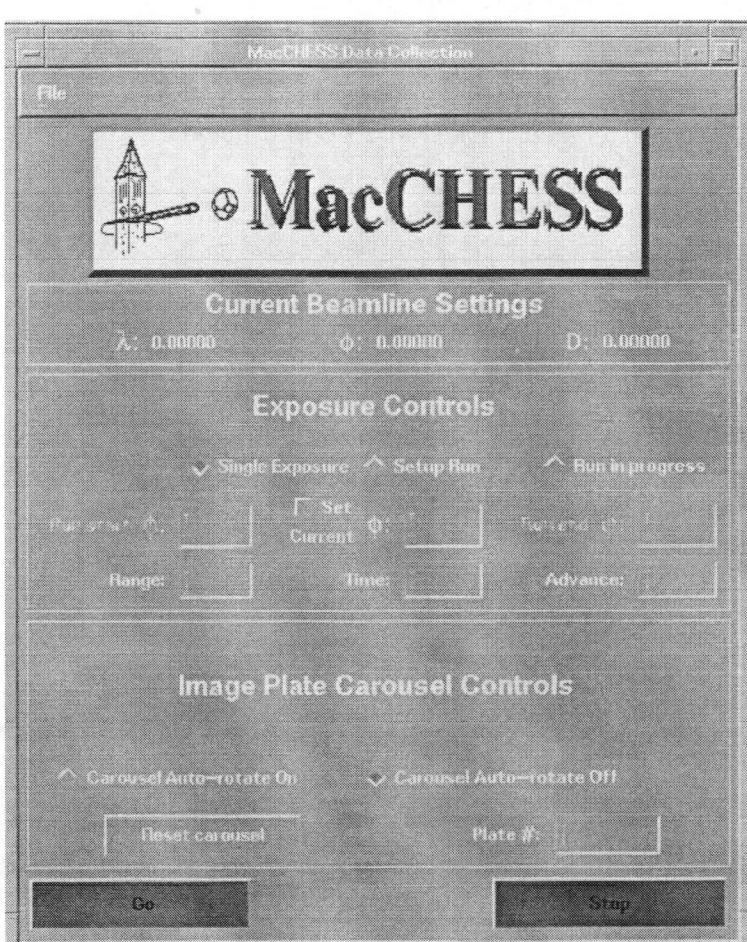

FIGURE 3. Main window of **IMPspec**, shown as it appears on startup of the program.

When Fuji image plates are used to collect data, the "detector" heading covers the plates themselves and their scanner, which is run by the **BAS2000** program (from Fuji) running on a SPARCstation. There is no direct connection between the scanner and the station equipment, but scanned files are immediately available for processing, *via* NFS-mounted disks. "Station equipment" includes a carousel which holds several image plates. "Processing software" includes the **getbeam** program for locating direct beam spots. "Equipment control" and "status display" are now provided by the **IMPspec** GUI by J. LaIuppa, which runs on an H-P. **IMPspec** features include:

- The well-organized graphical interface is easy to understand and use.
- Setting up and initiating a single exposure or a data series is very straightforward. The user is automatically notified when image plate cassettes need to be reloaded.
- **IMPspec** communicates with **rspec** by means of a socket connection, for direct control of spindle and carousel rotation when a data series is not in progress.
- A means is provided for manually setting parameters such as spindle motor speed, which are normally determined automatically.
- The current spindle angle, carousel position, and x-ray wavelength are always displayed in the main window.
- Error messages appear in pop-up windows, making them hard to miss.

The few image plate users who have been at CHESS since completion of the GUI have been very pleased with it.

CONCLUSION

Crystallographic users at CHESS now have convenient, user-friendly graphical interfaces available for almost all routine operations needed to collect and process diffraction data. These interfaces increase the rate of production of good data by speeding up the user training process, reducing input errors, and making it easier to check data quality in a timely fashion. Developments are continuing, particularly in the fields of error handling, provision for unusual experiments, and integration of crystal evaluation, data collection, and data processing functions.

ACKNOWLEDGEMENTS

We are grateful to Andy Arvai and Chris Nielsen of Area Detector Systems Corp. for their customization of the ADX suite of programs for CHESS; Andy was also kind enough to expand the capabilities of **adxv** to handle images from many different detectors. Eric Eikenberry provided assistance in recent developments involving **TV6**. **Rspec** uses the **kblib** library developed by Keith Brister. CHESS staff and users have been helpful in testing interfaces.

REFERENCES

1. Sweet, R., and Skinner, J. M., Acta Cryst. **52**, C-27 (1996).
2. Kinder, S. H., McSweeney, S. M., and Duke, E. M. H., J. Synchrotron Rad. **3**, 296-300 (1996).
3. Szebenyi, D. M. E., Arvai, A., Ealick, S., LaIuppa, J. M., and Nielsen, C., J. Synchrotron Rad. **4**, 128-135 (1997).

Synchrotron Beam Stabilization Techniques at CHESS

J. White, E. Fontes, S. Peck*

CHESS and
*Laboratory of Nuclear Studies
Cornell University
Ithaca NY 14853

The reproducibility of operating conditions from experimental run to run, and during the run, have long been of major concern to the CHESS staff and users. Assuring this in a colliding beam machine with counter-rotating particle beams has long been a problem. Recently, there has been a great deal of progress made on this front. A software feedback system using magnetic bumps in the storage ring, CESR, in coordination with X-ray beam position monitors from CHESS, has been implemented during the High Energy Physics running to minimize the perturbations to the x-ray user community inherent in a colliding beam experimental facility. This has been done in a joint effort between a number of individuals from both the CHESS and CESR staffs.

Recent results show that the variation of the vertical position of the x-ray beam at the position monitors from one experimental run to another has been decreased from ~50 to 100 microns, in the best of conditions, to ~20 microns. The stability of the beam during an experimental run has increased as well, such that position drift has been reduced by a factor of two under normal circumstances. Due to the complexities involved in running a machine for high energy physics and the goals associated with it, there are conditions in which instabilities will arise. However, the feedback as been shown to effectively minimize the duration of many of these occurrences. As the CHESS staff works to increase the monitoring capacity of the laboratory, and thus the knowledge base of how to minimize excursions from optimum running for x-ray experimentalists, the CESR staff is working on increasing reliability of the beam conditions of the storage ring on many fronts.

REAL-TIME: <u>IN-SITU</u> MEASUREMENTS AND MATERIALS CHARACTERIZATION

Real-time x-ray scattering studies of thin-film growth and processing

R.L. Headrick

Cornell High Energy Synchrotron Source
Cornell University Ithaca, NY 14853

Recent advances in synchrotron sources and x-ray detectors allow processes such as thin-film growth, surface reactions, and strain-relaxation mechanisms to be probed in detail in real-time. Development of surface and interface scattering techniques, and instrumentation to perform these processes in-situ have also progressed significantly over the last 10-15 years. I will briefly review these developments, with emphasis on thin-film growth, and describe the current efforts at CHESS to develop a materials growth and processing user facility.

X-ray Studies of Annealing in Thin-film Semiconductors

Roy Clarke

*MHATT-CAT Center for Real-time X-ray
Studies at the Advanced Photon Source and*

*University of Michigan
Ann Arbor, MI 48109-1120*

One of the most important issues in thin-film heterostructures is the nature of the interfaces. In general, their structure is likely to involve some degree of lattice strain, especially when the film is deposited on a mismatched substrate. The evolution of the film stress and its behavior during post-deposition processing is critical for many device applications, in particular where extreme operating conditions are encountered. In this presentation we illustrate the use of real-time x-ray scattering for in-situ annealing studies of semiconductor films, emphasizing the advantages of appropriate x-ray optics and fast area detectors.

Low-Energy X-ray Dosimetry Studies (6 to 16 keV) at SSRL Beamline 1-5

N. E. Ipe, S. Chatterji, A. Fassò, K. R. Kase, R. Seefred*; P. Olko and P. Bilski[†]; and C. Soares[‡]

Stanford Linear Accelerator Center, Stanford University, Stanford, CA 94070; [†]Health Physics Lab, Institute Of Nuclear Physics, Krakow, Poland; and [‡]United States Department of Commerce, NIST Ionizing Radiation Division, Gaithersburg, MD 20899

Abstract. Synchrotron radiation facilities provide a unique opportunity for low-energy x-ray dosimetry studies because of the availability of monochromatic x-ray beams. Results of such studies performed at the Stanford Synchrotron Radiation Laboratory (SSRL) are described. Polish lithium fluoride thermoluminescent dosemeters (TLDs), MTS-N(LiF:Mg, Ti- 0.4 mm thick), MCP-N (LiF:Mg, Cu, P - 0.4 mm thick) were exposed free in air to monochromatic x-rays (6 -16 keV). These exposures were monitored with an SSRL ionization chamber. The responses (counts /Gy) of MTS-N and MCP-N were generally found to increase with increasing energy. The response at 16 keV is about 3 and 4 times higher than the response at 6 keV for MTS-N and MCP-N, respectively. Irradiation at 6 keV indicates a fairly linear dose response for both type of TLDs over a dose range of 0.01 to 0.4 Gy. In addition there appears to be no significant difference in responses between irradiating the TLDs from the front and the back sides. The energy response of the PTW ionization chamber type 23342 relative to the SSRL ionization chamber is within ±4.5 % between 6 and 16 keV. Both the TLDs and the PTW ionization chamber can also be used for beam dosimetry.

I. INTRODUCTION

There are currently over thirty synchrotron radiation (SR) facilities in operation around the world, with another twenty-eight in the construction, design, or proposal stage. As SR facilities are rapidly being built all over the world, they introduce the need for low-energy x-ray dosemeters because of the potential radiation exposure to experimenters who work in close proximity to the SR beamlines and experimental enclosures. However, they also are an important resource for providing monochromatic x-rays that can be used for research and calibration of dosemeters.

Most of the commercially available dosemeters are not designed to respond well to low-energy x-rays (< 30 keV) and frequently their responses at these low energies are not well known. Further, the holder or dosemeter packaging significantly attenuates the low-energy photons (< 10 keV). Hence there is a critical need to develop a low-energy x-ray dosemeter. With this as the ultimate goal, a series of studies have been undertaken at SSRL, in which the low-energy responses of both active and passive detectors such as thermoluminescent dosemeters (TLDs), are being determined using monochromatic x-rays.

In principle purely monoenergetic beams are necessary to determine the response function of any dosemeter. The availability of monochromatic x-rays at any discrete energy below 30 keV, at synchrotron radiation (SR) facilities provides unique opportunities for low-energy x-ray dosimetry studies. Calibration facilities normally make use of filtered techniques with x-ray machines that produce continuous spectra to generate "nearly" monoenergetic photon beams. These photon beams have a spectrum with a peak at the desired energy. Only a few energies in the desired energy range (<30 keV) are available.

In addition, low-energy x-ray dosemeters are useful in both radiotherapy and radiodiagnostic techniques. In radiotherapy, x-rays with energies between 5 and 20 keV are used to treat certain types of skin disorders. The spatial distribution of the dose over the treatment area can be determined with TLDs. X-rays with energies of 10 keV and above are frequently used in diagnostic medicine, for example, mammography at 17.5 keV.

In order to do accurate dosimetry, detectors that have been calibrated against primary or secondary standards are required. Since the ionization chambers that were used in the experiments at SSRL had not been calibrated against any standards, the response of the PTW ionization chamber Type 23342[1], in the energy range of 6 to 16 keV was investigated. This type of chamber is used routinely for absolute dosimetry in radiotherapy, and is also used in secondary standards laboratories.

Thermoluminescence describes the process of emission of optical radiation from a material upon heating. When a crystal is exposed to ionizing radiation, free electrons and holes are produced. The electrons and holes may become trapped at defect sites as they migrate through the crystal. The trapped electrons will remain in their traps provided that they do not acquire sufficient energy to escape. If the temperature of the material is raised trapped electrons may acquire sufficient thermal energy to be released. The energy gap between the trap and the conduction band determines the temperature

---

[1] PTW-Freiburg, Lorracher Strasse 7, D-79115 Freiburg, Germany

required to release the electron. Released electrons may then combine with holes at luminescent centers and the excess energy is radiated as visible or ultraviolet light. An annealing procedure is necessary for TLDs to empty all the shallow and deep traps.

The plot of light intensity as a function of temperature is referred to as a glow curve. Since there are many trapping levels present, each trapping level will give rise to an associated glow peak maximum, which may or may not be resolved during the readout. Most TLD materials have low temperature peaks (i.e., peaks occuring at temperatures lower than the main dosimetric peak). Since these peaks are not stable at room temperature and particularly below 100°C their fading may impact the results of measurements. Hence, a partial annealing before the readout known as a pre-readout annealing is used to remove the low temperature peaks. Thus, both accuracy and precision of the readout is improved.

The area under the glow curve and therefore the response of the TLD is proportional to the energy absorbed in the TLD for a given type of radiation. Preliminary studies on the low-energy (7 - 17.5 keV) response of the Harshaw/Bicron LiF TLDs indicate that for thicker TLDs (> 0.2 mm), the predicted response per unit absorbed dose in air, based on energy absorbed in the TLDs is significantly greater than the measured response (1). The differences between measured and predicted responses at low energies can be attributed to the attenuation of photons and the attenuation of light in the TLD. Once the low-energy response of various TLDs is known, additional characteristics such as overall TLD efficiency, effective energy attenuation coefficient, and light attenuation coefficient can be determined as a function of energy. The effective energy attenuation coefficient can be determined from Monte Carlo simulations of energy deposition in the TLD. Using a combination of measured responses obtained by irradiating the TLDs from the front and back sides, and a theoretical TLD response model, the light attenuation coefficient and overall TLD efficiency can be determined. These parameters are vital for the understanding of basic TL mechanisms. The energy response is particularly important in assessing the potential use of different TLD materials for low-energy x-ray dosimetry.

In this paper we report the results of experiments to determine the low-energy (6-16 keV) response of the Polish TLDs (MCP-N, MTS-N), and the PTW ionization chamber.

II. THEORETICAL MODEL FOR RESPONSE OF TLD

It can be shown that the responses R(k) and R'(k) of the TLD when read with the irradiated side facing the photomultiplier (PM) tube and away from the PM tube, respectively, are given by equations 1 and 2 (Appendix A).

$$R(k) = \frac{CN\mu(k)k}{\mu(k)+f}\left\{1 - e^{-[\mu(k)+f]t}\right\}$$

(1)

$$R'(k) = \frac{C'N\mu(k)k}{\mu(k)-f} e^{-ft}\left\{1 - e^{-[\mu(k)-f]t}\right\}$$

(2)

where:
R(k) = TLD response for N photons each of energy k when read with irradiated side facing PM tube
R'(k) = TLD response when read with irradiated side away from PM tube
C = $\eta\varepsilon$ = Overall TLD efficiency when read with irradiated side facing PM tube
C' = Overall TLD efficiency when read with irradiated side away from PM tube
η = Intrinsic luminous efficiency
ε = Reader efficiency
$\mu(k)$ = Effective energy attenuation coefficient
f = Effective light attenuation coefficient
t = TLD thickness

Dividing equation 1 by equation 2 and assuming C = C' we obtain equation 3:

$$\frac{R'(k)}{R(k)} = \frac{e^{-ft}\left\{1 - e^{-[\mu(k)-f]t}\right\}(\mu(k)+f)}{\left\{1 - e^{-[\mu(k)+f]t}\right\}(\mu(k)-f)}$$

(3)

$\mu(k)$ can be determined from Monte Carlo simulations of energy deposition studies in the TLD. Once $\mu(k)$ is known, f can be determined numerically from equation 3. Using $\mu(k)$ and f in equation 1, C can be determined. Energy deposition studies in the Polish TLDs, MCP-N (^{nat}LiF: Mg, Cu, P) and MTS-N (^{nat}LiF: Mg, Ti) were performed using FLUKA (2), a Monte

Carlo code. The energy deposited in thin layers, each of thickness about 0.1 mm and radius 2.25 mm was scored in ~5 cm-thick disks of MCP-N and MTS-N TLDs for photon energies of 7, 9, 12 and 15 keV. μ(k) was obtained from the energy vs depth in TLD curves. Table 1 lists the effective energy attenuation coefficients obtained from FLUKA and the theoretical energy absorption coefficients (provided by the manufacturer) of both MCP-N and MTS-N. The errors on the FLUKA values are about 0.2%. Thus it can be seen that the effective energy attenuation coefficient is the same as the energy absorption coefficient. The differences (up to 10%) can be attributed to the different cross sections used in the theoretical and FLUKA calculations.

TABLE 1. Energy Attenuation Coefficients For TLD's

Energy (keV)	Energy Attenuation Coefficient (cm^{-1})			
	MTS-N		MCP-N	
	FLUKA	Theoretical	FLUKA	Theoretical
7	43.29	44.25	47.47	48.00
9	20.25	20.90	22.87	23.45
12	8.42	8.99	9.45	10.13
15	4.27	4.71	4.86	5.32

III. EXPERIMENTAL CONDITIONS

Experiments were performed at the Stanford Synchrotron Radiation Laboratory (SSRL) Beamline 1-5 using synchrotron radiation from the 3 GeV electron storage ring SPEAR (a schematic of the beamline is shown in Fig. 1). The experimental layout is described in detail in Reference 1. The synchrotron radiation enters the beamline through a pair of beryllium windows, a pair of vertical slits and a pair of horizontal slits. The beam is then incident on a downward reflecting monochromator comprised of 2 silicon (111) crystals. A white beam stop in the monochromator ensures that white light does not enter the downstream experimental enclosure. A piezoelectric crystal adjustment is used to detune the monochromators to eliminate the higher energy harmonics. The x-ray beam is then transported into the experimental enclosures (not shown in the figure).

Two parallel-plate ionization chambers are mounted downstream of the mono beam shutters. The portion of the beam line downstream of the mono shutters is mounted on a motorized table that is capable of both horizontal and vertical motion. A shutter and a series of filters are mounted in a holder between the ionization chambers. The shutter regulates the x-ray dose to the sample. The filters are used to calibrate the energy of the mono beam by scanning across an absorption edge, while measuring the beam intensity with the ionization chambers both before and after the filter. The ionization chambers are operated at a high voltage of 310 volts and are filled with slowly flowing nitrogen at atmospheric pressure. A set of horizontal and vertical Huber slits define the beam size at the dosemeter position.. A third ionization chamber placed downstream of the manual slits is used to monitor the dose to the dosemeters or detectors. A collimator after the third ionization chamber acts as a guard slit. Dosemeters were mounted in specially designed plexiglass holders. The entire system is under the control of a computer that drives the monochromator to the desired energy and exposes the sample to the selected radiation dose.

The lithium fluoride Polish TLDs obtained from the Institute of Nuclear Physics, Poland, were MCP-N- 0.4 mm thick (3) and MTS-N -0.4 mm thick. The TLDs are solid circular sintered pellets of diameter 4.5 mm. According to the manufacturer, the detection threshold is about 10000 and 50 nGy, respectively and the response is linear between 5×10^{-5} and 3, and 10^{-7} and 10 Gy, respectively. The photon energy dependence is within 30 and 20 %, respectively between 30 keV and 1.3 MeV. The thermal fading at room temperature is about 5% per year for both types of TLDs and the fluorescent light effect on fading and zero reading is negligible.

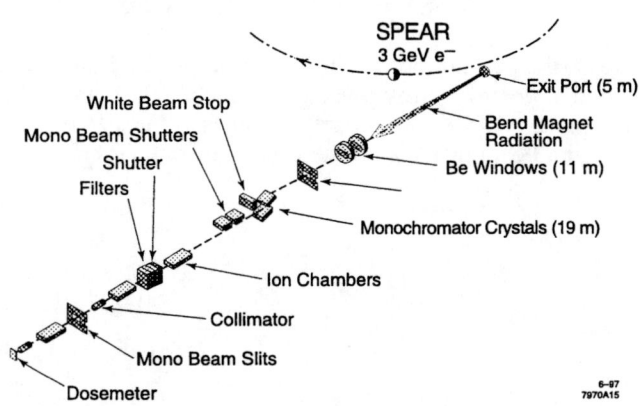

FIGURE 1. Schematic of SSRL Synchrotron Radiation Beam Line 1-5.

The TLDs were read out at the Institute of Nuclear Physics using an RA '94*[2] reader equipped with a photomultiplier tube (EMI 9789QB) with a bialkali photocathode and BG-12 optical filter. The pre-readout annealing consists of heating the dosemeters at 100 °C for 10 minutes.

The readout cycle consists of a linear ramp at 5 °C/s. The maximum temperature is typically 350 °C for MTS-N and 280°C for MCP-N. The annealing cycle is as follows:

MTS-N - 400 °C for 1 hour and 100 °C for 2 hours.

MCP-N - 240 °C for 10 min followed by fast cooling on an aluminum plate.

The "GaFChromic™ Dosimetry Media" film MD-55 is a radiochromic film which was used for imaging because of its high spatial resolution (4). It is colorless before irradiation and turns progressively blue with increasing absorbed dose.

The PTW ionization chamber type 23342 consists of a rectangular block 61 mm in length, 22m in width, and 14.4 mm in height. The cylindrical air-filled cavity inside has a diameter of 5.2 mm and is covered by a thin membrane window. The sensitive measuring volume is defined by the electrode which has a diameter of 3 mm and the distance (1mm) between the two electrodes. Additional information regarding the chamber can be obtained from the instrument manual available from the manufacturer. The chamber was connected to a Keithley electrometer 602. A variable battery source was used for voltage supply. The GaFChromic film was used to center the beam on the window. According to the manufacturer, the energy dependence of the chamber is within ±2 % in the energy range of 20 keV to 70 keV. The chamber is open to air and hence requires a temperature correction. At an operating voltage of 300 volts the chamber collects more than 99.5% of all the charges produced by continuous radiation with dose rates of up to 175 Gy/s and pulsed radiation with pulsed dose rates of up to 1.8 mGy/s.

IV. EXPERIMENTAL PROCEDURE

The horizontal and vertical slits upstream of the third ionization chamber were adjusted so that the beam size at the sample position was 1.4 mm × 1.4 mm. This was verified by measuring the size of the beam on the GaFChromic film with a microscope. The monochromator was calibrated using an iron filter. The monochromator was then set at various energies ranging from 6 to 16 keV. At each energy, the harmonics were monitored by studying the energy spectrum scattered by a thin capillary tube using a Bicron sodium iodide (NaI) scintillator (Model 1XMP040B-X)[3]. The NaI scintillator was mounted vertically at 90° to the incident beam direction and connected to a Trump 8K/2K multichannel analyzer (MCA). The scintillator/MCA was calibrated using a ^{241}Am source with various filters (Cu, Mo, Ba, Rb, Tb, Ag, Fe). The monochromator was detuned so that the intensity of the mono beam was reduced by 20%. Spectral measurements confirmed that this amount of detuning was sufficient to eliminate completely the higher harmonics at each energy (the harmonics were most significant at energies between 6 and 8 keV). At each energy the beam was centered on the dosemeter, by adjusting the beamline so that the current in the ionization chamber that monitors the dose to the sample was maximized. The GaFChromic film was used to obtain an image of the beam at each energy.

The dosemeters were mounted in specially designed sample holders. The temperature, pressure and ionization chamber integrated currents were noted at each irradiation. The ionization chamber readings were converted to dose in air and corrected for attenuation in the nitrogen and air paths to obtain the dose in air at the sample position. At each energy, 6 TLDs were exposed to doses varying from 0.01 to 0.03 Gy with the front (numbered) side facing the beam, and 6 with the front side away from the beam. The TLDs were also exposed to integrated doses varying from about 0.01 to 0.4 Gy at 6 keV. Sixteen TLDs were used as controls.

The PTW ionization chamber was exposed at each energy and the dose rate was corrected for temperature and pressure. The ionization chamber was operated at a voltage of 280 V.

V. RESULTS AND DISCUSSION

Figure 2 shows the response (response per unit absorbed dose in air at the dosemeter location) in counts/Gy for the MTS-N and MCP-N TLDs as a function of energy for integrated doses of about 0.01 to 0.03 Gy for a beam size of 1.4 mm x 1.4 mm. The average responses are shown. The error bars are smaller than the size of the symbols. The open and closed symbols represent the data for TLDs with beam incident on the front and back surfaces, respectively. The response of the TLDs to Cs-137 is also shown.

---

[2] Mikrolab, Krakow, Poland

[3] Harshaw/Bicron, 6753-I Cochran Road, Solon, Ohio 44139

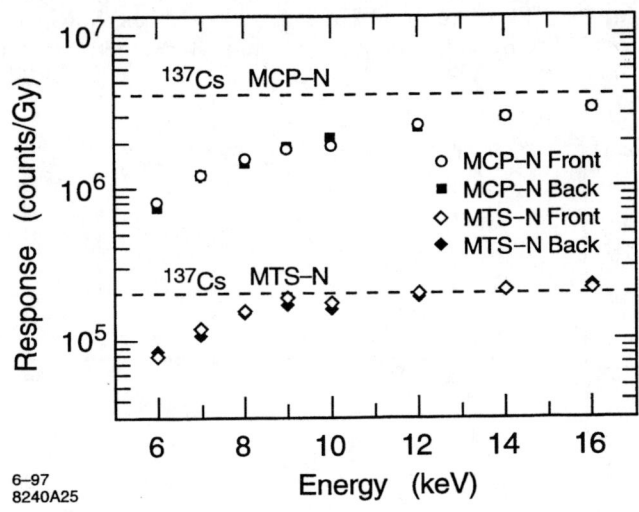

FIGURE 2. Response of Polish TLDs as a Function of Energy (0.01- 0.03 Gy)

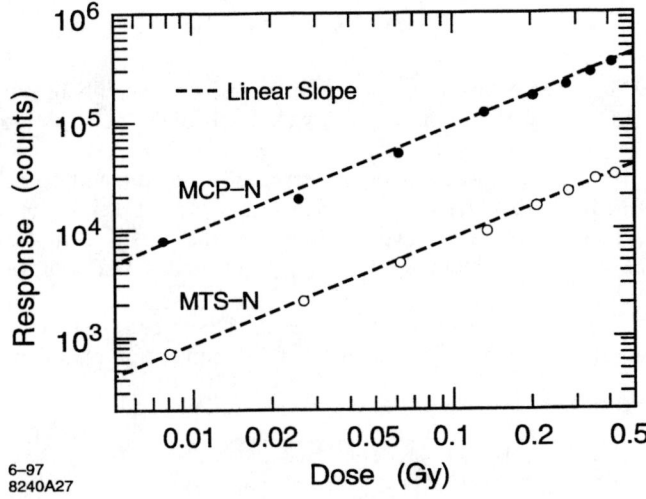

FIGURE 3. Response of Polish TLDs as a Function of Dose at 6 keV

The responses of both MTS-N and MCP-N increase with increasing energy. No differences in the shape of the glow curves were observed for x-ray energies between 6 and 16 keV. The responses of MTS-N and MCP-N at 16 keV are about 3 and 4 times higher than the response at 6 keV, respectively.

The response of MCP-N is about 10 to 15 times higher than the response of MTS-N at the low energies. For both TLD types the difference between the front and back irradiations is within the variability between the dosemeter. Hence, it was not possible to determine the light attenuation coefficient as we had hoped to. The differences in response between front and back irradiations may be more significant in thicker TLDs. This will be the thrust of future studies.

Figure 3 shows the response (counts) as a function of absorbed dose in air for 6 keV x-rays, for both MTS-N and MCP-N. The response is fairly linear over the dose range shown. Future studies will include the dose response over a wider range of doses, and at several energies.

Figure 4 shows the response of the PTW ionization chamber relative to the SSRL ionization chamber as a function of energy. The energy response of the PTW ionization chamber is within ±4.5 % between 6 and 16 keV for dose rates varying between 0.03 and 0.3 Gy/s.

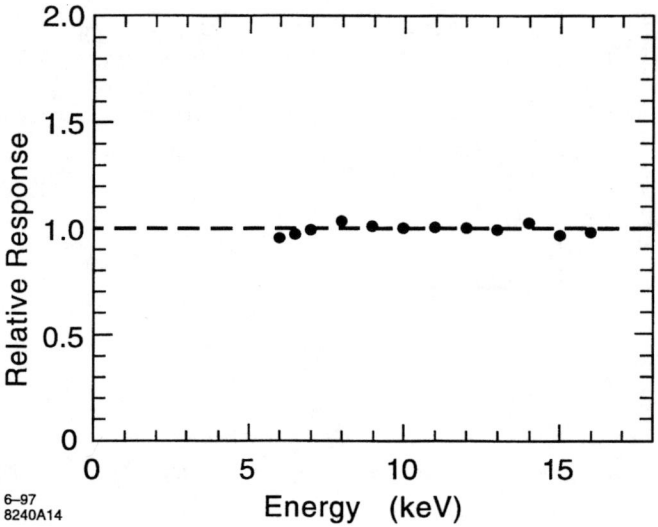

FIGURE 4. Response of PTW Ionization Chamber Relative to SSRL Ionization Chamber vs Energy

VI. CONCLUSIONS

The need for low-energy x-ray dosemeters has increased because of the increasing number of synchrotron radiation facilities being built around the world. The SSRL Beamline 1-5 was used for low-energy x-ray dosimetry studies (6 to 16 keV).

The measured responses of the Polish TLDs were found to increase between 6 and 16 keV, with the response of MCP-N being 10 times higher than that of MTS-N. The response at 16 keV is 3 and 4 times higher than the response at 6 keV for MTS-N, and MCP-N, respectively. The differences between irradiating the TLDs from the front and back is within the variability between the TLDs. Future studies will include the response of thicker TLDs, as well as the dose response over a greater range of doses at each energy.

The energy response of the PTW ionization chamber is within ±4.5 % between 6 and 16 keV. It is therefore suitable for low energy x-ray dose rate measurements. Both the TLDs and the PTW ionization chamber can be also be used for beam dosimetry.

VI. APPENDIX A

Proposed Model for Response of TLD

Under "ideal" narrow beam conditions, exponential attenuation (described by a linear attenuation coefficient μ) will be observed for a monoenergetic beam of photons incident on a material. The photons are "ideal" in the sense that they are absorbed without producing scattered or secondary radiation. Real photon beams interact with matter producing both secondary and scattered radiation. Monte Carlo simulations show that the energy deposited per unit thickness, dE/dx, is exponentially distributed in the TLD along the beam direction x:

$$\frac{dE(k)}{dx} = N\mu(k)ke^{-\mu(k)x} \tag{1}$$

where N is the number of incident photons, μ(k) is an effective energy attenuation coefficient and k is the photon beam energy.

Assuming a simple exponential model also for light absorption (ignoring reflection of light from the planchet) an elemental layer of thickness dx, located at a distance x from the face of the TLD, will contribute an amount dR to the response R. dR is proportional to the energy dE = (dE/dx)dx deposited in that layer, weighted by a light absorption factor e^{-fx} and is given by :

$$dR(k) = C\frac{dE(k)}{dx}e^{-fx}dx \qquad (2)$$

where C=ηε is the overall TLD efficiency, η is the intrinsic luminous efficiency, ε is the reader efficiency and includes geometry dependence, sensitivity of the photomulitplier (PM) tube, etc. and f is the effective light attenuation coefficient of the TLD material.

Combining equations 1) and 2), we obtain:

$$dR(k) = CN\mu(k)ke^{-[\mu(k)+f]x} \qquad (3)$$

if the TLD is read out with the irradiated side facing the PM tube.

Equation 4) is obtained by integrating the response over the whole TLD thickness t:

$$R(k) = \frac{CN\mu(k)k}{\mu(k)+f}\left\{1-e^{-[\mu(k)+f]t}\right\} \qquad (4)$$

In a similar way, we can obtain equation 5) for the case in which the detector is read out with the irradiated side facing away from the PM tube:

$$dR'(k) = C'N\mu(k)ke^{-[\mu(k)x-f(t-x)]}dx \qquad (5)$$

where C' is the overall TLD efficiency when the TLD is read out with the irradiated side facing away from the PM tube. Integration of equation (5) results in equation 6):

$$R'(k) = \frac{C'N\mu(k)ke^{-ft}}{\mu(k)-f}\left\{1-e^{-[\mu(k)-f]t}\right\} \qquad (6)$$

ACKNOWLEDGEMENTS

The authors acknowledge Henry Bellamy of SSRL for his assistance. Gratitude is also expressed to Sherry Oppenheim and the SLAC Technical Publications Department for their help in preparing this manuscript.

REFERENCES

1. Ipe, N. E., Bellamy, H., Flood, J. R., Kase, K. R., Velbeck, K. J., and Tawil, R., "Low-Energy X-ray Dosimetry Studies (7 to 17.5 keV) with Synchrotron Radiation," in *Proceedings of the 11th International Conference on Solid State Dosimetry*, 1995, pp. 69-74.
2. Fassò, A., Ferrari, A., Ranft, J., Sala, P.R., Stevenson, G.R., and Zazula, J.M. "FLUKA92" in *Proceedings of the First Workshop on Simulating Accelerator Radiation Environments (SARE)*, 1994, pp. 134-144.
3. Bilski, P., Olko, P., Burgkhardt, B., Piesch, P., and Waligorski, M. P. R., *Radiat. Prot. Dosim.*, **55**, 31-38 (1994).
4. McLaughlin, W. L., Yun-Dong, C., Soares, C. G., Miller, A., Van Dyk, G., and Lewis, D. F., *Nucl. Instrum. Methods*, **A302**, 165 - 176 (1991).

Ultrafast X-ray Diffraction of Laser-irradiated Crystals

P.A. Heimann,[1] J. Larsson,[2] Z. Chang,[3] A. Lindenberg,[2] P.J. Schuck,[2] E. Judd,[2] H.A. Padmore,[1] P.H. Bucksbaum,[3] R.W. Lee,[4] M. Murnane,[3] H. Kapteyn,[3] J.S. Wark[5] and R.W. Falcone[2]

[1]Advanced Light Source, Accelerator and Fusion Research Division, Lawrence Berkeley National Laboratory, Berkeley, CA 94720, [2]Physics Department, University of California, Berkeley, CA 94720, [3]Center for Ultrafast Optical Science, University of Michigan, Ann Arbor, MI 48109, [4]Lawrence Livermore National Laboratory, Livermore, CA 94551, [5]Department of Physics, Clarendon Laboratory, University of Oxford, Oxford OX1 3PU, U.K.

An apparatus has been developed for measuring time-dependent x-ray diffraction. X-ray pulses from an Advanced Light Source bend magnet are diffracted by a sagittally-focusing Si (111) crystal and then by a sample crystal, presently InSb (111). Laser pulses with 100 fs duration and a repetition rate of 1 KHz irradiate the sample inducing a phase transition. Two types of detectors are being employed: an x-ray streak camera and an avalanche photodiode. The streak camera is driven by a photoconductive switch and has a 2 ps temporal resolution determined by trigger jitter. The avalanche photodiode has high quantum efficiency and sufficient time resolution to detect single x-ray pulses in ALS two bunch or 'camshaft' operation. A beamline is under construction dedicated for time resolved and micro-diffraction experiments. In the new beamline a toroidal mirror collects 3 mrad horizontally and makes a 1:1 image of the bend magnet source in the x-ray hutch. A laser induced phase transition has been observed in InSb occuring within 70 ps.

1. INTRODUCTION

Laser sources currently produce ultrashort (< 1 ps) pulses at infra-red, visible, and ultraviolet wavelengths which can be used to study the dynamics of valence electrons in atoms, molecules, and solids. Studies of laser-induced phase transitions have involved time-resolved reflectivity and second-harmonic generation of optical light pulses. However, to directly probe structural properties of molecules and solids it is advantageous to use x-ray radiation. For example, conformational changes were recently studied in the ns time domain at the European Synchrotron Radiation Facility, using Laue diffraction from a rapidly evolving myoglobin sample. (1) Laue diffraction images were obtained from a single pulse of x-rays from the storage ring. With a resolution in the range of 1 ps, x-ray diffraction techniques could be applied to the study of ultrafast processes such as laser-induced phase transitions and transient excited modes in solids.

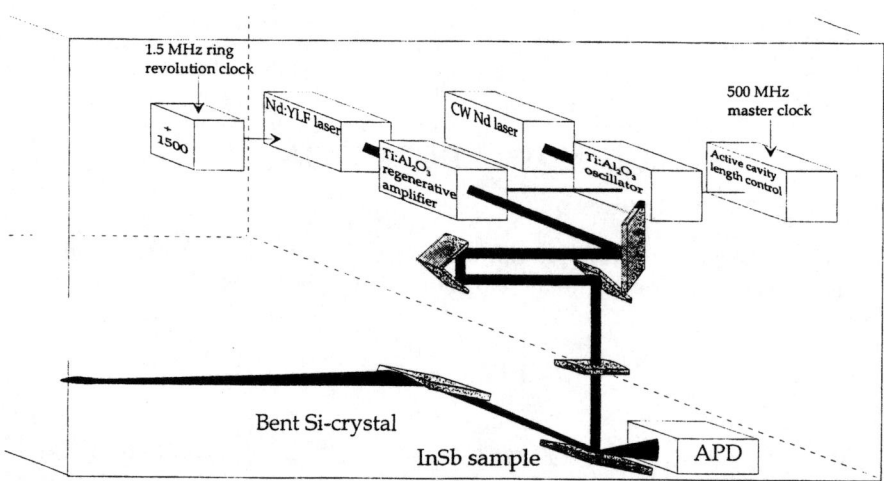

FIGURE 1. Experimental set-up showing the x-ray radiation from the synchrotron, two crystals, laser system, and avalanche photodiode detector.

We have established an x-ray diffraction experiment at the Advanced Light Source (ALS) synchrotron, which will enable the study of ultrafast structural dynamics. These dynamics are initiated by 100 fs laser pulses from a Ti:Al$_2$O$_3$ laser system, which is synchronized to the electron bunches in the storage ring. Radiation from a bend magnet beamline diffracts from two crystals into either an x-ray streak camera or an avalanche photodiode. This apparatus has enabled us to conduct pump-probe x-ray diffraction experiments in which we irradiate an InSb crystal with a 100 fs laser pulse, and probe the interaction with a 70 ps x-ray pulse from the ALS. The different techniques are described in the next section. The results of measurements demonstrating the capability of the instrumentation are presented in section 3.

2. TECHNIQUES

The experimental apparatus, including synchrotron radiation beamline, crystal chamber, detectors and laser system is shown in Fig. 1. The initial form of this apparatus has been described previously in Ref. 2. The ALS storage ring is operated with different timing modes. Some experiments were performed with the ALS in a double-bunch mode, for which the maximum beam current is 40 mA and the pulse repetition rate is 3 MHz. Other measurements were conducted during 'camshaft' operation, in which an isolated bucket, having a 66 ns gap before and a 16 ns gap afterwards, is filled to 20 mA. The lifetime of this isolated electron bunch is however shorter than the lifetime in double-bunch mode. The duration of x-ray pulses from the ring varies with the current in the electron bunches, $\Delta t \sim i^{-1/3}$, but is typically 30 - 80 ps. (3)

To date the beamline 10.3.2 has been used for these experiments. This beamline delivers the unfocused, white radiation through a Be window into a x-ray hutch. The available photon energy range is from about 3 keV where the Be window transmission cuts off to about 12 keV where the bend magnet spectrum decreases at 1.9 GeV electron beam energy. The angular acceptance of the x-ray radiation is determined by apertures in the endstation, typically 0.3 mrad horizontally by 0.03 mrad vertically. These apertures are set to provide a 1 mm^2 footprint on the sample crystal not larger than the footprint of the laser beam.

The ALS beamline 7.3.3 is under construction and is specifically designed for time resolved x-ray diffraction and micro-diffraction experiments. Figure 2 shows a layout of this beamline. A toroidal mirror provides a 1:1 double focus of the bend magnet source in the endstation hutch. The angular acceptance of the mirror, 3 mrad horizontally x 0.2 mrad vertically, results in a 40-fold increase in the useable flux for time-resolved diffraction experiments compared with the other beamline without focusing. The energy range extends from about 2 keV determined by the thickness of the C filter and Be window to about 12 keV resulting from both the bend magnet spectrum and the mirror reflectivity.

The toroidal mirror produces a 1:1 image of the ALS electron beam in a center bend magnet, whose dimensions are 250 μm horizontally x 50 μm vertically (fwhm). The mirror has a grazing incidence angle of 5.4 mrad and equal source and image distances of 16 m. Consequently, the sagittal radius is polished to be 87 mm, while the tangential radius is bent to 2.94 km. The mirror substrate has dimensions of 700 x 75 x 75 mm^3 and is manufactured from single crystal silicon. The mirror is bent with equal couples applied by a leaf spring bender. The bender mechanism is attached to the mirror through metal blocks glued onto the ends of the mirror. The mirror, which absorbs 5 W, is not actively cooled. Analysis of the thermal and gravitational deformations of the mirror shows that these errors can be compensated by adjusting the bending of the mirror to a residual error of ~ 1 μrad rms. (4)

Some additional components of beamline 7.3.3 are shown in figure 2. A 10 μm thick C filter reduces the power absorbed in the Be window. Horizontal and vertical apertures downstream of the mirror chamber adjust the angular acceptance

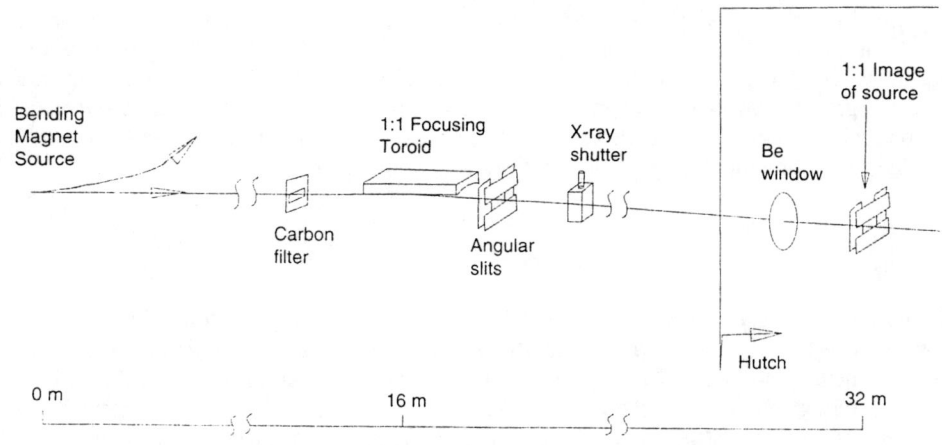

FIGURE 2. Layout of ALS beamline 7.3.3.

for the x-ray micro-diffraction optics, which have a limited collection aperture. A fixed bremsstrahlung backstop blocks the high energy radiation. In addition, there is an x-ray shutter, which allows personnel to enter the hutch. The Be window, 50 µm thick and spherical in shape, separates the UHV of the beamline from the inferior vacuum or He atmosphere of the experiments. A hutch contains the endstations, while an adjoining room encloses the laser system.

In the present sychrotron beamline 10.3.2, the white bend magnet radiation is monochromatized by a bent Si (111) crystal, and a sample crystal is illuminated by the laser. The crystals diffract in the vertical plane, corresponding to s-polarization. The Bragg angle of the Si (111) crystal is fixed at 22.5° which results in the reflection of 5 keV x-rays. In order to increase the x-ray intensity within the laser illuminated area on the sample crystal, the first crystal is sagittally curved. The thin Si crystal is bent into a cylinder with R ~ 0.75 m corresponding to a focal distance of 1 m. Two rectangular springs bend the rectangular Si crystal, which has a length / width ratio ~ 4 in order to reduce anticlastic bending. (5) The crystal bender is actuated by a motor driven slide allowing the line focus to be positioned either at the sample or the detector. The width of the focal line was measured with the streak camera to be about 100 µm.

In the experiment described below, the sample is InSb (111) chosen to have comparable penetration of laser and x-ray radiation into the crystal. At high laser fluence the pump-laser penetration depth is in general limited by free-carrier absorption to about 100 nm. Because InSb has L-shell binding energies between 3.7 and 4.7 keV, the penetration of 5 keV x-rays is reduced by photoabsorption. An alternative approach is to use thin crystals grown on substrates.

The laser system consists of a Ti:Al$_2$O$_3$ laser oscillator pumped by a frequency-doubled Nd-laser (Spectra Physics: Millenia), and a Ti:Al$_2$O$_3$ regenerative amplifier system (Positive Light: Spitfire/Merlin) operating at a repetition rate of 1 kHz. The pulse duration is approximately 100 fs and the pulse energy is about 1 mJ. Synchronization of the laser pulses with the synchrotron pulses is achieved by locking the laser-oscillator repetition frequency (set by the oscillator cavity length) to the master RF clock of the ALS. (6) From the error signal observed in the stabilization electronics, we estimate the jitter between the laser and the synchrotron pulses to be about 5 ps. The delay between the laser and the x-ray pulse can be varied using an optical delay line in the laser beam path. The laser beam is focused on the InSb (111) sample at a fluence of about 30 mJ/cm^2, below the threshold for rapid and visible damage.

Two types of detectors have been employed for time resolved x-ray diffraction: x-ray streak cameras and avalanche photodiodes. Using a Si photodiode 1500 diffracted photons / pulse were collected. Because of the low x-ray flux / shot and the kHz repetition rate of the laser, nearly all measurements have been done by averaging over multiple pulses. Multiple shot data acquisition is possible because at the fluence used in these studies irreversable damage of the InSb is not observed even after exposure times as long as 15 minutes.

The x-ray streak camera has been described in Ref. 7 and 8. Briefly, x-rays strike a CsI photocathode generating electrons, which are accelerated through a slit and deflected by sweep plates. The electrons are then focused with a magnetic lens onto an amplifying pair of microchannel plates and a phosphor screen. This x-ray streak camera has the capability to operate at a high repetition rate (1 kHz), high time resolution (540 fs in single shot mode), and low jitter triggering using a photoconductive switch. On the other hand, streak cameras have a low x-ray quantum efficiency, about 1 %, because of the 100 nm thick CsI photocathode.

Jitter in the streak camera timing with respect to the laser pulse was measured using a test setup. About 20 mJ of 800 nm light from the Ti:Al$_2$O$_3$ laser was used to trigger a GaAs photoconductive switch, (9) which drives the high-voltage pulse to the sweep plates of the streak camera. UV light, with a wavelength of 266 nm, was obtained by frequency mixing the 800 nm light in two KDP crystals. The UV light was split into two pulses separated by 20 ps, using a fused silica window of known thickness inserted halfway into the beam. This UV double-pulse was then incident on the photocathode and used to calibrate the time response and relative jitter of the streak camera. The resulting effective temporal resolution of the streak camera system was 2 ps in this averaging mode (here about 5000 pulses were summed).

Alternatively, the diffracted x-rays are detected by an avalanche photodiode (Advanced Photonix, Inc.). This detector with an active area of 1 cm^2 has near unity quantum efficiency and a gain of 200. (10) The response time is 10 ns, sufficient to resolve individual ALS x-ray pulses in either two bunch or 'camshaft' mode. The signal is recorded using a digital oscilloscope or a gated-integrator connected to a computer. In all the pump probe experiments, the gated-integrator was used, since that allowed us to record data at 1 KHz, as determined by the laser repetition rate. When using the avalanche photodiode, the time resolution is determined by the ALS x-ray pulse duration, about 70 ps.

3. MEASUREMENTS

As a test of our ability to observe time-resolved events, we measured the duration of the diffracted ALS pulses. For this test the sample crystal was Si(111) and no laser irradiation was applied. The x-ray streak camera was used as the detector. The result of the measurement is shown in Fig. 3. Data was obtained by averaging over 2000 pulses. The observed pulse duration of 70 ps is in good agreement with previous measurements of the ALS pulse duration. (3) This measurement confirms that the laser and x-ray streak camera are synchronized to the ALS time structure to within a fraction of 70 ps.

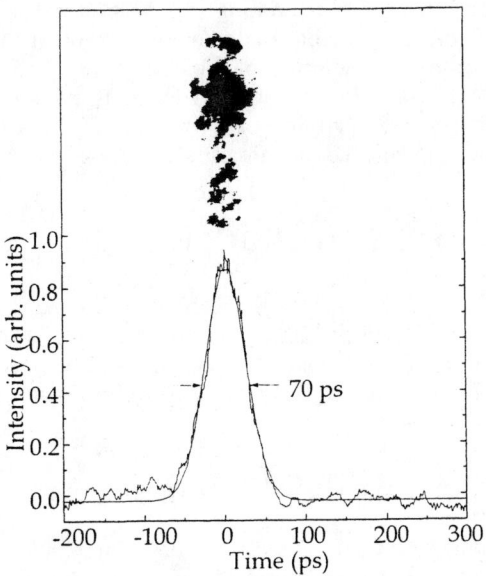

FIGURE 3. The pulse duration of the ALS was measured to be 70 ps with a streak camera operating in an averaging mode. The sample was Si(111). Both the camera output and a lineout of this data are shown.

In the following measurement a pump-probe technique is employed. In Fig. 4 the diffracted x-ray intensity is shown as a function of delay between the heating pulse and the x-ray probe. The length of the laser beam path is varied using the optical delay line. The detector was an avalanche photodiode. The drop in diffraction intensity has a time-resolution limited by the ALS-pulse duration. By differentiating the signal, the ALS pulse shape is retrieved as shown as the dashed trace in Fig. 4. About 10 ps broadening is attributed to timing jitter and the relatively course time steps were used. The data in Fig. 4 comprises about 10^5 laser pulses at the repetition rate of 1 KHz.

The drop in diffracted intensity is interpreted as resulting from the melting of a layer at the surface of the InSb crystal. The melted, disordered layer absorbs a fraction of the x-ray intensity. Before the next laser shot, the crystal lattice has regrown, and the crystal has equilibrated at an elevated temperature, in this case 50 °C above ambient temperature. It is concluded that a phase transition in InSb occurs with a characteristic time less than the ALS pulse duration, i.e. < 70 ps.

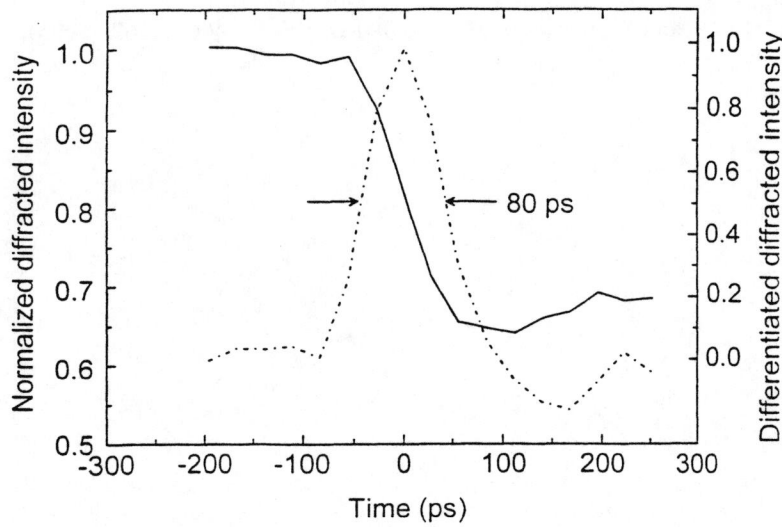

FIGURE 4. X-ray diffraction efficiency as a function of delay between the heating laser pulse and the x-ray pulse (solid line) The sample was InSb(111). The differential of the x-ray diffraction efficiency yields the expected pulse shape for the ALS pulse (dashed line).

In conclusion, we have demonstrated an apparatus combing synchrotron radiation with an ultrafast laser system, a crystal diffractometer and time resolved detectors. A beamline with focusing toroidal mirror is under construction for these experiments. The x-ray streak camera has temporal resolution of 2 ps operating in an averaging mode, for events that are triggered by ultrafast laser pulses. Our set-up will allow us to pursue laser-pump / x-ray-probe experiments to monitor structural changes in materials with ultrafast time resolution. In a first set of experiments, we show evidence of a laser-induced phase transition in InSb, where disorder is induced on a timescale < 70 ps.

ACKNOWLEDGMENTS

The contributions of Alistair MacDowell, Keith Franck, Robert Patton, Neal Hartman, Andrew Grieshop, Robert Duarte and Ted Lauritzen to the design and construction of ALS beamline 7.3.3 are acknowledged. This work was supported by an Academic Research Infrastructure Instrumentation Grant from the National Science Foundation, and the Department of Energy through Lawrence Berkeley National Laboratory and Lawrence Livermore National Laboratory.

REFERENCES

1. Bourgeois, D., Ursby, T., Wulff, M., Pradervand, C., Legrand, A., Schildkamp, W., Laboure, S., Srajer, V., Teng, T. Y., Roth, M., and Moffat, K., *J. Synchrotron Rad.* **3**, 65-74 (1996); Sanjer, V., Teng, T. Y., Ursby, T., Pradervand, C., Ren, Z., Adachi, S. I., Schildkamp, W., Bourgeois, D., Wulff, M., and Moffat, K., *Science* **274**, 1726-1729 (1996).
2. Larsson, J., Chang, Z., Judd, E., Schuck, P. J., Falcone, R. W., Heimann, P. A., Padmore, H. A., Kapteyn, H. C., Bucksbaum, P.H., Murnane, M. M., Lee, R. W., Machacek, A., Wark, J. S., Liu, X., and Shan, B., *Optics Lett.* **22**, 1012-1014 (1997).
3. Keller, R., Renner, T., and Massoletti, D. J., *AIP Conference Proceedings* **390**, 240-247 (1996).
4. Hartman, N., and Grieshop, A., *LSME* **712.**
5. Ferrer, S., Krisch, M., de Bergevin, F. and Zontone, F., *Nucl. Instrum. & Methods A* **311**, 444-447 (1992).
6. Schoenlein, R.W., Leemans, W.P., Chin, A.H., Volfbeyn, P., Glover, T.E., Balling, P., Zolotorev, M., Kim, K.-J., Chattopdhyay, S., and Shank, C.V., *Science* **274**, 236-238 (1996).
7. Chang, Z., Rundquist, A., Zhou, J., Murnane, M. M., Kapteyn, H. C., Liu, X., Shan, B., Liu, J., Niu, L., Gong, M., and Zhang, X., *Appl. Phys. Lett.* **69**, 133-135 (1996).
8. Chang, Z., Rundquist, A., Wang, H., Kapteyn, H. C., Murnane, M. M., Liu, X., Shan, B., Liu, J., Niu, L., Gong, M., Zhang, X., and Lee, R., *SPIE* **2869**, to be published (1996).
9. Knox, W., and Mourou, G., *Appl. Phys. Lett.* **36**, 623-626 (1980); Yen, R., Downey, M. M., Shank, C.V., and Auston, D.H, *Appl. Phys. Lett.* **44**, 718-720 (1984); Maksimchuk, A., Kim, M., Workman, J., Korn, G., Squier, J., Du, D., Umstadter, D., and Bouvier, M., *Rev. Sci. Instrum.* **67**, 697-699 (1996).
10. Gullikson, E.M., Gramsch, E., and Szawlowski, M., *Applied Optics* **34**, 4662-4668 (1995).

Information Stored in High-Q Space: Role of High Energy Scattering

T. Egami[1], S. J. L. Billinge[2], S. Kycia[3], W. Dmowski[1], and A. S. Eberhardt[4]

[1] *Department of Materials Science and Engineering, University of Pennsylvania, Philadelphia, PA 19104-6272,*
[2] *Department of Physics and Astronomy and Center for Fundamental Materials Research,*
Michigan State University, East Lansing, MI 48824-1116
[3] *CHESS, Cornell University, Ithaca, NY 14853*
[4] *Los Alamos National Laboratory, Los Alamos, NM 87545*

Abstract Much of crystallographic diffraction measurements are focused on obtaining information with Q ($=4\pi\sin\theta/\lambda$) below 17 Å^{-1} or d > 0.35 Å, with the implicit assumption that no useful information is stored in the Q space above. However, this assumption is valid only with respect to the periodic lattice structure. Actually, high-Q space is full of information on the local atomic structure that could be of major importance in some cases. We discuss high energy x-ray or neutron scattering as the methods of obtaining the data from the high-Q space, and the atomic pair-distribution function (PDF) analysis as the means of extracting information from such data. Preliminary data of our recent high-energy x-ray scattering measurement on a MX compound are shown for which this type of analysis is likely to play a significant role in understanding the properties.

INTRODUCTION

Powder x-ray diffraction with Cu-K$_\alpha$ radiation is a standard method of materials characterization in laboratories. By using the Cu-K$_\alpha$ radiation with the wavelength λ of 1.54 Å, the range of Q ($=4\pi\sin\theta/\lambda$) space the diffraction measurement covers is limited to about 8 Å^{-1}, which corresponds to the d-spacings greater than 0.8 Å. Most of simple structures are easily identified by such a study. When the structure is more complex and more information is needed the use of Mo-K$_\alpha$ radiation with λ = 0.71 Å extends the Q range up to 17 Å^{-1}, d-spacings down to 0.35 Å. Rarely structural determination requires diffraction data beyond these limits. In fact the Rietveld powder refinement is usually terminated around Q = 12 Å^{-1}, or d = 0.5 Å, since usually the overlap of the Bragg peaks becomes more severe beyond this point. Thus in carrying out these measurements we consciously or unconsciously make an assumption that the Q space beyond these limits carries no relevant information, even though the diffraction intensity continues to vary with Q, up to large values of Q as shown in Fig. 1. Such a luxury of discarding the data is usually permitted, since the crystal structure is specified only by a limited number of parameters, and if the position and the intensity of a certain number of independent Bragg peaks are determined, we have sufficient information to determine the crystal structure.

However, in a number of cases the crystal structure determined from the Bragg peaks does not accurately describe the true structure of the solid, but represents only the periodic average. They are the following cases:

1. When the crystal contains some randomness that affect the properties.

Even when the average structure is crystalline, in the alloys or mixed ion systems in which more than one kind of elements occupy the crystallographically equivalent sites, the local structure around each element is different. Such a local variation in the structure can be important for some of the properties. In this case the crystal structure only describes the average periodic structure. Local variation is reflected in the structure merely as large thermal amplitudes.

2. When the lattice dynamics is so anharmonic that atoms become locally displaced.

Strictly speaking no crystal has a truly periodic structure, because of the lattice vibrations that introduce randomness. Even at T = 0 K, zero-point lattice vibration keeps all the atoms moving around the lattice sites. However, these atomic vibra-

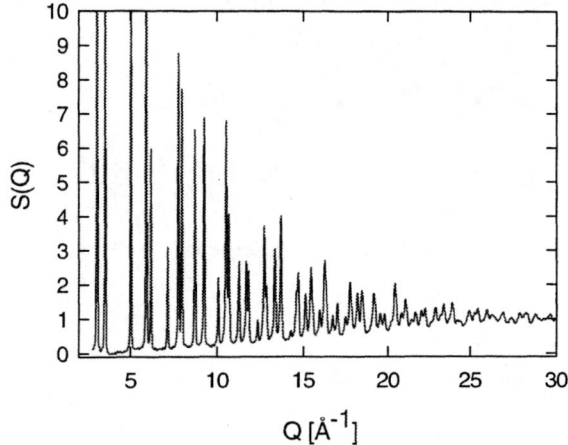

Fig. 1 S(Q) of crystalline Ni powder determined by pulsed neutron scattering.

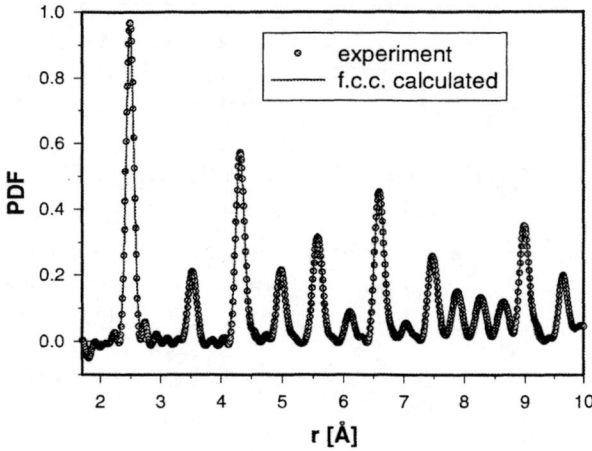

Fig. 2 Pulsed neutron PDF of crystalline Ni obtained by the Fourier-transformation of S(Q) (circles) compared to the PDF calculated for the f.c.c. structure (line).

tions can be described well by the Debye-Waller factor, as long as the vibrations are harmonic. But if the lattice vibration is strongly anharmonic and the local potential for an atom becomes double-well, Debye-Waller factor can no longer describe these local modes of vibrations. An example is the local structural fluctuation associated with the structural transition. Even before the whole structure undergoes a transition the local structure can already show changes in the short range order. For instance local lattice polarization may occur before the ferroelectric transition takes place (1,2). When the coupling between the lattice and charge is very strong, local states such as polarons are formed, breaking the local symmetry before the total symmetry is broken (3,4).

3. When the unit cell is so large and contains a large number of atoms.

Even when the structure is fully periodic, if the unit cell is so large and contains a large number of atoms the reliability of the atomic positions determined by crystallographic analysis becomes questionable. Compared to the high accuracy of *the lattice constants* determined by diffraction the atomic positions are much less accurately known. Diffraction data from high-Q space often provide crucial information on the local structure that complements the crystallographic analysis.

The range of the Q-space in which useful information is stored can be estimated by the Debye-Waller approximation. A rule of thumb is that if Q is beyond 3 σ of the Debye-Waller envelope variations in the diffraction intensity are suppressed by the Debye-Waller factor (5). This means,

$$Q_{max} \approx 3/<2u^2>^{1/2} \qquad (1)$$

where u is the r.m.s. amplitude of lattice vibration. For u = 0.04 Å that corresponds to the zero-point amplitude of a relatively strong covalent bond, Q_{max} is 53 Å^{-1}. Therefore, while Q_{max} of 30 ~ 40 Å^{-1} may be sufficient in describing soft metals or Van der Waals bonds, a larger Q space up to 50 ~ 70 Å^{-1} needs to be explored in order to describe covalently bonded materials with sufficient accuracy. In the discussions to follow we define the Q range of 20 ~ 70 Å^{-1} as high-Q space.

METHODS OF OBTAINING HIGH-Q DATA

Principal methods of obtaining the diffraction data in high-Q space are the following:
1. High energy (pulsed) neutron scattering.
2. XAFS.
3. Electron diffraction.
4. High-energy x-ray scattering.

* *High energy (pulsed) neutron scattering*

Since thermal neutrons of room temperature have the wavelength of about 2 Å, the Q space probed using room temperature neutrons is limited to about 6 Å^{-1}. In order to probe high-Q space neutrons with higher energy are necessary. Such high energy (hot) neutrons can be attained from a high temperature moderator (presently available only at the Institut-Laue-Langevin, Grenoble) or as epithermal neutrons from a radial source or pulsed neutron source. The data shown in Fig. 1 were obtained at the Intense Pulsed Neutron Source (IPNS) of the Argonne National Laboratory. A similar facility is available at the Los Alamos National Laboratory (LANSCE).

Since the neutron scattering length is not proportional to the atomic number Z, neutron scattering is a useful tool to study systems containing light elements such as oxides. In the last decade the Egami group at Penn carried out a large number of diffraction measurements on oxides using pulsed neutron sources as we describe later. For a typical powder diffraction measurement a powder weighing about 15 gram is needed. While a pulsed neutron source provides high energy neutrons sufficient to determine the diffraction intensity up to 40 Å^{-1} or so, the counting statistics becomes worse quickly beyond this with the spectrometers currently available at the IPNS or LANSCE. Since the energy of neutrons is proportional to Q^2 at a constant scattering angle, the spectral intensity goes down quickly if one tries to go beyond 40 Å^{-1}.

* *XAFS*

X-ray absorption fine structure (XAFS) spectroscopy is a very popular method of local structural study. The high-Q diffraction data are obtained as the EXAFS, oscillations in the absorption coefficient with energy above the absorption edge. While this is a relatively easy method of measurement with a synchrotron source, the data processing is complex because of the phase-shifts, multiple scattering and inelastic scattering. The method gives only the nearest neighbor distance with some accuracy, and perhaps the second neighbor distance, but intermediate range order cannot be probed by this method. The coordination number and the amplitude of lattice vibration can also be deduced, but they are strongly model dependent. In addition, the Q range obtainable by XAFS is not sufficiently large. Just as the neutron scattering the energy of the photoelectrons emitted by x-rays is proportional to Q^2, and the period of oscillation in x-ray energy is proportional to Q. Thus beyond k ~ 15 Å^{-1}, or Q ~ 30 Å^{-1} the data become too noisy to analyze.

* *Electron diffraction*

In principle high-Q data can be obtained also by electron diffraction. However, a large aperture needs to be used, and since the Ewald-sphere deviates from the plane at high-Q the analysis is not simple. Furthermore quantitative accuracy of electron diffraction does not compare well against neutron and x-ray scattering measurements.

* *High-energy x-ray scattering*

High energy synchrotron sources such as CHESS and APS provide intense high-energy x-rays (20 ~ 100 keV) suitable for high-Q diffraction measurement. Unlike neutron scattering (< 40 Å^{-1}) or XAFS (< 30 Å^{-1}) x-ray scattering can easily reach high-Q space even beyond 100 Å^{-1} without significant degradation of the data quality. Such a high Q cut-off translates to a very high real space resolution. As we show below spatial resolution of 0.01 Å or better can be achieved by high energy x-ray diffraction. Also, since absorption becomes very small at such high x-ray energies, it is easier to create a special environment, such as high and low temperature and high pressure. While this technique is at its infancy its future is quite bright.

METHOD OF EXTRACTING INFORMATION

The data obtained from such a high-Q space cannot be readily processed by the conventional crystallographic methods of analysis. In powder diffraction the Bragg peaks severely overlap at high-Q values, making indexing impossible. In order to collect data over a wide Q space the Q resolution has to be sacrificed, so that a detailed Q dependence study is difficult. An easiest, while certainly not exclusive, way of extracting information from the vast amount of high-Q data is to apply a Fourier-transformation and obtain the atom-atom correlation function. The atomic pair-density function (PDF) can be obtained by

$$\rho_0 g(r) = \frac{1}{2\pi^2 r} \int_0^\infty [S(Q) - 1] \sin(Qr) Q dQ \qquad (2)$$

where S(Q) is the normalized diffraction intensity (structure function) and ρ_0 is the number density of atoms. Since S(Q) can be determined only up to less than $4\pi/\lambda$ as we mentioned, if a long wave probe is used the integration above has to be terminated prematurely, resulting in spurious oscillations called termination error. The Q range we discussed above corresponds to

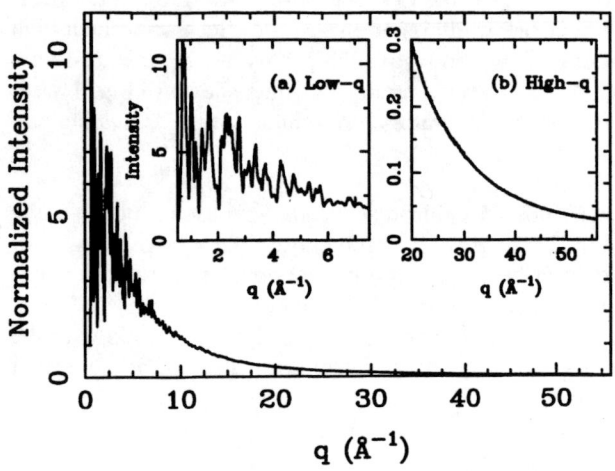

Fig. 3 Intensity of x-ray elastic scattering from PtI MX chain measured with 60 keV incident x-rays.

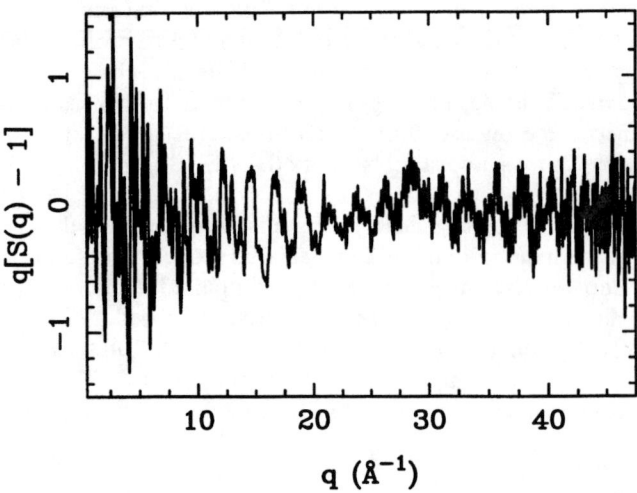

Fig. 4 Approximate interference function for the PtI chain compound, $i(Q) = Q[S(Q) - 1]$, obtained by a simple method of background smoothing.

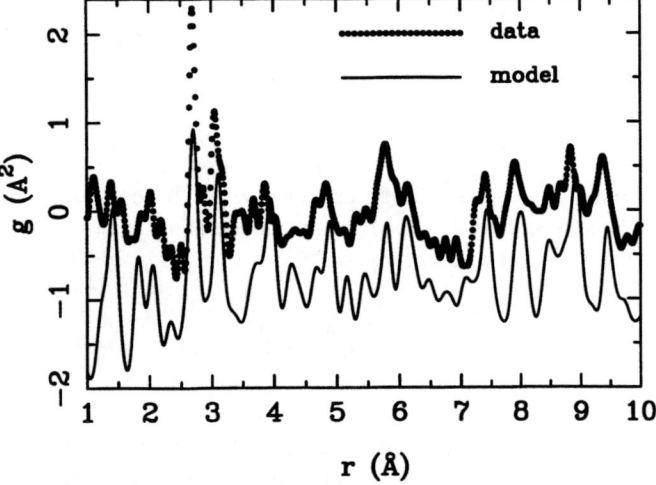

Fig. 5 Preliminary PDF of PtI MX chain compound obtained by a simplified data treatment (see text) compared to the crystallographic model. The split of the Pt-I distances at 2.7 Å and 3.1 Å due to Peierls distortion is clearly seen, in spite of the approximate nature of the data treatment.

the range of integration in (2). The criterion is that at Q_{max} the integrand of (2) is very small compared to unity so that the termination there does not introduce errors in integration.

Experimentally S(Q) is determined by a powder diffraction measurement, subtracting the measured background rather than the fitted background as in many crystallographic analysis, and making necessary corrections for absorption and multiple-scattering. In the case of x-ray scattering Compton scattering has to be separated or subtracted, and the coherent scattering intensity has to be divided by the square of the atomic scattering factor, $f(Q)^2$. Thus S(Q) includes both Bragg peaks and diffuse scattering, and therefore the PDF can describe periodic as well as aperiodic structures. For this reason the PDF analysis has widely been used in the structural studies of liquids and glasses. It is also quite useful in describing the atomic correlations in crystals that include aperiodic atomic displacements. Pulsed neutron PDF analysis has been very effective in observing local oxygen displacements in high-T_C cuprates (6,7), lattice polaron formation in manganites that show colossal magnetoresistance (CMR) (3,4), local lattice distortion in relaxor ferroelectrics (8,9), and nano-scale structure in catalytic support oxides (10).

HIGH ENERGY X-RAY SCATTERING

High intensity high-energy x-rays can be obtained from a high-energy synchrotron or a wiggler insertion device. The setup is the same as the regular powder diffraction experiment, but because of low absorption very heavy shielding is necessary everywhere to eliminate stray radiation, making the alignment procedure non-trivial. An intrinsic Ge solid state detector can be used in separating the Compton scattering intensity. Fig. 3 shows an example of the elastic scattering intensity I(Q) determined over a wide range of Q. The measurement was recently carried out at the A-2 wiggler beamline of CHESS of Cornell University. The sample studied was a one-dimensional organic compound $[Pt(en)_2I_2][Pt(en)_2](ClO_4)_4$, where en is 1,2-diaminoethane. This compound is one of the so-called MX solids that shows a strong charge density wave (CDW) due to Peierls distortion (11). Since Pt and I are the only heavy elements in the system x-ray scattering is presumed to be a powerful tool to describe the CDW set up by Pt and I. At high-Q the structure in I(Q) dies down quickly (inset) and it appears that no information is stored in the data.

However, from the point of view of the PDF analysis the relevant quantity that carry information is the integrand of eq. (2), $i(Q) = Q[S(Q) - 1]$. In order to obtain $i(Q)$, I(Q) has to be corrected for absorption and multiple-scattering, divided through $f(Q)^2$ and multiplied by Q. Since the intensity of multiple-scattering is high and the calculated values of f(Q) may not be so reliable at high Q, this is a non-trivial process. Instead, as a preliminary process we just smoothed the data over a wide range, and normalized the data by the smoothed curve. The result, shown in Fig. 4, clearly demonstrates that the oscillations persist up to large values of Q. This approximate $i(Q)$ was used in eq. (2) to obtain the preliminary PDF shown in Fig. 5. The double peaks at 2.7 and 3.1 Å are due to Pt-I distances, split into two due to the CDW. High resolution of the PDF, of the order of 0.01 Å or better, is obvious. Thus even at this preliminary stage the PDF gives a clear picture of the local structure. The study of temperature dependence is expected to provide insights regarding the nature of phase transitions in this complex system. This example demonstrates that high-energy x-ray scattering is a powerful and unique tool to probe a portion of the Q-space with high Q values.

CONCLUSION

Most of the diffraction measurements are carried out in the Q-space below 12 $Å^{-1}$, with an implicit assumption that the region of the Q-space beyond carries no relevant information. This assumption is valid for perfect crystals with a relatively simple structure, but it breaks down when the crystal contains some structural disorder. In such a case information concerning the short range atomic structure is contained in the high-Q space. High-energy x-ray or neutron scattering and XAFS measurements are capable of extracting information stored in high-Q space. Among them high-energy x-ray scattering is most promising in reaching very high Q values. Neutron scattering and XAFS are in practice limited to Q limits of 40 and 30 $Å^{-1}$, respectively, while x-ray scattering can probe the Q-space even beyond 100 $Å^{-1}$. In spite of some difficulties associated with working with high energy x-rays, high-energy x-ray scattering is likely to become a powerful and unique tool in the study of local structure.

ACKNOWLEDGMENTS

This research at the University of Pennsylvania was supported by the National Science Foundation, DMR96-28134. SJLB was supported in part by the Alfred P. Sloan Foundation. The high energy x-ray measurements were made at the Cornell High Energy Synchrotron Source at Cornell University which is supported by the National Science Foundation through grant DMR-9311772.

REFERENCES

1. Billinge, S. J. L. and Kwei, G. H., *J. Phys. Chem. Solids* **57**, 1457 (1996).
2. Kwei, G. H., Billinge, S. J. L., Cheong, S.-W. and Saxton, J. G., *Ferroelectrics*, **164**, 57 (1995).
3. Billinge, S. J. L. *et al.*, *Phys. Rev. Lett.* **77**, 715 (1996).
4. Louca, D. and Egami, T., *J. Appl. Phys.* **81**, 5484 (1997).
5. Toby, B. H. and Egami, T., *Acta Cryst. A* **48**, 336 (1992).
6. Toby, B. H., *et al.*, *Phys. Rev. Lett.* **64**, 2414 (1990).
7. Egami, T. and Billinge, S. J. L., in *Physical Properties of High Temperature Superconductors V*, ed. Ginsberg, D. M. (World Scientific, 1996) p. 265.
8. Rosenfeld, H. D. and Egami, T., *Ferroelectrics*, **150**, 183 (1993).
9. Teslic, S. and Egami, T., *Acta Cryst B*, submitted.
10. Egami, T., Dmowski, W. and Brezny, R., *SAE Publication* **970461** (1997).
11. Scott, B. *et al.*, *J. Molecular Structure*, **356**, 207 (1995).

Measurement of EUV optical constants of *in-situ* deposited films

C. Tarrio, R. N. Watts[†], T. B. Lucatorto

Electron and Optical Physics Division
National Institute of Standards and Technology
Gaithersburg MD 20899

J. M. Slaughter[*] and Charles M. Falco

Optical Sciences Center and Department of Physics
The University of Arizona
Tucson, AZ 85721

Abstract We describe a DC sputtering thin-film deposition chamber and load-lock system recently installed on an extreme ultraviolet beamline at the NIST Synchrotron Ultraviolet Radiation Facility (SURF II). The system allows for the measurement of reflectance vs. angle, and, from those measurements enables the determination of the optical constants of pristine films. We present preliminary results.

Until recently, extreme ultraviolet (EUV) studies focused primarily on elucidating the energetics of atoms, molecules, and solids. However, since the advent of EUV multilayer mirrors capable of providing relatively high normal incidence reflectivities in the wavelength region between 3 nm and 40 nm, a growing interest has developed in "optical" applications such as high resolution EUV astronomy, lithography and microscopy. Optimization of the design of the EUV optics for such applications has increased the demand for more accurate optical constant data, particularly for materials used in multilayer reflective coatings. A lack of reliable optical constant measurements in this region has meant that the most commonly used constants are derived from calculated atomic scattering factors.[1]

Traditionally EUV optical constants have been obtained from Kramers-Kronig analyses of either reflectivity or absorptance data.[2] These data must of course be extrapolated to infinite photon energy, and the particular form assumed for the extrapolated region often affects the data over a more significant energy range than one might assume.[3] Thus to obtain accurate optical constants by this method, one needs to make measurements over a rather extensive energy range that is often hard to produce in a given experimental arrangement. (Inelastic electron scattering [4] can cover from a few eV to a few hundred eV on a single instrument; however this technique requires very thin, self-supporting films.)

The recent interest in more accurate EUV optical constants has led to the increased use of angle-dependent reflectance measurements[5] that eliminate the need for Kramers-Kronig transforms. In this technique the reflectance is measured at several angles of incidence including the region around the critical angle. The data are then fit to the Fresnel equations:

$$r_s(\phi) = \frac{\sin\phi - \sqrt{N^2 - \cos^2\phi}}{\sin\phi + \sqrt{N^2 - \cos^2\phi}} \qquad (1a)$$

$$r_p(\phi) = \frac{N^2 \sin\phi - \sqrt{N^2 - \cos^2\phi}}{N^2 \sin\phi + \sqrt{N^2 - \cos^2\phi}} \tag{1b}$$

where r_s is the reflected amplitude with polarization perpendicular to the plane of reflection, r_p is for parallel polarization, f is the grazing angle of incidence and N is the complex index of refraction, $N=1-\delta+ik$., where δ represents the scattering contribution to the index and k, the extinction coefficient, represents absorption.

Because EUV radiation penetrates a very small distance into a sample, the typical surface layer (oxide, water, and/or hydrocarbon) that a sample develops in air can have a significant impact on measurements, especially at grazing angles where the penetration may be of the same order as the surface layer thickness. Since in a multilayer mirror most of the reflected amplitude is generated by reflections from interfaces that have never been exposed to air, accurate modeling of the reflectivity requires accurate data on the optical constants of pristine films. With the system described here, we are able to acquire the most reliable data possible for the determination of EUV optical constants.

Several years ago we constructed a new monochromator [6] on the SURF II storage ring. This beamline is dedicated primarily to making reflectance measurements in the 3 nm to 40 nm region. The measurement chamber is 25 cm in diameter and contains a goniometer arm with 50×25×25 mm³ x-y-z motion and a full 360° of motion in the reflection plane. A photodiode detector also has 360° of motion in the plane of reflection and rotates about a disc opposite the feedthrough of the goniometer. We have recently added a thin-film deposition chamber onto the measurement chamber as shown in Fig. 1.

We have chosen DC magnetron sputtering as our method of sample preparation to closely match the materials currently used in multilayer mirror fabrication by the major developers. The substrate to be coated is placed into a holder which attaches to a transfer rod that allows both rotation and translation. The rod is used to hold the sample 11.4 cm above the sputtering target during deposition and then transfer it, immediately afterward, into a sample dock on the reflectometer's goniometer arm. The 3.8 cm diameter sputtering source is typically operated in a 250 Pa Ar environment at bias voltages of <500V and deposition rates of 0.1-0.3 nm/s. The deposition rate is measured with a quartz crystal thickness monitor

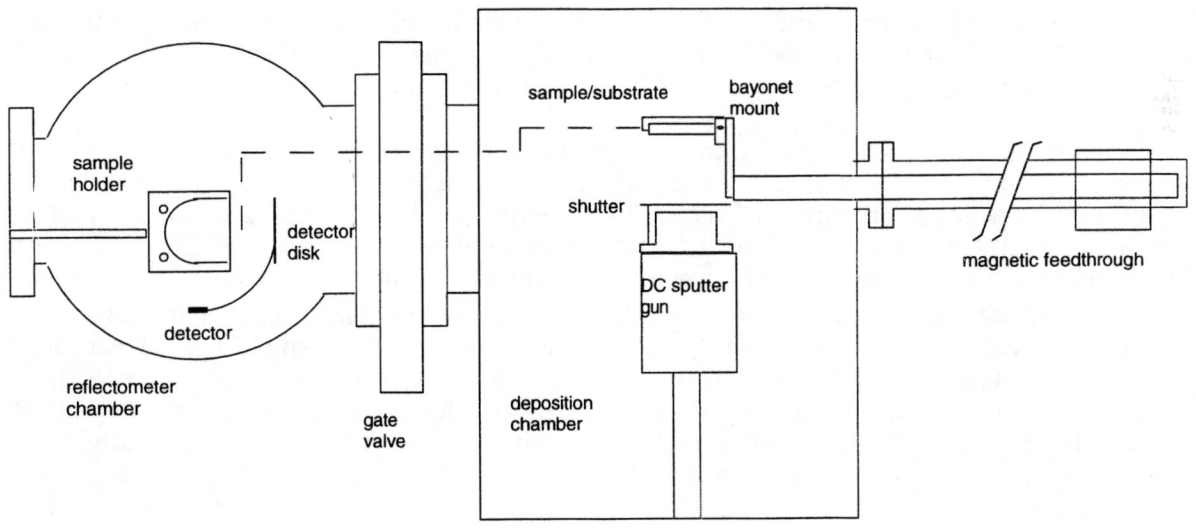

Figure 1. Schematic drawing of the in-situ deposition system.

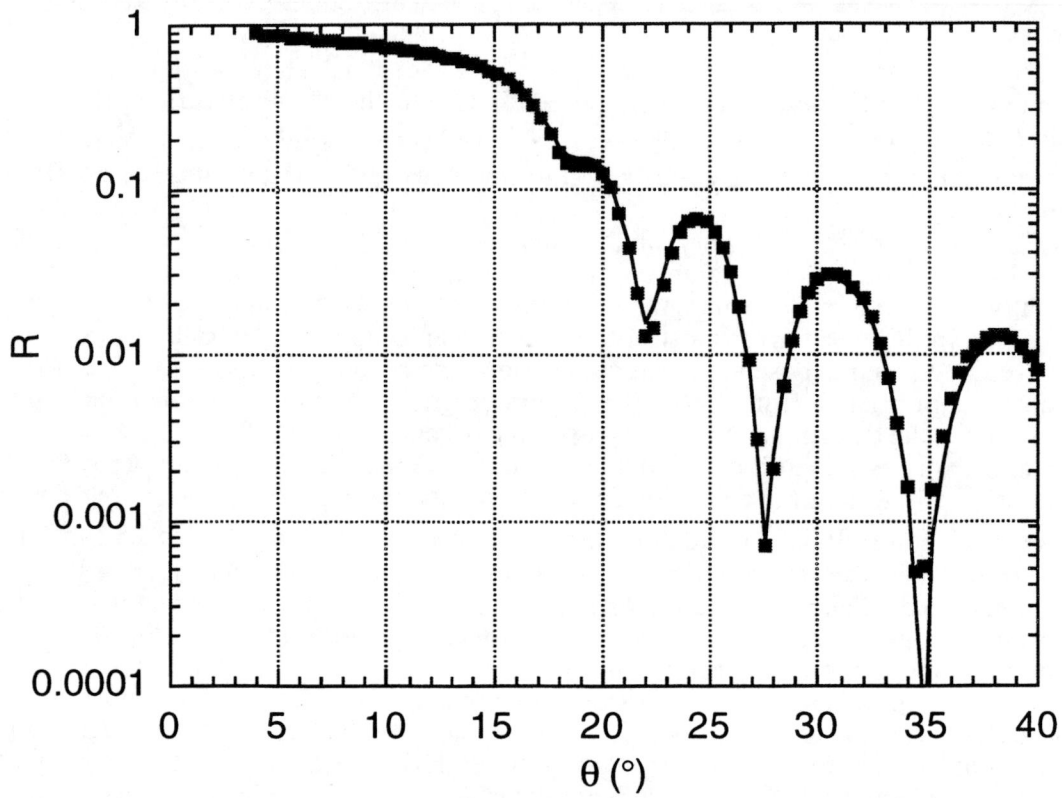

Figure 2: Reflectance of a 92 nm thick Si film on float glass as measured at 50 eV (squares) and fit using the parameters in the text (solid line).

and controlled by providing constant power to the sputtering source while controlling the Ar gas flow with a mass flow controller to keep the chamber pressure constant. The chamber is 30 cm in diameter by 46 cm tall and is pumped by an eight-inch cryopump which is throttled during deposition to set the desired pressure.

Samples are introduced and removed from the system through a quick entry port on the sputtering chamber while an isolation gate valve allows the ion-pumped measurement chamber to remain under UHV. Preserving the vacuum integrety of the measurement chamber is critical; it allows us the time necessary to make the measurements without significant accumulation of impurities. The high pumping speed of the cryopump allows quick pumpdown of the relatively small deposition chamber to high vacuum. After deposition, the chamber recovers to the low 10^{-8} torr range in approximately one minute.

For EUV measurements, substrates are required that are both flat and smooth. We have found that float glass is very well suited for such use. Off-the-shelf float glass is very smooth (<1 nm RMS), and the 1 cm thick samples we use are flat enough to allow for grazing-incidence x-ray reflectivity measurements, a desirable possibility that allows us to fully characterize the film morphology subsequent to the EUV reflectometry. The size of the sample we can introduce to our system is limited to a 5 cm diameter disk. This allows for measurements down to 5° grazing at short wavelengths and 2.5° at longer wavelengths where the beam divergence is smaller.

Some early results are presented in Fig. 2 which shows a measurement of a Si film on float glass at 50 eV energy (24.8 nm wavelength). The solid curve is a least-squares fit to the reflectance from a film of thickness t[7]:

$$R = \left| \frac{r_t + r_b \exp(2i\varphi)}{1 + r_t r_b \exp(2i\varphi)} \right|^2 \tag{2}$$

where r_t and r_b are the reflected amplitudes from the top and bottom surfaces of the film, and $\varphi = 2\pi Nt\sin\phi$. The optical constants determined are $\delta=0.0430$ and $k=0.00763$. The 1σ root-sum-square (rss) uncertainty in the determination of δ is ± 0.0012 (2.7%) with a 2.5% component arising from angle metrology and 1% from fitting uncertainty. The rss uncertainty in k is ± 0.0004 (5%) with components of angle metrology (1%), fitting errors (2.7%), and measurements of roughness (1%), incident beam polarization (1%), and substrate index (4%) For comparison other measurements have found values of $\delta=0.0437$ and $k=0.00627$ for a HF-dipped silicon wafer[8], $\delta=0.0433$ and $k=0.0123$ for an evaporated film[9], and $\delta=0.046$ and $k=0.0124$ for a sputtered film[10]. Small differences in the optical properties between prepared films and single crystals have been observed[9], however, the large differences in k between the current work and Refs. 9 and 10 is most likely due to oxide layers since these samples were handled in air. This illustrates the importance of *in-situ* sample preparation. We plan to make measurements on all the major materials used in multilayer EUV optical coatings.

†deceased

* Current address: Motorola, Phoenix Corporate Research Laboratories, MD EL308, 2100 East Elliot Road, Tempe, AZ 85284

References

[1] B.L. Henke, E.M. Gullikson, and J.C. Davis, Atomic Data and Nuclear Data Tables, **54**, 181 (1993).
[2] D. Y. Smith, "Dispersion Theory, Sum Rules, and Their Application to the Analysis of Optical Data," in *The Handbook of Optical Constants*, edited by E. D. Palik, (Academic Press, Orlando, 1985).
[3] M. L. Bortz and R. H. French, Appl. Spect. **43**, 1498 (1989).
[4] S. E. Schnatterly, "Inelastic Electron Scattering," in Solid State Physics, edited by F. Seitz and D. Turnbull, vol. 34, p 275.
[5] M. L. Scott, "Measurement of n and k in the XUV by the angle-of-incidence total-external-reflectance method," in *The Handbook of Optical Constants, II*, edited by E. D. Palik, (Academic Press, Orlando, 1991).
[6] C. Tarrio *et al*, J. X-ray Sci. Tech. **4**, 96 (1994).
[7] E. Spiller, *Soft X-ray Optics* (SPIE Press, Bellingham, WA, 1994) p. 104.
[8] R. Soufli and E. M. Gullikson, to appear in Appl. Optics.
[9] C. Tarrio and S. E. Schnatterly, J. Opt. Sci. Am. B**10**, 952 (1993).
[10] D. L. Windt, Appl. Optics **30**, 15 (1991).

Real-Time X-ray Diffraction Measurements of GaN Growth on Sapphire(0001)

A. R. Woll, R. L. Headrick*, S. Kycia*, J. D. Brock

*Department of Applied Physics
and*

Cornell High Energy Synchrotron Source

Cornell University
Ithaca NY 14853

Real-Time X-ray diffraction techniques have been used to examine the growth of GaN on Sapphire (0001) using Metal-Organic Molecular Beam Epitaxy (MOMBE). The formation of an AlN buffer layer upon high temperature exposure of sapphire substrates to NH_3 is indicated, at 900C, by a sudden change in specular reflectivity as well as the appearance of the AlN (1,0,-1,0) bragg peak. Subsequent growth of GaN shows a gradual shift of this bragg peak from the AlN position towards that of GaN. The time-dependence of this shift is inconsistent with a previous measurement (Kim *et al*, Appl. Phys. Lett, **69** (16), 14 October 1996) predicting a critical thickness of 29 Å for a GaN grown on a 32 Å AlN buffer layer on Sapphire (0001). Possible explanations for this discrepancy, including the role of buffer layer thickness, will be discussed.

RADIATION DOSES TO INSERTION DEVICES AT THE ADVANCED PHOTON SOURCE

E.R. Moog, P.K. Den Hartog, E.J. Semones, and P.K. Job

Advanced Photon Source, Argonne National Lab, Argonne, IL 60439

Abstract

Dose measurements made on and around the insertion devices (IDs) at the Advanced Photon Source are reported. Attempts are made to compare these dose rates to dose rates that have been reported to cause radiation-induced demagnetization, but comparisons are complicated by such factors as the particular magnet material and the techniques used in its manufacture, the spectrum and type of radiation, and the demagnetizing field seen by the magnet. The spectrum of radiation at the IDs has been measured and found to include a large high-energy (7 GeV) component, at least during some runs. Lead shielding installed immediately upstream of the IDs has been found to decrease the dose to the upstream ends of the IDs. It has almost no effect on the dose to the downstream ends of the IDs, however, since much of the radiation travels through the ID vacuum chamber and cannot be readily shielded. Opening the gaps of the IDs during injection and at other times also helps decrease the radiation exposure.

INTRODUCTION

The insertion devices (IDs) at the Advanced Photon Source (APS) use Nd-Fe-B permanent magnets to produce their magnetic field. Although NdFeB magnets are known to be sensitive to radiation damage (1-10), no radiation damage has yet been observed in ID magnets at the APS. We seek to anticipate whether the dose levels presently being observed at the IDs are high enough to cause demagnetization within the desired 20-year lifespan of the IDs. The dose received by an ID at the European Synchrotron Radiation Facility (ESRF) after being installed for only one year was high enough to cause partial demagnetization of the ID magnets (1), so there is cause for concern at the APS as well.

These questions give rise to other questions: What is the dose required to damage the magnets and how does that depend on the spectrum of the radiation? What is the dose actually received by the ID magnets and what is its spectrum? How effective are the measures that have been and are being taken to reduce the dose to the ID magnets? These different questions will be examined below.

WHAT DOSE IS REQUIRED TO DAMAGE THE ND-FE-B MAGNETS?

Others have exposed magnets to various types of radiation and determined the damage threshold. Some of this work is not directly applicable to insertion device magnets, however, because the type of radiation used in the study (e.g., neutrons rather than electrons or photons) is not what is expected to cause radiation damage in a storage ring (11).

Some flux loss vs. dose results relevant to storage rings include work by Luna et al. (2) who exposed magnets to an 82 MeV direct electron beam. A 1.5% remanence loss was measured after only a 36 krad exposure. When the radiation exposure was to bremsstrahlung from an 85-MeV electron beam, a 14% remanence loss was seen after exposure of one sample to 450 Mrad, whereas another magnet from a different manufacturer showed only 2% remanence loss after 1370 Mrad. At ESRF, the dose received by the ID whose magnets were partially demagnetized (1) was estimated to be 6.7 Mrad for the first upper magnet and 5.1 Mrad for the first lower magnet (12). The peak field loss at the upstream end of the ID was nearly 8%, but magnet regions immediately above or below the particle beam showed greater demagnetization than the rest of the magnet (13). Colomp and Bräuer (3) exposed some magnets to the direct 200 MeV electron beam from the ESRF linac. They observed demagnetization of from 1.9% to 2.7% after an exposure of 300 krad. These experiments were troubled, however, by a spatial variation of the dose by a factor of at least 200. The observed demagnetization could have actually consisted of higher levels of demagnetization that were localized to small regions of the magnet block where the actual dose was much higher. Measurements made of an ID at HASYLAB showed no radiation-induced effects despite an estimated exposure of 7.2 Mrad directly above the beam and 3.3 Mrad and 12 Mrad at positions horizontally displaced by 4 cm (14). Another study irradiated magnet blocks with 17 MeV electron beams and found a 9% flux loss after an exposure of 260 Mrad (4).

A consideration in attempting to use published results to determine lifetimes of APS ID magnets is that wide variation in radiation sensitivity between magnets from different vendors has been reported in a number of studies (2, 3, 5 - 8). Some differences can be attributed to the manufacturing process of the magnets (5), and some differences can be attributed to the presence of small amounts of other materials in the magnet mix (7). Nd magnet technology has developed rapidly in recent years so that the magnet material used in the APS insertion devices was not available eight years ago; it would be expected that these advances in magnet technology might make a significant difference in the magnets' radiation sensitivity.

In addition to effects based on the magnet material itself, there has been found to be an effect due to the strength of the demagnetizing field in which the magnet is placed while it is irradiated (6,7). If the demagnetizing field is stronger, the magnet will more readily demagnetize. This would mean that the probability of radiation-induced demagnetization in a particular ID would be a function of the magnetic design of that ID.

Any stabilization that the magnets may have undergone, either by exposure to a reverse field or to elevated temperature, may also influence the likelihood of radiation-induced demagnetization. If the small regions of a magnet block that will change their magnetization easily have already had their magnetization changed by the stabilization procedure, the block should be more resistant to further flux loss. The APS magnets were all stabilized before the IDs were assembled, so that temperatures up to 60°C would not cause any demagnetization, nor would a demagnetizing field up to 1.2 H_C.

It is interesting that magnet blocks demagnetized due to radiation damage can be remagnetized to full strength (4, 8-10). While this would mean that ID magnets that have been partially demagnetized by radiation do not need to be replaced, it would not eliminate the need for a complete disassembly and reassembly of the ID magnetic structure. Also, the remagnetized and restabilized magnets might need to be sorted differently for the best overall magnetic results, and the ID would need a full magnetic tuning procedure. Therefore, although it would be possible to recover from radiation demagnetization of the ID magnets without purchasing new magnets, the recovery would not be painless.

The published results that are probably most applicable to the APS IDs are those from the study of Okuda et al. (4). The magnet blocks that were used in that study were manufactured by the Shin-Etsu Chemical Corp. (2-1-5 Kitago, Takefu, Fukui 915, Japan), as were many of the magnet blocks used in the APS IDs. The study was published in 1994; if the magnets were not manufactured much before that time then they would probably have used similar technology to that of the APS magnets from Shin-Etsu, which were purchased in 1995. The 9% flux loss observed after a 260 Mrad exposure is probably overly optimistic for APS ID magnets, however, since the study was performed with single magnet blocks rather than with an assembled ID magnetic structure in which a demagnetizing field is imposed on the blocks. A different demagnetizing field can change the dose required for a particular flux loss by well over one order of magnitude (7).

WHAT DOSE LEVELS ARE OBSERVED AT APS?

The first running period with IDs installed and with the 8-mm aperture vacuum chamber was in the late fall of 1995 through January 1996. Dose measurements made then alerted us to the need for some radiation shielding for the insertion devices. Injection efficiency was quite low during much of the run, and for long periods of time beam was being injected although it was not being successfully stored. When the dosimeters were removed and read out, two of the three installed IDs were found to have been exposed to extremely high radiation doses. TLD response saturates at exposures over about 300 krad, so it is difficult to know the actual dose to within a factor of 2, but estimates place it as high as 5.4 Mrad on one ID and 1.1 Mrad on the other. The gap on the third ID had been kept open during the entire run, so its dose was only 52 krad on the first pole. However, a second dosimeter that was placed to measure what the dose would have been for that ID if it had been at minimum gap for the entire run gave an estimated dose of 3 Mrad. The ID with the highest dose has since accumulated another 120 krad at the upstream end and 260 krad at the downstream end. The magnetic fields of the undulators that have received the highest doses were rechecked most recently in March 1997, and no demagnetization was found.

After these surprisingly high dose rates and given the experience at ESRF, it was decided that measures to reduce the exposures encountered by the installed insertion devices were warranted. As a result, insertion devices are now installed in the downstream part of the ID straight section whenever possible, and Pb shielding is installed in the open space upstream of the ID. This has been done since Feb. 1996. The effect of the Pb shielding will be discussed below.

The doses to the upstream and downstream ends of the ID magnetic structures have been monitored during storage ring operation. Doses have also been monitored on the ID vacuum chambers and in the vicinity of the IDs. The dosimeters used have been TLDs and radiochromic (15) and GafChromic (16) films (17). Doses are not uniform from sector to sector,

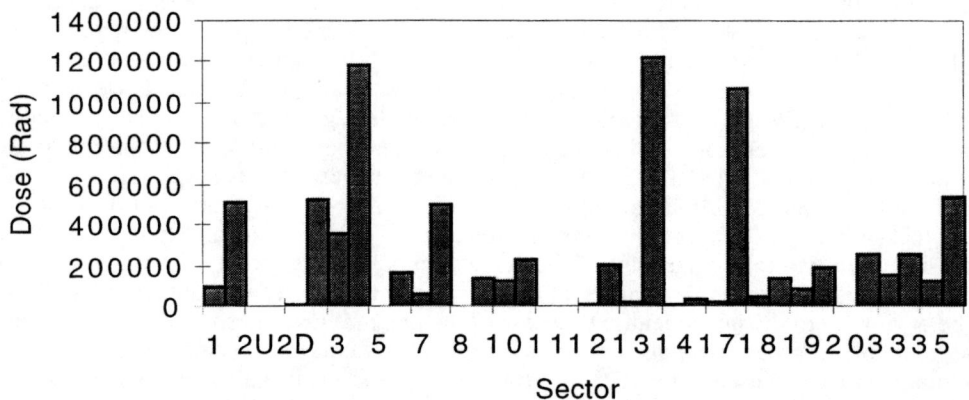

Figure 1. Total accumulated dose recorded by dosimeters on the upstream (first bar for each sector) and downstream (second bar) ends of the ID magnetic structure near the gap. These dosimeters open and close with the gap. This dose was accumulated during the approximately 21 weeks of total running time between when a run began on 10 Sept. 96 and when a run ended on 4 May 97. Note that there are 2 IDs installed in sector 2, 2U is upstream.

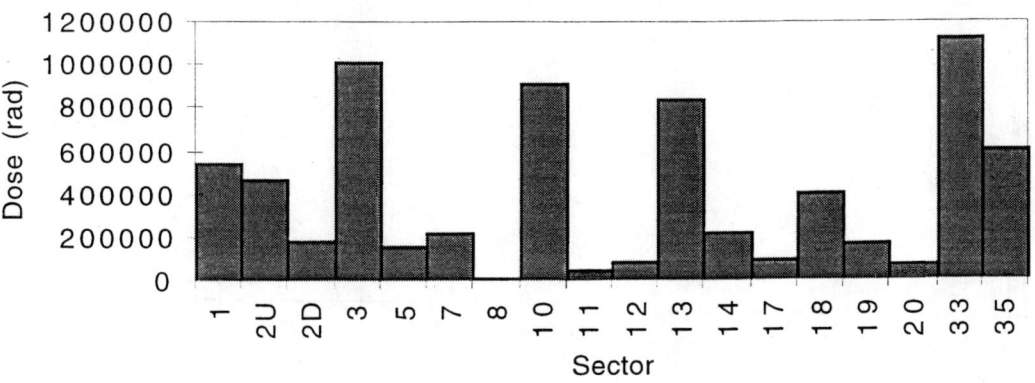

Figure 2. Total accumulated dose recorded by dosimeters on the ID vacuum chamber immediately upstream of the ID, at the same distance from the vacuum chamber as the dosimeters from Fig. 1 would be at minimum gap. These dosimeters do not move with the gap. The time span is the same as for Fig. 1.

even for those sectors that are far from the injection point. In the early runs, there would usually be a sector where the dose was much higher than other sectors, but this high-dose sector would vary from run to run. Now that more operational experience has been gained so that the operation of the storage ring is more routine, some systematics in where the dose levels are higher are beginning to be seen. Unexplained incidences of a high dose somewhere that only occurs during one run are still found, however.

It is of interest to look at the doses that have been accumulated during the approximately 21 weeks of total running time that occurred from when a run began on 10 Sept. 1996 until a run ended on 4 May 1997. Dosimeters are mounted on the upstream and downstream ends of the ID magnetic structures, fastened to the outside of the last pole, near the ID gap. The doses recorded here will be a measure of the exposure to the magnetic structure because they open and close with the gap. The accumulated dose by sector from these dosimeters is shown in Fig. 1. (This accumulated dose may not all have been to one particular ID, however, because some IDs were exchanged between runs for mechanical upgrades.). Pb shielding was present in all sectors but 14 and 35 and, for only the last 4.5 weeks of running time, sector 2. The lower dose at the upstream ends of the IDs is due to the Pb, as will be explained. In sectors 14 and 35, and in the upstream position in sector 2 (an ID was only installed there for those 4.5 weeks), there is no space for Pb between the upstream end of the ID and the transition region where the vacuum chamber aperture narrows down to the smaller size used for the IDs. If much of the radiation is created at this transition, the cone of radiation may not have grown large enough at the position of the upstream ends of the magnetic structure to reach the dosimeters that are mounted there.

Dosimeters have also been mounted on the ID vacuum chambers immediately upstream of the ID and at the same height above the vacuum chamber as the dosimeter on the magnetic structure when the ID is at minimum gap. The total dose in this location, shown in Fig. 2, reflects the sector-to-sector variation in the dose levels. The sector-to-sector variation is not the same for each run, however. Injection into the ring is in sector 39 (of 40), 2 sectors upstream of the sector 1 ID.

As can be seen in Fig. 1, the IDs in three separate sectors have accumulated a total dose of 1.1 or 1.2 Mrad during this time period, and IDs in three other sectors have accumulated 0.5 Mrad. In yet another sector (sector 35), the total of 0.5 Mrad was accumulated in the approximately 8 total weeks of running time between 2/18/97 and 5/4/97 (no ID was installed there for any earlier runs).

WHAT DIFFERENCE DOES THE TYPE OF RADIATION MAKE?

The type of radiation to which the magnets are exposed has been found to be significant. A number of studies that exposed magnets to 1.17 MeV ^{60}Co γ-rays (2, 4, 8) found no radiation-induced demagnetization, despite total doses as high as 280 Mrad (4). This, combined with the demagnetization seen with lower doses of higher-energy radiation quanta, suggests that the spectrum of the radiation is very important in determining whether there will be damage. It is probably not important whether the energy quanta in the incident radiation are electrons or photons as long as the energy of the incident quanta is high enough to cause a radiation shower, because the shower will consist of both electrons and photons no matter what the incident radiation.

Clearly, then, because 1 MeV gammas do not cause demagnetization whereas 17 MeV electrons do (4), one needs to know the spectrum of radiation to which the magnets are exposed in order to predict the likelihood of damage. An experiment was carried out at the APS to determine whether the dose rates being measured at the IDs were from high or low energy quanta. A multi-layered sandwich of alternating Pb and film dosimeters was placed so the dosimeters were approximately 30 mm directly above the positron beam, as shown in Fig. 3. The dose as a function of depth of Pb was measured, and the results are shown in Fig. 4a. For comparison, the absorbed energy as a function of depth in Pb due to a 6.3-GeV cascade is shown in Fig. 4b (18). The depth at which the peak dose occurs is a function of the energy of the incident quanta.

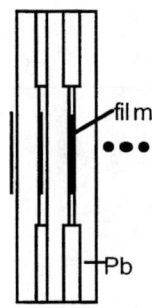

Figure 3. A schematic of the arrangement for a dose-depth experiment. The stack continues for many more layers than shown here.

The similarity between these curves suggests that 7 GeV quanta made up a large fraction of the incident radiation in the APS test. This radiation is high enough in energy to demagnetize the magnets.

The depth-dose experiment was repeated during the subsequent run. In that experiment, the peak dose occurred at a shallower depth, suggesting that some of the cascade had already taken place before the radiation reached the experiment. It may be that the spectrum of the dose reaching the ID varies strongly with events or beam characteristics that are specific to the particular run. More experiments will be conducted to further characterize the exposure.

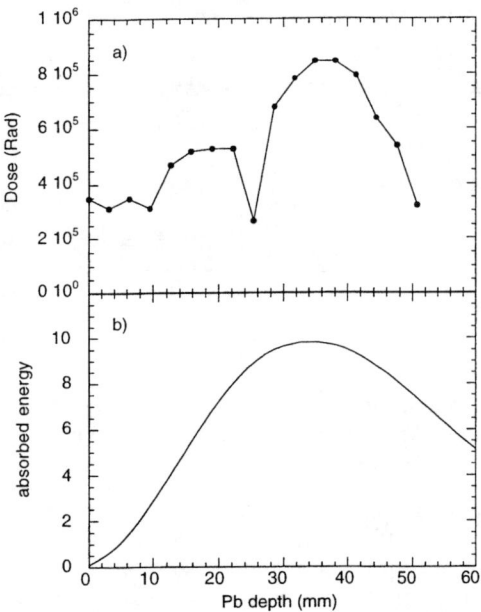

Figure 4. a) Dose measured as a function of Pb depth using the setup of Fig. 3. b) For comparison, absorbed energy vs. depth of Pb from a 6.3 GeV electromagnetic cascade. (data taken from 18). The similarity of these curves suggests that the incident radiation in the current measurement was at an energy near 7 GeV.

WHAT IS THE EFFECT OF THE MEASURES TAKEN TO REDUCE RADIATION DOSE TO THE IDS?

The Pb shielding mentioned above that has been installed immediately upstream of the IDs is more than 30 radiation lengths thick and should reduce the radiation levels to something too small to measure. The measured effect of the Pb gives us insights into the spatial distribution of the radiation. Fig. 5 shows the dose measured during a 3.5-week run by film dosimeters laid flat on the upper and lower faces of the (10-mm outside dimension) vacuum chamber and laid flat on the pole faces of the ID. This ID was kept at 100 mm gap throughout the run. The position of the Pb shielding is marked on the graph; note that it has no effect on the distribution of doses measured at the face of the vacuum chamber, despite the fact that there is no gap between the Pb shielding and the vacuum chamber. The Pb does dramatically reduce the dose at the upstream face of the first pole where the unshielded doses were highest, however; that point is not shown on the graph because the dose there was too small to measure with the film dosimeters that were used. The Pb shielding has essentially no effect on the dose at the downstream end of the ID. This shows that, while there is radiation traveling through the air above the vacuum chamber, there is also a non-negligible amount of radiation that will reach the magnets of an ID at closed gap

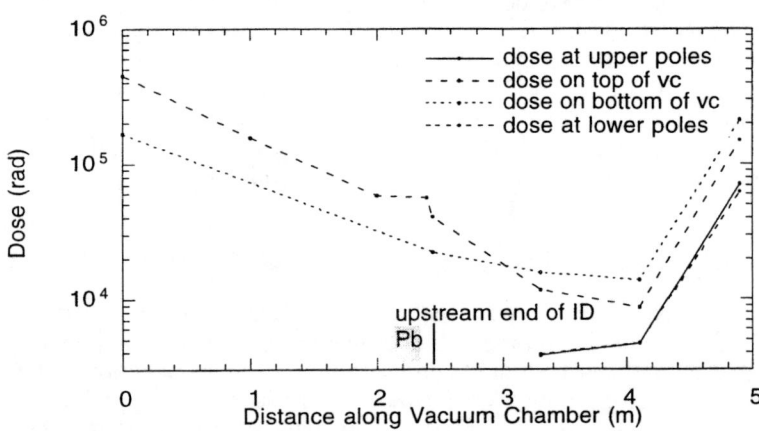

Figure 5. Effect of the Pb shield on the dose to the ID. The shielding has no effect on the dosimeters that were placed flat on the top and bottom faces of the ID vacuum chamber, showing that much of the radiation travels through the vacuum chamber rather than through the air outside the chamber. The dose at the pole faces at the upstream end of the ID was too low to measure; it has been markedly decreased by the Pb. The Pb has no effect on the dose level at the downstream end of the ID. The ID was kept at 100 mm gap during this run. The increase in dose at the downstream end of the ID straight section is not yet understood, although it is often observed.

by traveling through the vacuum chamber. So although the shielding dramatically reduces the radiation dose at the upstream end of the ID, where it would otherwise usually be highest, it has essentially no effect on the lower dose levels at the downstream end. The other drawback of the Pb shielding is that it cannot be placed in every sector. In some sectors, the total ID length is nearly as long as the small-gap region of the vacuum chamber, so that no Pb can be installed with a small enough gap to protect the ID.

Another means that is used to reduce the radiation dose to the magnets is to open the gaps for injection. This is also done at the Advanced Light Source (ALS) in Berkeley where it is found to be effective (19). Figure 6 shows the total dose measured during a 6-week running period at the first poles of each of the installed insertion devices. Dosimeters were also mounted immediately upstream of the first pole of each ID, at the same distance from the vacuum chamber as the dosimeter on the first pole would be if the ID were at minimum gap. Thus, the difference in dose rates indicates the amount of dose the upstream end of the ID was spared by having the gap open when the ID was not being used.

SUMMARY

The rate of radiation dose accumulation for the APS IDs has been and will continue to be measured. No radiation-induced demagnetization of the APS IDs has been observed to date. Attempts to compare these dose levels to dose levels where damage has been reported by others are complicated by factors such as the energy spectrum of the radiation and type of radiation, the magnet manufacturing technique and the demagnetizing field at the magnet blocks. The radiation spectrum has been measured and found to be high enough in energy to cause demagnetization. Lead shielding placed upstream of the IDs has helped reduce the dose to the upstream ends of the IDs but has much less effect at the downstream ends. Opening the ID gaps during injection and when the ID is not in use is also effective in reducing the dose.

ACKNOWLEDGMENTS

The authors would like to thank John Grimmer, Tim Roberts, John Attig, and others for their help in placing the dosimeters. We also thank Elwyn Dolecek and the ANL dosimetry group for all their unstinting help in preparing and reading out the TLDs. Thanks, too, to Jenny Erdmann and Jerry Moore who helped with the film dosimeter readings.

This work was supported by the U.S. D.O.E., BES-Materials Sciences, under Contract #W-31-109-Eng-38.

REFERENCES

(1) Chavanne, J., Elleaume, P., and Van Vaerenbergh, P., "Partial demagnetization of ID6 and dose measurements on certain IDs," ESRF Machine Technical Note 1-1996/ID, (Jan. 1996).
(2) Luna, H.B., Maruyama, X.K., Colella, N.J., Hobbs, J.S., Hornady, R.S., Kulke, B., and Palomar, J.V., *Nucl. Instrum. Meth. Phys. Res.* **A285**, 349 (1989).
(3) Colomp, P., and Bräuer, E., "Measurements on demagnetization of the ID magnets by irradiation with the linear accelerator," ESRF Report 143/93/EB (30 Mar 1993).
(4) Okuda, S., Ohashi, K., and Kobayashi, N., *Nucl. Instrum. Meth. in Phys. Res.* **B 94**, 227(1994).
(5) Brown, R.D., Cost, J.R., Meisner, G.P., and Brewer, E.G., *J. Appl. Phys* **64**, 5305 (1988).
(6) Brown, R.D., and Cost, J.R., *J. Appl. Phys* **63**, 3537 (1988).
(7) Brown, R.D., and Cost, J.R., *IEEE Trans. Magn.* **25**, 3117 (1989).
(8) Cost, J.R., Brown, R.D., Giorgi, A.L., and Stanley, J.T., *Mat. Res. Soc. Symp. Proc.* **Vol. 96**, 321 (1987).
(9) Blackmore, E.W., *IEEE Trans. on Nucl. Sci.* **NS-32**, 3669 (1985).
(10) Cost J., Brown, R., Giorgi, A., and Stanley, J., *IEEE Trans. Magn.* **24**, 2016 (1988).
(11) Some neutrons are produced in the storage ring environment, but in the experiments of Luna et al. [2], no loss of remanence was seen that would have been attributable to the neutrons produced by using an 85 MeV electron beam to produce bremsstrahlung.
(12) Van Vaerenbergh, P., personal communication.
(13) Elleaume, P., personal communication.
(14) Pflüger, J., Heintze, G., and Vasserman, I., *Rev. Sci. Instrum.* **66**, 1946 (1995).
(15) Mincher, B.J., and Zaidi, M.K., *Rad. Protection Dosimetry* **47** (1993) 571.
(16) Mincher, B.J., Zaidi, M.K., Arbon, R.E., McLaughlin, W.L., and Schwendiman, G.L., *Rad. Protection Dosimetry* **66** (1996) 233.
(17) Dosimeter films and reader were from Far West Technology, Goleta, CA, USA.
(18) Bathow, G., Freytag, E., and Tesch, K., *Nucl. Phys.* **B2** (1967) 669.
(19) Krebs, G.F., and Holmes, M., to be published in the proceedings of the 1997 Particle Accelerator Conference, held 12-16 May 1997 in Vancouver, B.C., Canada.

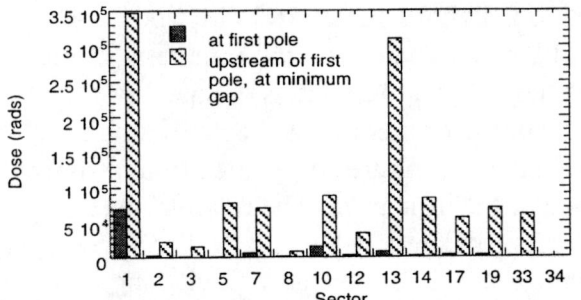

Figure 6. Doses measured by dosimeters fastened to the upstream ends of the ID magnetic structures and by dosimeters mounted on the ID vacuum chambers immediately upstream of the ID, at the same height above the vacuum chamber as the ones on the magnetic structure would be at minimum gap. The difference between the two shows the dose the ID was spared by not always being at minimum gap.

Application of electroreflectance to Stark spectroscopy.

A.K. Gaigalas, T. Ruzgas, and G. Niaura
Biotechnology Division, NIST, Gaithersburg, MD 20899

Abstract:

Electroreflectance(ER) was measured for immobilized tryptamine on polycrystaline gold electrode. The measurements were performed in the wavelength region 200-300nm using the NIST synchrotron radiation source. The ER response was interpreted as a Stark shift in the tryptamine absorption line at 218 nm. The large ER response suggests that the method could be applied to the study of the tryptophan environment in adsorbed proteins.

Introduction:

Tryptamine is similar to the amino acid tryptophan; both have an indole-like part comprised of two coupled rings with very strong electronic absorption lines in the spectral region between 200 and 250 nm. The electronic transitions are mainly between states of the π manifold and are sensitive to the solvation shell of the rings. In the following, we interpret the ER measurements in terms of potential induced shift in the tryptamine absorption line where the shift is due to the difference in the static dipole moments of the ground and excited states of the solvated tryptamine molecule.

Discussion:

The ER measurements were performed at the NIST synchrotron radiation source(SURF). Figure 1 shows a schematic of the apparatus. The beam was passed through a monochromator and focused on a gold electrode whose surface was modified with a layer of biomolecules. The angle of incidence was approximately 10^0 which is sufficiently close to normal to permit the simplification of the analysis of the ER response and to minimize distortion in the shape of the absorption line of the biomolecules(1). The polarization of the incident light was parallel to the electrode surface. For most measurements, the potential on the electrode was modulated with an amplitude of 0.2V and the DC level was varied between -0.1V and +0.1V relative to Ag(AgCl) reference electrode The photons reflected from the electrode were collected and detected by a photo multiplier(PM). The output from the PM was send to a digital voltmeter(DVM) and a lock-in amplifier driven at same frequency as the modulation of the electrode potential. The ratio of the outputs of the lock-in amplifier and the DVM constitutes the ER response and is denoted by $\Delta R/R$. Figure 2 and Figure 3 compare the measured ER response of immobilized cytochrome c using the SURF and Xe arc lamp sources. The ER signal in the two figures originates from a shift of the heme absorption line due to oxidation and reduction of the iron atom. The relative noise of the SURF measurement is 10% while the Xe arc lamp results indicate a value of 1%. The synchrotron source has great utility for these measurements because of the ease of switching between the deep UV and visible spectral regions.

Figure 4 shows an example of the ER response from an electrode modified with tryptamine. The shape of the ER response is very similar to the derivative of the observed absorptivity in solution and there is no redox process associated with the immobilized tryptamine. We assumed that the ER response from an electrode modified with a layer of biomolecules could be described by the response of a stack of noninteracting thin layers(2). The absorptivity of the biomolecular layer and the dependence of the absorptivity on electric field was modeled in accordance to Ref (3). The analysis showed that the measured tryptamine ER response could be represented by a product of the derivative of the measured tryptamine absorptivity in solution and the ER amplitude which is related to the Stark shift of the absorption line. The *slope* of the dependence of the ER amplitude on the potential modulation amplitude is:

$$slope = -1.84 \frac{\sqrt{\varepsilon'_T}\, \Gamma_T}{hca} (\mu_e^* - \mu_g^*) \beta \qquad Eq\ 1$$

where h is Planck constant, c is the speed of light, ε_T' is the real part of the tryptamine layer dielectric constant, a is a unit conversion factor(100cm/m), $\mu_e^* - \mu_g^*$ is the difference of the dipole moments of the ground and excited states, Γ_T is the tryptamine surface concentration, and β is the proportionality coefficient between the electric field at the site of the molecule and the applied electrode potential. For the case of 0.01M phosphate buffer(PB), (solid circles in Figure 4) we obtain a measured slope~-22±5*10^{-5} mole/(cm m^2 V^{-1}). Assuming that the tryptamine molecule occupies an area of approximately 1nm^2, a 20% coverage (estimated from the reduction peak of DSP, an immobilizing agent, and a quess of 80% efficiency for the immobilization reaction) of the geometric area of the electrode leads to a surface concentration of 3.3 10^{-7} moles/m^2. The modified Gouy-Chapman theory of the double layer (Eqs. 2.11 and 2.13 of Ref. 4)gives an estimate of $\beta = 2.4*10^8$ m^{-1} for a solution with ionic strength of 0.01M and tryptamine located 1 nm from the electrode surface. Using the above estimates for β and Γ_T, Eq [1] then gives $(\mu_e^* - \mu_g^*) = 21*10^{-30}$ Cm(~6 D). The continuous line in Figure 4 shows the ER response of an electrode with immobilized tryptamine in 1.0M $NaClO_4$ +0.01M PB. There is still the strong spectral feature at 215nm which shifts by about 1nm to shorter wavelengths at high ionic strength. Measurement of absorption in a solution of 0.01M tryptamine + 1.0M $NaClO_4$ +0.01M PB yielded a result identical to that of tryptamine in 0.01M PB indicating that there is no specific interaction between the per chlorate ions and tryptamine. Measurement of ER on a modified electrode(no tryptamine) in a solution of 0.1M $NaClO_4$ +0.01M PB also yielded no spectral features. Therefore the properties of the ER spectra shown in Fig 4(solid line) most likely depend on the Stark shift of the solvated tryptamine states similar to those in low ionic strength buffers. The primary difference would be the change in double layer potential spatial gradients. In the context of the Gouy-Chapman theory of the double layer, the value of β at 1 nm (the putative location of the indole ring) from the electrode surface in the 1M $NaClO_4$ solution should be $1.8*10^8$ m^{-1}. Since the surface density should remain the same for covalently immobilized species, Eq [1] together with the measured slope from -12±4*10^{-5} mole/(cm m^2 V^{-1})) leads to an estimate of the difference of the static dipole moments of $16*10^{-30}$ Cm(~ 11 D). This is of the same order as the value determined in the low ionic strength.

We used PM3 semi empirical calculations(contained in a commercial quantum chemistry software package) to obtain preliminary estimates of the expected dipole moment difference. The calculation of tryptamine in vacuum with configuration interaction using three occupied and unoccupied molecular orbitals gave electronic transitions at 224.8nm and 237.8nm with a net increase in the static dipole moments of $4.6*10^{-30}$ Cm and $3.26*10^{-30}$ respectively. The calculation results are in qualitative agreement with the observed electronic absorption spectrum and the estimated change in the static dipole moment. PM3 code with three state configuration mixing was used to calculate the electronic spectrum for solvated tryptamine molecule. Solvation was approximated by periodic box boundary conditions with 18 water molecules in the box placed randomly around the tryptamine centered in the box. The resulting electronic spectrum had two transitions at 225.02 nm and 237.9 nm, which have small red shifts relative to the vacuum values. The ground state dipole moment of the tryptamine molecule was estimated from the point charges at each atomic position of tryptamine. The result was about $10*10^{-30}$ Cm compared to $5.4*10^{-30}$ Cm for the molecule in vacuum. Assuming that the excited state also experiences an increase in dipole moment on solvation, the calculation suggests that the observed differences in the dipole moment may be a sensitive measure of solvation.

Conclusion:

The measured ER response of immobilized tryptamine in low ionic solutions is dominated by the absorption line at 218 nm. The dependence of the ER magnitude on modulation amplitude was used to obtain an estimate of the difference in static dipole moments of the excited and ground state of $21*10^{-30}$ Cm. The ER signal in 1.0M $NaClO_4$ displayed a small blue shift relative to the signal in low ionic strength solution and the dependence on modulation amplitude was of the same order as in low ionic strength solutions. The estimated difference in static dipole moments was $16\ 10^{-30}$ Cm which suggests a lack of sensitivity to the ionic composition of the solvation shell. After additional efforts to obtain an independent measurement of the surface concentration Γ_T and a more rigorous evaluation of the ratio of electric field to potential, the ER technique could be a of great utility as a probe of solvation at the interface.

The large ER response suggests that the technique could be applied to tryptophan located inside proteins where the ER response would be influenced by the local protein structure surrounding the indole ring of the tryptophan residue. Figure 5 shows the ER response of cytochrome c in the 200 nm spectral region(0.02V dc 0.2V ac 91 Hz) at an electrode potential where electron transfer (ET) is taking place between the heme in cytochrome c and the electrode. The signal is much smaller than that shown in Figure 2, however, a narrow feature is discernible at the location of the tryptophan absorption line(the interpretation of the broad feature is uncertain). In addition to the Stark shift, the absorption line of tryptophan could be shifted due to protein structural changes induced by ET. Determining the nature of the ER response is the next challenge.

References:
1. Kim, S., Wang, Z., and Scherson, D.A., *J. Phys. Chem.*B, **101**(1997) pp 2735-2740
2. Kolb, Dieter M., in "Spectroelectrochemistry: Theory and Practice"(R.J. Gale, Ed.), Plenum Press, New York, N.Y., 1988, p.87
3. Liptay, Wolfgang, "Excited States" **1**, (Edward C. Lim, Ed.), Academic Press, N.Y., 1974, p. 129-229
4. Schmickler, W., and Henderson, D., *Progress in Surface Science* **22**, 323 (1986)

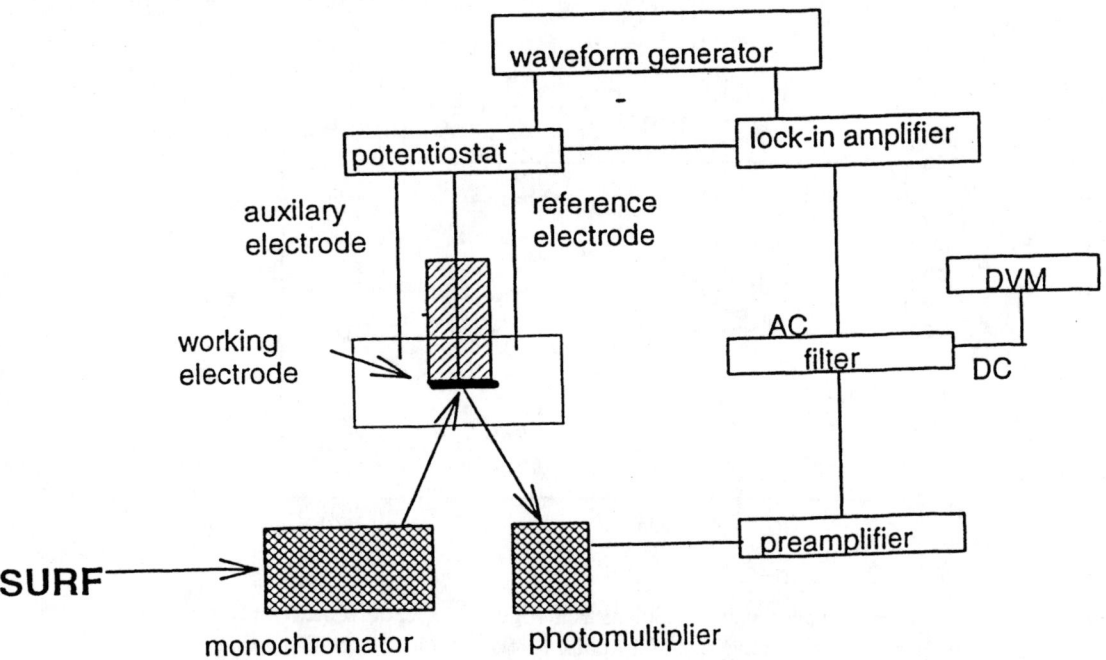

Figure 1. The schematic of the experimental apparatus. The monochromator was used to select the spectral region. A lens and a mirror(not shown) were used to guide the light to the electrode surface. The angle of incidence was ~10^0. The reflected light was detected by a photo multiplier(PM) and processed by low-pass and high-pass filters whose outputs was detected by a voltmeter and a lock-in amplifier respectively. The ratio of the detected outputs was the primary measured quantity.

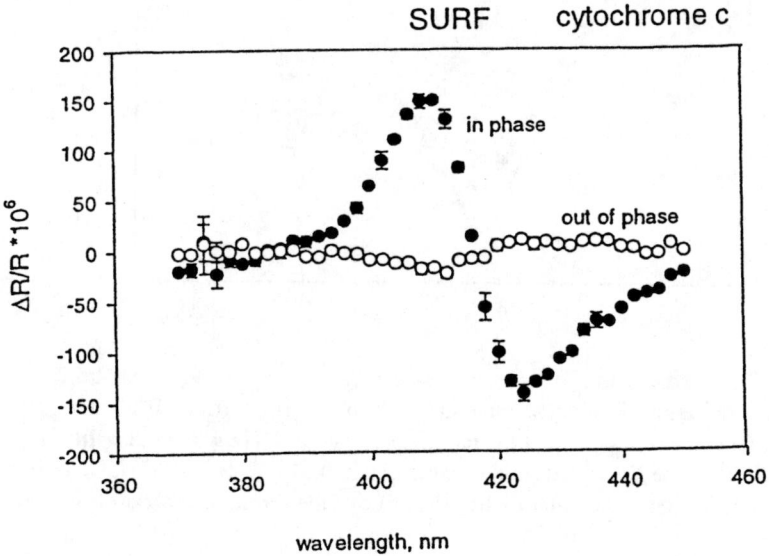

Figure 2. The wavelength dependence of the electroreflectance signal from a gold electrode with immobilized cytochrome c. The source of the light was SURF. The constant electrode potential was set to 0.01V(vs Ag/AgCl) and the modulation was 41Hz with an amplitude of 0.02V. The signal corresponds to the difference in absorption of the reduced and oxidized states of the metal site in cytochrome c.

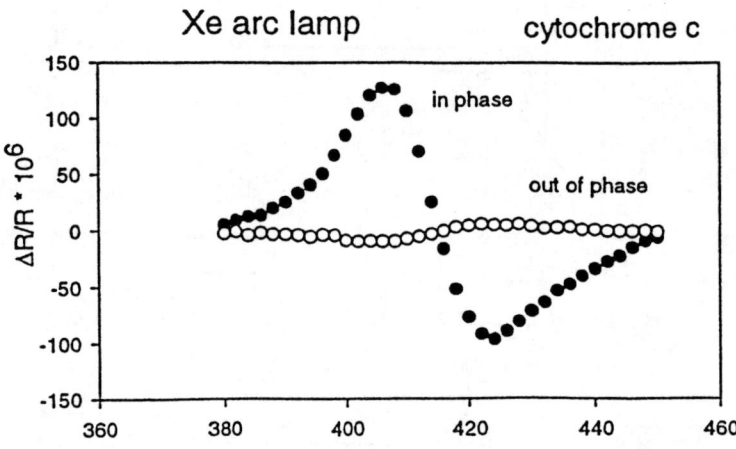

Figure 3. Same experimental conditions as in Figure 2 except that the light source was a 70W Xenon lamp. Comparison of the noise levels in Figures 2 and 3 gives an approximate relative stability of the two light sources.

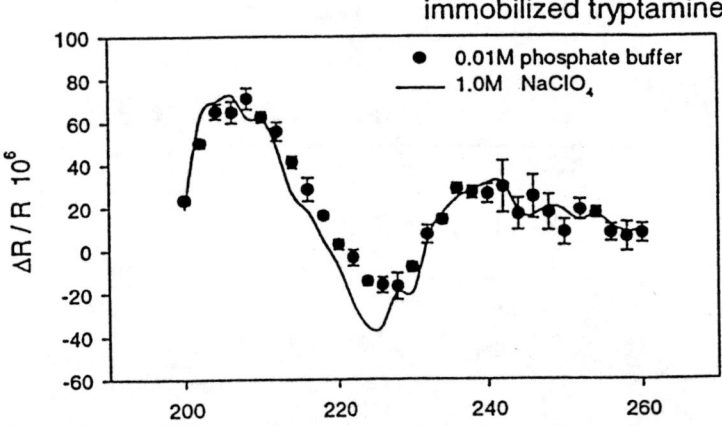

Figure 4. The solid circles show the in-phase electroreflectance(ER) signal from a gold electrode with immobilized tryptamine in 0.01M phosphate buffer(PB). The constant potential was set to 0.0V and the modulation was at 91Hz with an amplitude of 0.2V. The solid line is the ER signal from an electrode in 0.01M PB+ 1M of NaCl buffer. In both cases, the ER signal is similar to the derivative of the tryptamine absorption line at 218 nm.

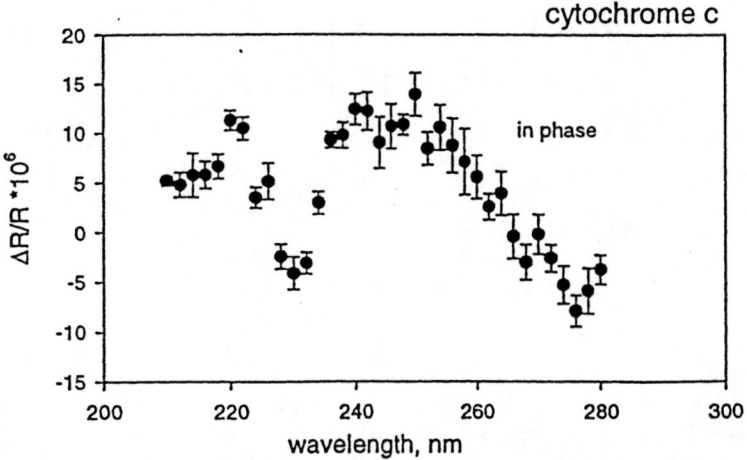

Figure 5. The electroreflectance signal from an electrode with immobilized cytochrome c in 0.01M phosphate buffer. The potential of the electrode was set to 0.01V. Most likely the response at ~225nm originates from the tryptophan residue in the protein.

High-Energy X-Ray Experiments at the APS Sector 1 Beamlines

D. R. Haeffner, S. D. Shastri, and D. M. Mills

Argonne National Laboratory
Advanced Photon Source
9700 S. Cass Ave., Argonne, IL 60439

The goal of the SRI-CAT high-energy x-ray program at Sector 1 of the Advanced Photon Source (APS) is to develop instrumentation and techniques that utilize the abundant potential of the APS as a source of photons in the 30 keV to 200 keV range. The initial efforts have been to characterize the high-energy x-rays from undulator A, to measure the scattering from several liquid and amorphous materials for comparison to neutron data, and to characterize optics for high-resolution Compton scattering. Results from these experiments will be presented along with a discussion of future Sector 1 high-energy x-ray development.

This work is supported by the U.S. Dept. of Energy, BES-Materials Sciences, under contract No. W-31-109-ENG-38.

DETECTORS: LARGE, SMALL, FAST, ENERGY-RESOLVING

X-ray imaging characteristics of a direct conversion detector using selenium and thin film transistor array

Brian Rodricks, Denny L. Lee, Lawrence K. Cheung,
Lothar S. Jeromin, Eugene F. Palecki

Sterling Diagnostic Imaging, Inc.
P.O. Box 6101
Newark, DE 19714-6101

Progress on the development of a semiconductor-based, direct-detection, flat-panel digital radiographic imaging device will be discussed. The device consists of a 500 mm thick amorphous selenium sensor coupled to an amorphous silicon thin-film-transistor (TFT) readout matrix. This detector has an active imaging area of 14"x17", 3072 x 2560 pixels with dimensions 139 mm x 139 mm and a geometrical fill factor of 86 percent. Charges generated primarily as a consequence of photoelectric interaction between the incoming x-rays and Se are integrated on storage capacitors that are located at each pixel. The high electric field applied across the Se minimizes the lateral spreading of the signal resulting in a significantly higher spatial resolution when compared to conventional film/screen and image plate systems. The sensor array is read out one pixel line at a time by manipulating the source and gate lines of the TFT matrix. Data are digitized to 14 bits.

This paper will discuss the statistical photon counting analysis performed on an early prototype device. Measurements will include modulation transfer function, detector quantum efficiency, linearity, and noise analysis.

Photoemission from Silicon Photodiodes and Induced Changes in the Detection Efficiency in the Far Ultraviolet

R. E. Vest and L. R. Canfield

Electron and Optical Physics Division, Physics Laboratory, National Institute of Standards and Technology, Technology Administration, U. S. Department of Commerce, Gaithersburg, Maryland, 20899

Abstract. Photoemission from semiconductor photodiodes exists as a spurious effect in the detection of radiation in the far ultraviolet. The magnitude of this photoemission effect in silicon transfer standards has been measured and found to be significant in the spectral region from 5 nm to 165 nm. The efficiency of these photodiodes in the more conventional semiconducting mode was altered by this photoemission as much as 14% under some conditions of operation. Application configurations which virtually eliminate the influences of photoemission have been identified and should be employed. The possibility of using these photodiodes in a photoemissive mode (as solar-blind detectors) was found to be limited by the temporal instability and spatial uniformity of the devices tested.

INTRODUCTION

Silicon n-on-p photodiodes fabricated with a nitrided passivating oxide layer (1) and used in the photovoltaic mode are issued as transfer standard detectors for the far ultraviolet by the National Institute of Standards and Technology (NIST) (2,3). Their high detection efficiency, spatial uniformity, and temporal stability make them useful detectors for many applications (1,4,5). These photodiodes are being used in solar physics, astronomy, plasma diagnostics, calibration of space-based instruments, and general-purpose radiation detection in many synchrotron radiation-based experiments. Depending on the external electronics configuration, one can measure either positive or negative electric current in response to radiation incident on the photodiode. From internal considerations, no difference in magnitude is expected. We will show that in some regions of the far ultraviolet this equality is disrupted by the inclusion of a spurious photoemission current in only one of the measurement configurations.

In this paper two types of radiation-induced currents are discussed. To avoid confusion, we will use the term "photocurrent" to mean the internal conversion current flowing across the junction of a semiconductor device, and the term "photoemission" to mean the flow of electrons from the surface of the device into vacuum.

The photoemission current may inadvertently be included in photocurrent measurements under some conditions of use in portions of the far ultraviolet spectral region. An electrometer whose input is connected to the front region of a silicon far ultraviolet transfer standard photodiode (position B of Fig. 1) will detect the photoemission as well as the conventional internal photocurrent. If the electrometer is connected to the rear region (position A of Fig. 1), the front region is essentially at ground potential, and only the internal photocurrent will be detected.

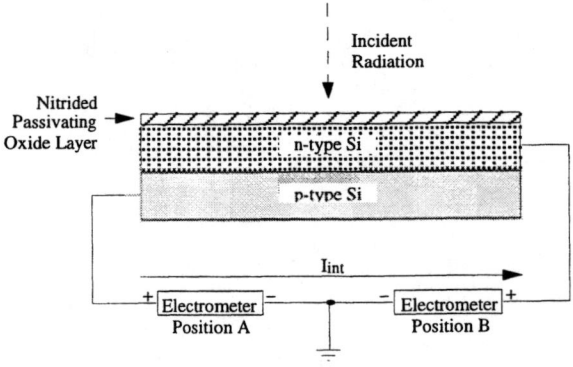

FIGURE 1. Schematic diagram of a simple detector circuit using a silicon transfer standard photodiode. I_{int} is the internal conversion current across the semiconductor junction. The electrometer in position A measures positive current and in position B measures negative current.

NIST calibrations of n-on-p devices in the far ultraviolet are always carried out with the photodiode operated in the positive photocurrent configuration. This method of calibration measures photocurrent in the back portion of the photodiode and ensures that the measured detection efficiency is not influenced by photoemission. We will show here that operation in a negative photocurrent configuration may compromise the NIST calibration of these silicon transfer standards in a portion of the far ultraviolet spectral region.

PHOTOEMISSION

The photoemission efficiency of a silicon photodiode (defined as the average number of electrons emitted from the surface per incident photon at a specific wavelength under specified electric field conditions) may be measured by connecting the two photodiode leads together, as shown in Fig. 2, to create a photocathode whose outer surface is nitrided SiO_2. (The photoelectric *yield*, a fundamental property of materials, is defined as the greatest photoemission efficiency possible under any field conditions at the specified wavelength.) A cylindrical external electrode is placed so that the photon beam passes through it and strikes the photocathode at normal incidence. The electrode may be biased relative to the photocathode to control the photoemission. The measured quantity is emission of electrons from the photocathode, rather than electrons collected at the external electrode, because the collection efficiency will, in general, be less than unity.

The magnitude of the electric field established by the voltage applied to the external electrode will determine whether all sufficiently energetic electrons leave the photocathode surface. A bias of +15 V or more (with the geometry used here) results in a relatively constant level of photoemission at any of the wavelengths shown in Fig. 3. However, at lower voltages the photoemission efficiency falls to zero as bias decreases. Figure 3 shows that there may be significant response at some wavelengths without any applied bias.

A bias of +100 V (5 nm to 17 nm) or +60 V (17 nm to 165 nm) was applied to the external electrode in the present investigation, giving saturated emission. The photoemission efficiency of the photodiode was measured at NIST's far ultraviolet calibration facilities by direct comparison to NIST working standard photodiodes (2,3). The relative expanded combined standard uncertainty (2-σ) of efficiency measurements ranges from 7% to 22%, depending on wavelength. Figure 4 shows the measured photoemission efficiency from 5 nm to 165 nm. The photoemission efficiency peaks near 82 nm, in the region of maximum oxide absorption, and falls with decreasing wavelength.

One can envision using photoemission from the front surface of a silicon photodiode, rather than the conventional internal conversion, to detect radiation in the 5 nm to 165 nm spectral region. Such a detector would have reasonably high efficiency, would have a response function which varies smoothly as a function of wavelength over much of its useful range, would be relatively insensitive to radiation of wavelengths longer than about 165 nm, and could be switched from photoemissive to semiconductive operation by means of external electronics. However, repeated measurements of the photoemission efficiency of a silicon photodiode have shown that the temporal stability of such a photoemissive detector is rather poor for radiometric use. The observed standard deviation of the photoemission efficiency for a data set of five measurements over a ten month period (2-σ) is 3% to 8%, depending on wavelength. Measurements made within a two day period exhibit a 2-σ variability of up to 5%. (Changes in the photoemission efficiency from a photodiode are measured with a precision (2-σ) of 1%.) Additionally, the local photoemission efficiency measured in 1 mm by 1 mm pixels over the central 8 mm square (the diode has a 10 mm by 10 mm active area) varies by $5 \pm 1\%$ (2-σ) of the mean value.

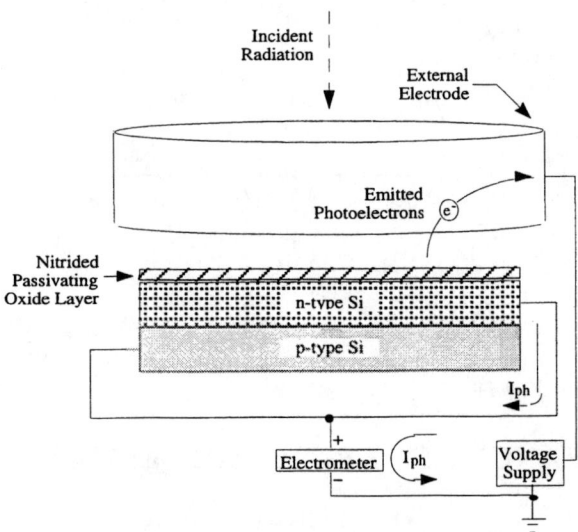

FIGURE 2. Schematic diagram for measurement of photoemission current I_{ph} from the front surface of a silicon transfer standard photodiode.

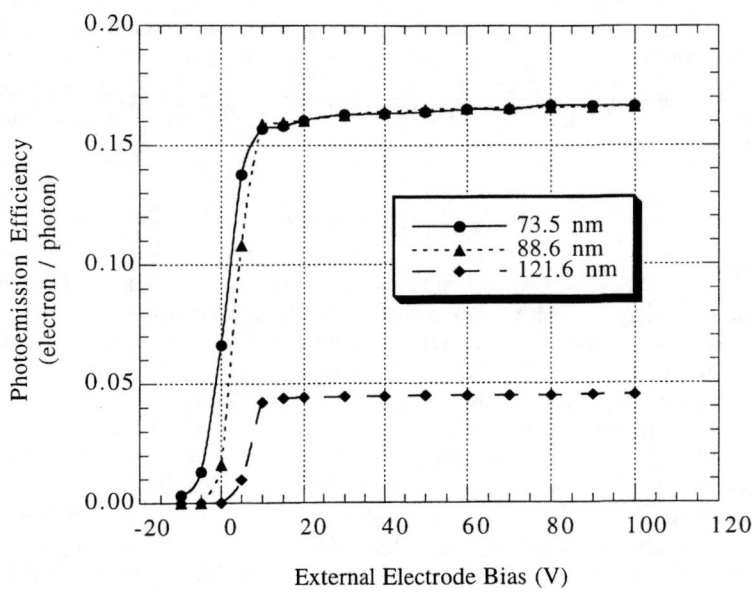

FIGURE 3. Photoemission efficiency from the front surface of a silicon transfer standard photodiode as a function of bias voltage at three wavelengths. At each wavelength, the detector response reaches a plateau by +15 V. The unbiased efficiency is not necessarily zero and rises with decreasing wavelength.

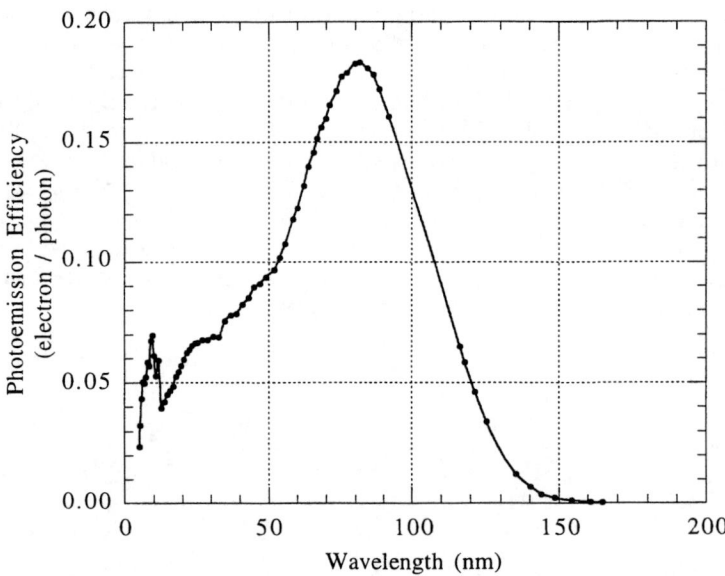

FIGURE 4. Photoemission efficiency from the front surface of a silicon transfer standard photodiode with the external electrode at +100 V (5 nm to 17 nm) or +60 V (17 nm to 165 nm). Photoemission becomes negligible at 165 nm, where the passivating oxide layer (the photocathode) becomes transparent.

CHANGES IN DETECTION EFFICIENCY

Theoretical Model

The photocurrent from an unbiased p/n junction semiconductor photodiode may be measured with either of two electrical configurations, which result in currents of opposite signs. It is expected that these photocurrents would be equal in magnitude. The currents generated and electronics used to measure them are indicated in Fig. 5. (All present NIST radiometric

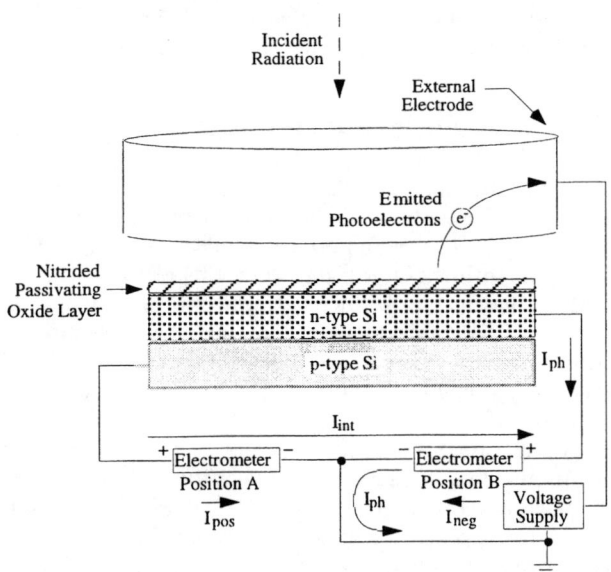

FIGURE 5. Schematic diagram of a silicon transfer standard photodiode with both current measurement configurations shown. I_{int} is the internal conversion current across the semiconductor junction and I_{ph} is the photoemission current. The electrometer in position A measures positive current I_{pos} and in position B measures negative current I_{neg}. The photoemission current is detected in position B, but not in position A.

silicon photodiodes calibrated in the far ultraviolet spectral region are n-on-p devices.) In the positive photocurrent configuration, the electrometer in position A measures the current I_{pos} flowing from the backside (anode) of the diode. In the negative photocurrent configuration, the electrometer in position B measures the current I_{neg} flowing from the front region (cathode) of the diode. When the diode is illuminated, the current measured in position B will usually be a negative quantity.

Incident photons within the 5 nm to 165 nm region studied here generate two electric currents in the photodiode. Photons transmitted through the oxide layer are absorbed in the silicon and produce a photocurrent I_{int} flowing across the semiconductor junction. This current flows through both the front and rear portions of the diode and is determined by the internal conversion efficiency (the average number of electron-hole pairs contributing to the measurable photocurrent per incident photon) and photon flux. Photons that are absorbed near the surface can also produce a photoemission current I_{ph} that flows through only the front portion of the photodiode and is determined by the photoemission efficiency and photon flux. The detection efficiency (the average number of electrons detected by external electronics per incident photon) is determined by measuring the electric current generated by a known photon flux. If the current is measured from the back portion of the device (position A of Fig. 5 – positive configuration) then the measured current I_{pos} is simply the internal conversion current I_{int}. Measurements made from the front region (position B of Fig. 5 – negative configuration) will detect both the internal conversion current I_{int} and the photoemission current I_{ph}. The measured current I_{neg} is the sum of these two currents. Using the sign conventions indicated in Fig. 5, we determine the relationship between the various currents:

$$I_{pos}(\lambda, \Phi) = I_{int}(\lambda, \Phi) \tag{1}$$

$$I_{neg}(\lambda, \Phi) = -I_{int}(\lambda, \Phi) + I_{ph}(\lambda, V, \Phi) \tag{2}$$

Combining these expressions:

$$I_{neg}(\lambda, \Phi) = -I_{pos}(\lambda, \Phi) + I_{ph}(\lambda, V, \Phi) \tag{3}$$

where I_{pos} is the positive configuration current measured in position A, I_{neg} is the negative configuration current measured in position B, I_{ph} is the photoemission current, λ is the wavelength, V is the voltage applied to the external electrode, and Φ is the incident photon flux. The negative configuration detection efficiency is less than the positive configuration detection efficiency, and the difference is the photoemission efficiency:

$$\varepsilon_{neg}(\lambda) = \varepsilon_{pos}(\lambda) - \varepsilon_{ph}(\lambda, V) \tag{4}$$

where ε_{neg} is the negative configuration detection efficiency, ε_{pos} is the positive configuration detection efficiency, ε_{ph} is the photoemission efficiency, λ is the wavelength, and V is the voltage applied to the external electrode. It should be noted that the internal conversion efficiency of the photodiode is never changed by photoemission. The negative configuration detection efficiency is determined by both the internal conversion efficiency and the photoemission efficiency.

Experimental Results

The detection efficiency of a NIST transfer standard silicon semiconductor photodiode was measured in the 52 nm to 200 nm region at NIST's far ultraviolet calibration facilities by comparison with NIST working standards (2,3). The relative expanded combined standard uncertainty (2-σ) of detection efficiency measurements in this region is 7% to 18%. However, most of the uncertainty is due to systematic influences, and measurements of the change in efficiency have a precision (2-σ) of 1%. Figure 6 shows the detection efficiency of a silicon photodiode measured at 88.6 nm as a function of external electrode voltage in both positive and negative configurations. The positive configuration efficiency, measured in the back region of the photodiode, is not affected by the external electrode voltage. However, the negative configuration efficiency drops by 14 ± 1% as the electric field strength established by the voltage applied to the external electrode allows photoemission to take place.

The reduction in detection efficiency is most pronounced around 90 nm, but is significant throughout the 52 nm to 145 nm spectral region. Figure 7 shows the measured detection efficiency of a silicon photodiode in the negative photocurrent configuration with the external electrode biased to 0 V and +20 V. The change in detection efficiency can be suppressed by applying a bias of –20 V to the external electrode. (Photoemission characteristics depend on the electric field strength at the surface. The voltages given here are appropriate for our geometry.) The change in these detection efficiency measurements, as well as similar measurements in the positive configuration, are shown in Fig. 8. The solid curve is calculated from Equation (4).

Since the change in the detection efficiency is equal to the photoemission efficiency, any factors that alter the photoemission will affect the change in negative configuration detection efficiency as well. Electric fields, spatial non-uniformity, and temporal instability all contribute to the variability of the photoemission efficiency and will all affect the change in negative configuration detection efficiency. Since photoemission may be a significant effect without any bias on the external electrode (See Fig. 3), the negative configuration detection efficiency may be less than the expected value even when no voltages are present. Photoemission, and therefore the change in detection efficiency, can be suppressed by a negative bias on the external electrode.

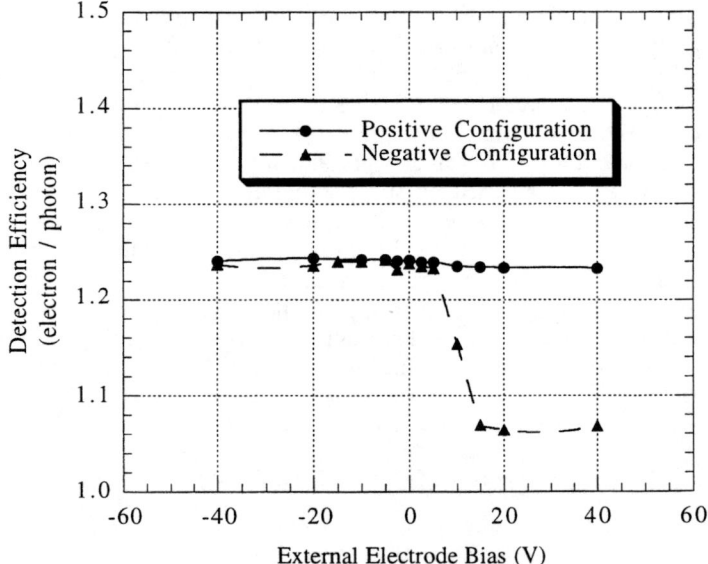

FIGURE 6. Detection efficiency of a silicon transfer standard photodiode at 88.6 nm as a function of external electrode voltage. In the positive photocurrent configuration, the detection efficiency is constant. In the negative configuration, the efficiency drops by 14% as the photoemission efficiency increases.

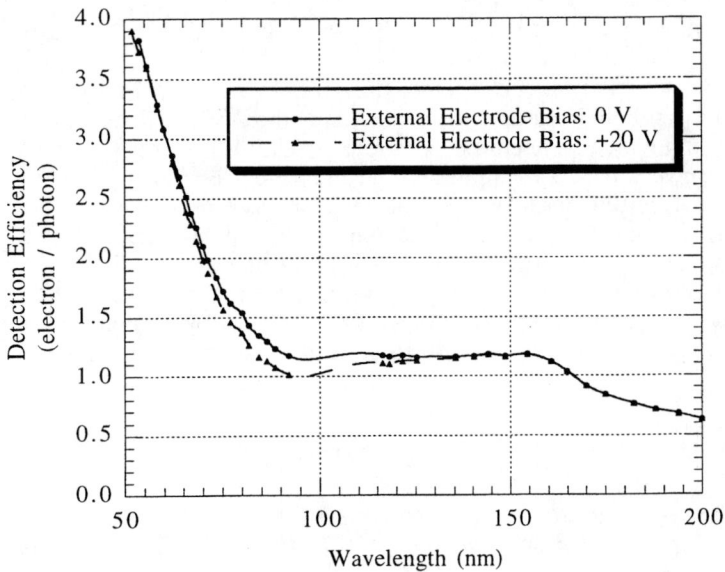

FIGURE 7. Detection efficiency of a silicon transfer standard photodiode in the negative photocurrent configuration with the external electrode biased to 0 V and +20 V.

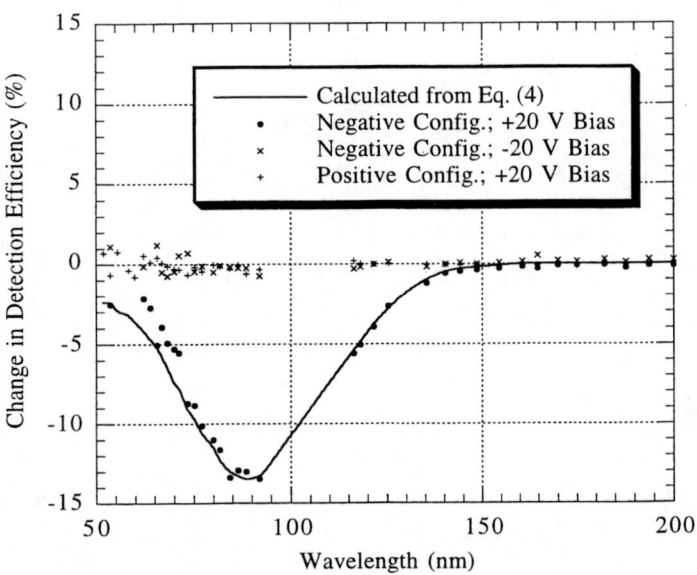

FIGURE 8. Change (relative to unbiased value) in detection efficiency of a transfer standard silicon photodiode. The change in the negative photocurrent configuration detection efficiency with the external electrode biased to +20 V agrees with the calculated change due to photoemission. There is no significant change in the detection efficiency with a negative bias on the external electrode (photoemission is suppressed) or when the photodiode is operated in the positive photocurrent configuration (photoemission is not detected).

CONCLUSIONS

The photoemission efficiency of NIST far ultraviolet transfer standard silicon semiconductor photodiodes has been measured. Photoemission is a significant process from 5 nm to 165 nm. The presence of a photoemission current modifies the detection efficiency of the silicon semiconductor n-on-p photodiode when the device is operated in a negative photocurrent configuration. The decrease in efficiency was found to be as much as 14% in the spectral region from 52 nm to 145 nm. At shorter wavelengths, the photoemission current is not a significant fraction of the total current, and at longer wavelengths photoemission is negligible.

Since NIST calibrates transfer standard silicon photodiodes (n-on-p devices) in this spectral region in the positive photocurrent configuration, end users should take care to avoid altering the device's detection efficiency due to operation in the negative photocurrent configuration with a photoemission current present. While it is possible to suppress photoemission by the application of a nearby negatively biased electrode, we recommend operating the device in the positive photocurrent configuration only. Negatively biased electrodes may accelerate residual gas ions into the photodiode, which would appear to the external electronics as a spurious electric current. The ion collection current could alter the negative photocurrent configuration detection efficiency through the same mechanism as photoemission.

Testing and modeling has been conducted only on NIST far ultraviolet transfer standard silicon photodiodes with an n-on-p construction. The recommended current-measuring configuration given here should be reversed if the device under consideration has a p-on-n construction. Photoemission effects can always be made negligible by measuring the current from the rear portion of the device.

REFERENCES

1. Korde, R., Cable, J. S., and Canfield, L. R., *IEEE Trans. Nucl. Sci.* **40**, 1655–1659 (1993).
2. Canfield, L. R. and Swanson, N., NBS Special Publication 250-2, Gaithersburg, MD: National Institute of Standards and Technology, 1987.
3. Furst, M. L., Graves, R. M., Canfield, L. R., and Vest, R. E., *Rev. Sci. Instrum.* **66**, 2257–2259 (1995).
4. Gullikson, E. M., Korde, R., Canfield, L. R., and Vest, R. E., *J. Elect. Spect. and Rel. Phenom.* **80**, 313–316 (1996).
5. Rabus, H., Scholze, F., Thornagel, R., and Ulm G., *Nucl. Instrum. Methods A* **377**, 209–216 (1996).

X-ray Imaging with Amorphous Silicon Active Matrix Flat-Panel Imagers (AMFPIs)

Youcef El-Mohri, Larry E. Antonuk, Kyung-Wook Jee, Manat Maolinbay, Xiujiang Rong, Jeffrey H. Siewerdsen, Manav Verma and Qihua Zhao

Department of Radiation Oncology, University of Michigan Medical Center, Ann Arbor, Michigan 48109

Abstract. Recent advances in thin-film electronics technology have opened the way for the use of flat-panel imagers in a number of medical imaging applications. These novel imagers offer real time digital readout capabilities (~30 frames per second), radiation hardness (>10^6 cGy), large area (30 x 40 cm^2) and compactness (~1 cm). Such qualities make them strong candidates for the replacement of conventional x-ray imaging technologies such as film-screen and image intensifier systems. In this report, qualities and potential of amorphous silicon based active matrix flat-panel imagers are outlined for various applications such as radiation therapy, radiography, fluoroscopy and mammography.

The imagers under development by our group generally consist of a two-dimensional array of imaging pixels coupled to a dedicated external electronic acquisition system (1). The array is a glass substrate onto which amorphous silicon circuits have been deposited. Each pixel consists of a photodiode sensor coupled to a field-effect-transistor (FET). Incident radiation is detected indirectly by means of an overlying scintillator (e.g. phosphor or CsI). X-rays interacting in the scintillator generate light photons which are detected in the photodiodes. (An alternative AMFPI approach involves direct detection of the incident x-rays by means of a thick converting material [e.g. a-Se] deposited over the FET matrix (2).) While the photodiodes act as storage capacitors for charge generated by the light photons, the FETs act as switches to enable the readout of this charge. During the course of our research dating back to 1987, a series of progressively larger arrays has been developed starting from small 64x40 pixel devices (3). This work has advanced to the point where it has become possible to construct large area devices up to 30 x 40 cm^2. Array designs with pixel pitches ranging from 100 μm to 508 μm have been developed for various medical imaging applications such as radiation therapy, radiography and fluoroscopy.

In radiation therapy, successful treatment of cancer using an external beam of megavoltage (>1 MeV) photons is contingent on accurate patient positioning prior to and during the treatment, so that maximum dose is delivered to tumor cells while minimizing the dose to surrounding healthy tissue. To achieve this goal, we have developed imagers (4,5) for this application which offer numerous advantages. Most recently, an 26x26 cm^2 imager, with a pixel pitch of 508 μm, has been developed (5,6). Such AMFPI devices are well suited to the cancer therapy environment where high radiation doses are expected (~1x10^6 cGy per year). Previous studies demonstrated that amorphous silicon devices are highly radiation resistant and suffer only slight, or no degradation with exposure to high levels of radiation (7,8). Also, the real-time readout offered by these imagers is ideal for rapid verification of patient positioning as compared to commonly used techniques involving film that require several minutes for film development. Recent observer-based studies have shown that performance achieved with these imagers is superior to that of conventional radiotherapy film systems (5).

In diagnostic imaging, where lower energy x-rays are used, the demands for higher spatial resolution and the associated smaller signal sizes present more of a challenge. For radiographic imaging, which is the production and viewing of single x-ray images, and for fluoroscopic imaging, which is the real time production and viewing of continuous sequences of images, we have developed a prototype imager. This imager has a format of 1920x1536 pixels with a pixel pitch of 127 μm giving a total area of 19.5 x 24.4 cm^2 (9). Detailed characterization and performance assessment of this imager was performed for radiographic and fluoroscopic operation (10,11) along with initial demonstrations of imaging with human subjects (10,12). Although the performance of this imager is far from optimal, it is providing valuable information which, along with detailed theoretical analysis of the system (11), will allow the development of more optimized imagers in the future. For example, higher performance devices are possible through further improvements in thin-film processing techniques such as the reduction of minimum feature sizes. Such improvements will make possible the creation of imagers suitable for the high resolution requirements of applications such as mammographic imaging. Overall, the potential advantages of AMFPIs for medical imaging (e.g. real-time digital readout, compactness, radiation resistance) are sufficiently numerous that there is very strong motivation to develop this technology to the point of matching or exceeding the imaging performance of established, highly optimized imaging technologies such as film-screen and image intensifiers.

In conclusion, our research suggests that active matrix, flat-panel imagers may well become a base technology in the 21st century, providing superior radiation therapy imaging devices and potentially offering an alternative to current radiographic, fluoroscopic and mammographic imaging technologies.

ACKNOWLEDGMENTS

This work is supported by National Institutes of Health grants no. R01-CA51397 and R01-CA56135

REFERENCES

1. Morton, E.J., Antonuk, L.E., Berry, J.E., Huang, W., Mody, P., and Yorkston, J., *IEEE Trans. Nucl. Sci.* **41**, 1150-1154, (1994).
2. Zhao, W., and Rowlands, J.A., "A large area solid state detector for radiology using amorphous selenium", *Proc. SPIE* **1651**, 134-143 (1992).
3. Street, R.A., Antonuk, L.E., and Perez-Mendez, V., "Amorphous silicon sensor arrays for radiation imaging", *Mat. Res. Soc. Proc.* **192**, 441-452 (1990).
4. Antonuk, L.E., Yorkston, J., Huang, W., Sandler, H., Siewerdsen, J.H., and El-Mohri, Y., *Int. J. Radiat. Onc. Biol. Phys.* **36**, 661-672 (1996).
5. Antonuk, L.E., El-Mohri, Y., Yorkston, J., Huang, W., Jee, K.W., Siewerdsen, J.H., Maolinbay, M., Scarpine, V.E., and Sandler, H., "Initial performance evaluation of an indirect-detection, active matrix flat-panel imager (AMFPI) prototype for megavoltage imaging", submitted to *International Journal of Radiation Oncology, Biology, Physics*.
6. Antonuk, L.E, Boudry, J., El-Mohri, Y., Huang, W., Siewerdsen, J.H., Yorkston, J., "Large area, flat-panel, amorphous silicon imagers", *Proc. SPIE* **2432**, 216-227 (1995).
7. Boudry, J.M., and Antonuk, L.E, *IEEE Trans. Nucl. Sci.* **41**, 703-707 (1994).
8. Boudry, J.M., and Antonuk, L.E., *Med. Phys.* **23**, 743-754 (1996).
9. Antonuk, L.E., El-Mohri, Y., Huang, W., Jee, K.W., Maolinbay, M., Scarpine, V.E., Siewerdsen, J.H., Verma, M., Yorkston, J., and Street, R.A., "Development of a high resolution, active-matrix, flat-panel imager with enhanced fill factor", *Proc. SPIE* **3032**, 2-13 (1997).
10. Antonuk, L.E., El-Mohri, Y., Siewerdsen, J.H., Yorkston, J., Huang, W., Scarpine, V.E., and Street, R.A., *Med. Phys.* **24**, 51-70 (1997).
11. Siewerdsen, J.H., Antonuk, L.E., El-Mohri, Y., Yorkston, J., Boudry, J.M., and Huang, W., *Med. Phys.* **24**, 71-89 (1997).
12. Antonuk, L.E., El-Mohri, Y., Huang, W., Sandler, H.M., Yorkston, J., Scarpine, V.E., "Real-time, Digital Radiography and Fluoroscopy of human subjects using active-matrix, flat-panel imagers at diagnostic energies." 1996 Meeting of Radiological Society of North America, Chicago, IL, December 1-6, 1996, supplement to *Radiology* **201(P)**, pp. 327 (abstract).

A new large area x-ray image sensor

Don Ouimette

Division of Imaging Research
Department of Radiology
University of Connecticut Health Center
Farmington CN 06030

A new high speed/high resolution x-ray detector call the XEBIT (X-ray sensitive Electron Beam Image Tube) is currently under development at the University of Connecticut Health Center. This large area (9 inch diameter) direct conversion detector is based on an x-ray photoconductor called thallium bromide. The device utilizes cathode ray tube technology to provide a 30 frame per second raster scanned electron beam to both charge and readout the photoconductor. Thallium bromide is a high Z material with a linear attenuation coefficient of 28.11 cm^{-1} at 60 keV. This high stopping power results in a quantum efficiency in excess of 50% at 60 keV for 300 micron thick layers.

Thallium bromide is a very good X-ray photoconductor which requires 6.5 ev to create an electron-hole pair. For 60 kev photons, this results in a gain of 9230 per absorbed photon. With a hole-mobility lifetime product of 1.5×10^{-6} cm^2/volt, good charge collection can be achieved at reasonable field strengths. Thallium bromide has a very high band gap of 2.7 ev and a dielectric constant of 33. Its resistivity, which is 5×10^9 ohm-cm at room temperature, is dominated by ionic conductivity. Fortunately, ionic conductivity has a strong temperature dependence and can be significantly reduced with moderate cooling to -25 degrees centigrade.

The XEBIT is based on using thallium bromide as a photoconductor in a vidicon type image tube. Its principles of operation are very similar to the standard light sensitive vidicon which were utilized extensively in the commercial television industry. A cathode generates an electron beam which is focused and accelerated down the tube to the target. A fine mesh in front of the target focuses and decelerates the beam resulting in low velocity landing of electrons on the insulating photoconductor surface. The front surface of the photoconductor is on a metallic window and is biased positively with respect to the cathode. Electromagnetics deflect the electron beam in a raster fashion to uniformly coat the surface of the photoconductor with electrons charging it down to cathode potential. This charged surface, with respect to the positively biased front surface, provides the necessary electric field across the photoconductor for charge transport. X-rays then penetrate the window and are absorbed by the thallium bromide. The absorbed photons generate large numbers of electron-hole pairs due to the high conversion gain. Electrons drift under the electric field to the positive bias electrode and the holes drift to the vacuum surface and annihilate stored charge. This results in a image dependent charge pattern on the vacuum surface of the photoconductor. A subsequent scan of the photoconductor generates the capacitively coupled signal by replacing the annihilated electrons.

The XEBIT is currently under development as a replacement for x-ray image intensifiers in medical imaging applications. The first devices are 9 inch prototypes designed to be no larger than standard intensifiers. The image quality is far superior to intensifiers with over 50% modulation at 5 line pairs per millimeter. The XEBIT is capable of full field imaging as well as under scanning to view smaller regions with higher detail. Although the XEBIT has been initially designed for medical imaging, it also has application to other areas such as non-destructive testing and x-ray diffraction.

SRI'97 National Conference — Tuesday, June 17

8:00 AM	Registration, Vendor and Poster setup
	Continental Breakfast (Noyes Community Center dining, second floor)
9:00	**E. Fontes** - Conference opening and welcome

Facilities: history, status, and upgrades

9:05	**B. Batterman** - History of synchrotron radiation at Cornell University
9:30	**R. Madden** - The NIST SURF II storage ring is upgraded
10:00	**S. Krinsky** - Brightness and stability advances at the NSLS
10:30	Coffee break
11:00	**E. Gluskin** - APS insertion devices: magnetic performance and radiation characteristics
11:30	**G. Rosenbaum** - The Structural Biology Center at the APS: an integrated user facility for macromolecular crystallography Facilities
12:00	Lunch (Noyes Community Center, second floor)
1:00 PM	Vendor and poster displays open for public viewing

Sources of x-ray and IR: insertion device, laser excited, and other

1:00	**R. Carr** - Design of a x-ray free electron laser undulator
1:30	**R. Bosch** - Long-wavelength edge radiation in an electron storage ring
2:00	**O. Makarov** - Control system for insertion devices at the Advanced Photon Source
2:15	**G. Fraser** - Characterization of the coherent microwave emission from the SURF II synchrotron storage ring (moved to Thursday program)
2:30	**J. Lang** - Characterization of the Elliptical Multipole Wiggler at the Advanced Photon Source
2:45	(MOVED TO POSTER)**R. Dejus** - Computation of Undulator Tuning Curves Sources
3:00	Vendor and Poster Exhibit Session (and refreshments)

SRI'97 National Conference — Wednesday, June 18

8:00 AM Registration (continued) and Continental Breakfast
Vendor and Poster Exhibit all day

Making beams: novel optics design, fabrication, testing and use

- 9:00 **S. Rogers** - Cryogenic high-heat-load optics at the APS
- 9:15 **K. Smolenski** - Silver bonded, internally water-cooled monochromators for CHESS wiggler beamlines
- 9:30 **L. Berman** - Performance of the double multilayer monochromator on the NSLS wiggler beamline X25
- 9:45 **K. Finkelstein** - Inelastic x-ray scattering at modest energy resolution
- 10:00 **T. Toellner** - X-ray monochromators with sub-meV resolution
- 10:15 **P. Fernandez** - Test results of a diamond double-crystal monochromator at the Advanced Photon Source
- 10:30 Coffee break (in area of Vendor and poster displays)
- 10:45 **Z. Zhong** - A tunable Laue bent-Laue monochromator with fixed second crystal for synchrotron radiation
- 11:00 **Z. Cai** - Beam size measurement of the stored electron beam at the APS storage ring using zone plate optics
- 11:15 **B. Ren** - Beam smiling in bent Laue monochromators
- 11:30 **J. Cross** - The performance of a wide band x-ray Bragg polarizer grown by molecular beam epitaxy
- 11:45 **S. Irick** - Measurement of diffraction gratings with a Long Trace Profiler with application for synchrotron beamline gratings
- 12:00 Lunch

Small Stuff: Making and imaging microstructures

- 1:00 **D. Bilderback** - Glass capillary optics for making x-ray beams of 0.1 µm to 50 µm diameter
- 1:30 **E. Johnson** - Beyond Sunshine: Hard x-rays for precision microfabrication
- 2:00 **Y. Vladimirsky** - X-ray lithography at CXrL - 3D nanostructures
- 2:30 **J. Wang** - X-ray fluorescence correlation spectroscopy for studying particle dynamics in condensed matter

- 3:00 Tour of CHESS
- 3:00 Vendor and Poster Exhibit Session (and refreshments)
- 5:00-9:00 Picnic outing at Taughannock State Park (register in advance)

SRI'97 National Conference — Thursday, June 19

8:00 AM   Continental Breakfast

Final day for Vendor Exhibits

High-tech: beam stability, beamline hardware and software control

9:00 AM	**D. Shu**	Progress of the APS high heat load x-ray beam position monitor development
9:15	**G. Rosenbaum**	High heat load fixed primary aperture for an undulator beamline with integral beam position monitors
9:30	**G. Fraser**	Characterization of the coherent microwave emission from the SURF II synchrotron storage ring (moved from Tuesday program)
9:45	**D. Shu**	Mirror mounts designed for the Advanced Photon Source SRI-CAT
10:00	**G. Rosenbaum**	Miniaturized Kappa goniometer for macromolecular crystallography

10:15   Coffee break (in area of Vendor and poster displays)

Real-time: *In-situ* measurements and materials characterization

11:00	**R. Headrick**	Real-time x-ray scattering studies of thin-film growth and processing
11:30	**R. Clarke**	X-ray Studies of Annealing in Thin-film Semiconductors

12:00   Lunch

1:00-4:00   Optional tour of Cayuga Lake Wineries (register in advance)

1:00	**N. Ipe**	Low energy x-ray dosimetry (6-16 keV) studies at SSRL B11-5
1:30	**P. Heimann**	Ultrafast x-ray diffraction of laser-irradiated crystals
2:00	**T. Egami**	Information stored in high-Q space: role of high-energy scattering
2:30	**C. Tarrio**	In-situ measurements of EUV optical constants of thin films
3:00	**A. Woll**	Real-time x-ray diffraction measurements of GaN growth on a sapphire (0001) surface

3:15   Vendor and Poster Exhibit Session (and refreshments)

5:00   Vendor Exhibit Closes

6:00   Conference Reception and Banquet at Statler Hotel (register in advance)

SRI'97 National Conference — Friday, June 20

8:00 AM Continental Breakfast

Detectors: large, small, fast, energy-resolving

9:00   **E. Eikenberry** - Overview on CCD-based area detectors for synchrotron applications

9:30   **B. Rodricks** - X-ray imaging characteristics of a direct conversion detector using selenium and thin film transistor array

10:00 **R. Vest** - Photoemission from silicon photodiodes and induced changes in the external quantum efficiency in the far ultraviolet

10:30 Coffee break

10:45 **C. Rossington** - Development of fully depleted MOS CCD's (moved to poster)

11:15 **Y. El-Mohri** - X-ray imaging with amorphous silicon active matrix flat-panel imagers (AMFPIs)

11:45 **D. Ouimette** - A new large area x-ray image sensor

12:15 **E. Fontes** - Close of SRI'97 National Conference

Attendee Address and E-mail Directory

Mr. Joel Anspach
Rocketdyne Albuquerque Operations
2511C Broadbent Pkwy NE
Albuquerque NM 87107
joel.e.anspach@boeing.com
505-345-2660 X680

Mr. Ravi Bains
Advanced Research Systems
1942 Riverbend Rd.
Allentown PA 18103
arscryo@aol.com
610-439-8022

Dr. Juan Barraza
Argonne National Laboratory
9700 S. Cass Ave.
Argonne IL 60439
barazza@aps.anl.gov
630-252-4661

Dr. Boris Batterman
CHESS
Cornell University
Wilson Lab.
Ithaca NY 14853
bwb1@cornell.edu
(607) 255-0917

Prof. Michael Bedzyk
Materials Sciences
Northwestern University
2225 N. Campus Dr.
Evanston IL 60208
bedzyk@nwu.edu
847-491-3570/630-252-7763

Dr. Lonny Berman
National Synchrotron Light Source
Brookhaven National Laboratory
Building 725D
Upton NY 11973
berman@bnl.gov
(516)344-5333

Dr. Don Bilderback
CHESS
Cornell University
281 Wilson Lab.
Ithaca NY 14853
dhb2@cornell.edu
(607) 255-0916

Mrs. Virginia Bizzell
CHESS
Cornell University
Wilson Lab.
Ithaca NY 14853
vfb2@cornell.edu
(607) 255-0922

Mr. Peter Blanchet
Advanced Research Systems
1942 Riverbend Rd.
Allentown PA 18103
arscryo@aol.com
610-439-8022

Dr. Nagel Bolding
Oxford Instrument
130A Baker Ave. Extension
Concord MA 01742
penn@oxford.usa.com
508-369-9933

Dr. Jeffrey Bonanno
Rockefeller University
1230 York Ave.
New York NY 10021
bonanno@rockvax.rockefeller.edu
212-327-7429

Mr. Russell Bonn
Engineering
Boeing North American
2511C Broadbent Pkwy
Albuquerque NM 87107
russell.h.bonn@boeing.com
505-345-2660x615

Dr. Robert Bosch
Synchrotron Radiation Center
3731 Schneider Dr.
Stoughton WI 53589-3097
bosch@src.wisc.edu
608-877-2197

Dr. Keith Brister
CHESS
Cornell University
Wilson Lab.
Ithaca NY 14853
kb22@cornell.edu
(607) 255-0920

Prof. H. Brumberger
Chemistry Department
Syracuse University
Syracuse NY 13244
315-443-5923

Dr. Robert Bubeck
Dow Chemical
Bldg. 1897
Midland MI 48674
bubeckra@dow.com
517-636-2957

Dr. Stephen Burley
The Rockefeller University/HHMI
1230 York Avenue
New York NY 10021
212-327-8336

Dr. Zhonghou Cai
Argonne National Laboratory
9700 S. Cass Ave. Bldg. 431/1300S
Argonne IL 60439
cai@aps.anl.gov
630-252-0144

Dr. Roger Carr
SLAC-SSRL
MS 69, PO Box 4349
Stanford CA 94309
carr@slac.stanford.edu
415-926-3965

Mr. Brian Carrol
CHESS
Cornell University
Wilson Lab.
Ithaca NY 14853
(607) 255-7163

Mr. Ben Clark
CHESS
Cornell University
Wilson Lab.
Ithaca NY 14853
bc19@cornell.edu
(607) 255-7163

Prof. Roberto Colella
Physics Department
Purdue University
West Lafayette IN 47907-1396
colella@physics.purdue.edu
765-494-3029

Mr. Chris Conolly
CHESS
Cornell University
Wilson Lab.
Ithaca NY 14853
cjc16@cornell.edu
(607) 255-7163

Dr. Julie Cross
Code 6680
Naval Research Laboratory
4555 Overlook Ave SW
Washington DC 20375
jox@ccf.nrl.navy.mil
202-404-4132

Dr. Mark Daly
X-ray Instrumentation Associates
2513 Charleston Rd. STE 207
Mountain View CA 94043-1607
bill@xai.com
415-903-9980

Dr. Paola De Cecco
SSRL
2575 SandHill Rd.
Menlo Park CA 94025
dececco@ssrl.slac.stanford.edu
415-926-3938

Dr. Ashley Deacon
MacCHESS
Cornell University
209 Biotechnology Building
Ithaca NY 14853
ash@bio.cornell.edu
607-255-2174

Mr. Hans DeJong
Johnsen Ultravac, Inc.
3470 Mainway
Burlington Ontario
juvinfo@ultrahivac.com
800-268-4980

Gregory Denbeaux
FEL Laboratory
Duke University
Lasalle St. Ext.
Durham NC 27708-0319
denbeaux@fel.duke.edu
919-660-2667

Mr. Park Doing
CHESS
Cornell University
Wilson Lab.
Ithaca NY 14853
pad9@cornell.edu
(607) 255-7163

Dr. Christopher Dorn
Electrofusion Products
Brush Wellman Inc.
44036 South Grimmer Blvd.
Fremont CA 94538
Electrofusion@BrushWellman.com
510-661-9747

Prof. Takeshi Egami
LRSM
Univ. Of Pennsylvania
3231 Walnut St.
Philadelphia PA 19104
egami@seas.upenn.edu
215-898-5138

Dr. Eric Eikenberry
Robert Wood Johnson Medical School
Dept. Of Pathology
Piscataway, NJ 08854
eric@viper.princeton.edu
908-235-4770

Dr. Youcef El-Mohri
Radiation Oncology
University of Michigan Medical Center
1500 E. Medical Ctr. Dr., Room UH-B2C432
Ann Arbor MI 48109
elmohri@umich.edu
313-936 9482

Dr. Yiping Feng
Experimental Facilities Division
Argonne National Laboratory
9700 S. Cass Ave.
Argonne IL 60439
yfeng@aps.anl.gov
630-252-1081

Dr. Patricia Fernandez
Experimental Facilities Division
Argonne National Laboratory
9700 S. Cass Ave., Bldg. 401
Argonne IL 60439
fernandez@aps.anl.gov
(630) 252-2901

Mr. George Ferrio
Micro Photonics, Inc.
PO Box 3129
Allentown PA 18106
surftest@aol.com
610-366-7103

Mr. Farshid Feyzi
Physical Sciences Laboratory
University of Wisconsin-Madison
3725 Schneider Drive
Stoughton WI 53589
ffeyzi@ps1.wisc.edu
608-877-2213

Dr. Ken Finkelstein
CHESS
Cornell University
285 Wilson Lab.
Ithaca NY 14853
kdf1@cornell.edu
(607) 255-7163

Dr. Ernest Fontes
CHESS
Cornell University
284 Wilson Lab.
Ithaca NY 14853
ef11@cornell.edu
(607) 255-2959

Dr. Gerald Fraser
Optical Technology Division
NIST
B208 - 221
Gaithersburg MD 20899
fraser@tiber.nist.gov
301-975-3797

Dr. Adolfas Gaigalas
NIST
222/A353 NIST
Gaitersburg MD 20899
gaigalas@nist.gov
301-975-2873

Mr. Scott Giancola
Newport Corporation
1791 Deere Ave.
Irvine CA 92606
ppaetzold@newport.com
603-891-2353

Dr. Efim Gluskin
Experimental Facilities Division
Argonne National Laboratory
9700 S. Cass Avenue, Bldg. 401/B3171
Argonne IL 60439
gluskin@aps.anl.gov
630-252-4788

Mr. Joseph Harkins
ALS Group
Lawrence Berkeley National Laboratory
1 Cyclotron Rd MS 80-101
Berkeley CA 94720
jpharkins@lbl.gov
510-486-7486

Dr. Michael Hart
NSLS
Brookhaven National Laboratory
Building 725B
Upton NY 11973
MHart@bnl.gov
516-344-5939

 Patric Hartog
Argonne/APS
Bldg. 401/B3177
9700 S. Cass Ave.
Argonne IL 60439
PDenHartog@anl.gov
630-252-3722

Dr. Randy Headrick
CHESS
Cornell University
Wilson Lab.
Ithaca NY 14853
rlh6@cornell.edu
(607) 255-0919

Mr. Christopher A. Heaton
MacCHESS
Wilson Laboratory
Ithaca NY 14853
cah19@cornell.edu
607-255-7163

Dr. Philip Heimann
ALS
Lawrence Berkeley Nat. Lab.
1 Cyclotron Rd.
Berkeley CA 94720
heimann@lbl.gov
510-486-7628

Dr. Jules Hendrix
MAR-USA
1840 Oak Ave.
Evanston IL 60201
info@mar-usa.com
847-869-1548

Mr. Don Holly
Physical Sciences Laboratory
Univ. of Wisconsin-Madison
3725 Schneider Drive
Stoughton WI 53589-3034
psl@psl.wisc.edu
608-877-2251

Mr. Gary Holzhausen
Applied Geomechanics Inc.
1336 Brommer St.
Santa Cruz CA 95062
applied@geomechanics.com
408-462-2801

Mr. Charles Howard
Russian Scientific Products
K-TEK International, Inc
7000 SW Hampton Street, Ste 101
Portland OR 97223
science@ktekintl.com
503 624 0315

Dr. Peter Ilinski
Argonne National Laboratory
9700 S. Cass Ave.
Argonne IL 60439
ilinski@aps.anl.gov
630-252-0145

Dr. Nisy Ipe
Radiation Physics
SLAC
P. O. Box 4349 m/s 48
Stanford CA 94309
ipe@slac.stanford.edu
415-926-4324

Mr. Steve Irick
Advanced Light Source
Lawrence Berkeley National Laboratory
MS 2-400
Berkeley CA 94720
SCIrick@lbl.gov
510-486-4077

Dr. De-Tong Jiang
Physics
PNC-CAT/Simon Fraser University
APS, Sector 20, Building 435E, Argonne
National Lab.
Argonne IL 60439
jiang@pnc.aps.anl.gov
630 252 0581

Dr. Erik Johnson
National Synchrotron Light Source
Brookhaven National Laboratory
Building 725D
Upton NY 11973
erik@bnl.gov
(516)344-4603

Mr. Glover Jones
CR&D S&E Labs
Dupont Company
EXSTA 228/320-D
Wilmington DE 19880-0228
gajones@esvax.dnet.dupont.com
302-695-3935

Ms. Beth Kohler
Advanced Research Systems
1942 Riverbend Rd.
Allentown PA 18103
arscryo@aol.com
610-439-8022

Dr. Sam Krinsky
NSLS
Brookhaven National Laboratory
Building 725B
Upton NY 11973
krinsky@bnl.gov
516-344-4740

Dr. Tim Kubala
Synchrotron Radiation Center
University of Wisconsin
3731 Schneider Drive
Stoughton WI 53589
kubala@src.wisc.edu
608 877-2144

Dr. Stefan Kycia
Cornell High Energy Synchrotron Source
Cornell University
Wilson Laboratory
Ithaca NY 14853
sk85@cornell.edu
(607) 255-3617

Dr. Barry Lai
Advanced Photon Source
Argonne National Laboratory
Bldg. 401
Argonne IL 60439
blai@aps.anl.gov
630-252-6405

Mr. Jim LaIuppa
CHESS
Cornell University
Wilson Lab.
Ithaca NY 14853
jml8@cornell.edu
(607) 255-1716

Dr. Jonathan Lang
Experimental Facilities Division
Argonne National Laboratory
9700 S. Cass Avenue, Bldg. 401/B3161
Argonne IL 60439
lang@aps.anl.gov
630-252-0122

Dr. Bennett Larson
Solid State Division
Oak Ridge National Laboratory
Bldg 3025 P.O. Box 2008
Oak Ridge TN 37831-6030
bcl@ornl.gov
423-574-5506

Dr. Heung-Rae Lee
XFD-ASP
Argonne National Laboratory
9700 S. Cass Ave. 431-B006
Argonne IL 60439
hrlee@aps.anl.gov
630-252-0150

Dr. Paul Lyman
MS&E
Northwestern University
LBNL MS 4-230
Berkeley CA 94720
plyman@bnl.gov
510-486-5804

Mr. Adrian Lyons
Micro Photonics, Inc.
PO Box 3129
Allentown PA 18106
surftest@aol.com
610-366-7103

Prof. Joselito Barbosa Maciel
Department of Physical Chemistry
Chemistry Institute
Federal University of Rio de Janeiro
CT Bl. A , Sala 408, Cidade Universitária -
Ilha do Fundão
Rio de Janeiro Rio de Janeiro 21949-900
jbm@iq.ufrj.br
(55) (21) 590 9890

Dr. Robert Madden
NIST
Far Ultraviolet Physics Group
Bldg. 245, Rm. B119
Gaithersburg MD 20899
robert.madden@nist.gov
301-975-3726

Claire Mainprize
Accelerator Technology Group
Oxford Instruments
Osney Mead
Oxford Oxon OX2 0DX
claire.mainprize@oxinst.co.uk
+44 (0) 1865 269500

Dr. Oleg Makarov
Experimental Facilities (XFD)
Advanced Photon Source,
Argonne National Laboratory
9700 South Cass Avenue
Argonne IL 60439-4800
makarov@aps.anl.gov
630 252-9138

Mr. Bill Mason
Physical Sciences Laboratory
Univ. of Wisconsin-Madison
3725 Schneider Drive
Stoughton WI 53589-3034
psl@psl.wisc.edu
608-877-2251
608-877-2201 FAX

Dr. Robert Mayer
SSRL
2575 Sand Hill Rd.
Menlo Park CA 94025
mayer@slac.stanford.edu
415-926-3842

Dr. William McHarque
Experimental Facilities Division
Argonne National Laboratory
9700 S. Cass Avenue, Bldg. 401/B3161
Argonne IL 60439
harque@aps.anl.gov
630-252-0124

Dr. Martin McMillan
3-Dimensional Pharmaceuticals, Inc.
665 Stockton Drive
Exton, PA 19341
mcmillan@3dp.com
610-458-8959

Dr. Dennis Mills
Experimental Facilities Division
Argonne National Laboratory
9700 S. Cass Avenue, Bldg. 401/B3161
Argonne IL 60439
dmm@aps.anl.gov
630-255-5680

Dr. Martin Molloy
Stanford Site Office (SLAC-SSRL)
U.S. Department of Energy
2575 Sand Hill Rd.
Menlo Park CA 94025
martin.molloy@oak.doe.gov
(415) 926-3774

Dr. Elizabeth Moog
Advanced Photon Source
Argonne National Laboratory
XFD-401
9700 S. Cass Ave.
Argonne IL 60439
moog@aps.anl.gov
630-252-5926

Dr. Tim Mooney
Advanced Photon Source (APS)
Argonne National Laboratory
9700 S. Cass Avenue
Argonne IL 60439
mooney@aps.anl.gov
(630)252-9976

Dr. Dean Morris
Oxford Instrument
130A Baker Ave. Extension
Concord MA 01742
morris@oxford.usa.com
508-369-9933

Mr. Joseph Navaie
MacCHESS
Wilson Lab
Ithaca NY 14853
jhn1@cornell.edu
607-255-7163

Mr. Kenneth Nelson
Blake Industries, Inc.
660 Jerusalem Road
Scotch Plains NJ 07076
blake4xray@worldnet.att.net
908-233-7240

Mr. Chris Nielsen
Area Detector Systems Corp.
12550 Stowe Drive
Poway CA 92064
cn@chem.ucsd.edu
619-486-0722

Mr. Peter Paetzold
Newport Corporation
1791 Deere Ave.
Irvine CA 92606
ppaetzold@newport.com
603-891-2353

Mr. Alan Pauling
CHESS
Cornell University
Wilson Lab.
Ithaca NY 14853
 (607) 255-7163

Mr. Nick Piscione
Kurt J. Lesker Co.
1515 Worthington Ave.
Clairton PA 15025-2700
sales@lesker.com
800-245-1656

Mr. Walt Protas
CHESS
Cornell University
Wilson Lab.
Ithaca NY 14853
 (607) 255-7163

Dr. Ruben Reininger
Optics Group
SRC/Univ. of Wisconsin
3731 Schneider Drive
Stoughton WI 53589
ryreinin@facstaff.wisc.edu
608-877-2158

Mr. Baorui Ren
Medical Department
BNL
30 Bell Ave.
Upton NY 11973
ren@bnlarm.bnl.gov
516-344-2045

Dr. David Rice
Cornell University
Wilson Laboratory
Ithaca NY 14853-8001
dhr1@cornell.edu
607-255-8786

Mr. Dana Richter
CHESS
Cornell University
Wilson Lab.
Ithaca NY 14853
der4@cornell.edu
(607) 255-7163

Dr. Kem Robinson
STI Optronics, Inc.
2755 Northup Way
Bellevue WA 98004
kem@stioptronics.com
206-827-0460x377

Dr. Brian Rodricks
Direct Radiography
Sterling Diagnostic Imaging
P. O. Box 6101, Mailbox 629
Newark DE 19714-6101
Rodrcib@strelingdi.com
302 451 3569

Dr. William Rodriques
Argonne National Laboratory
9700 S. Cass Ave.
Argonne IL 60439
william@nwu.edu

Dr. Shawn Rogers
Experimental Facilities Division
Advanced Photon Source
9700 S. Cass Avenue
Argonne IL 60439
rogers@aps.anl.gov
630-252-6613

Mr. David Rognlie
Blake Industries, Inc.
660 Jerusalem Road
Scotch Plains NJ 07076
blake4xray@worldnet.att.net
908-233-7240

Dr. Gerd Rosenbaum
Structural Biology Center
Argonne National Lab
9700 S. Cass Ave., Bldg 435D
Argonne IL 60439-4860
grosenbaum@anl.gov
630 252 3346

Dr. Carolyn Rossington
Lawrence Berkeley National Laboratory
1 Cyclotron Road, MS 70A-3363
Berkeley CA 94720
csrossington@lbl.gov
510-486-7827

Mr. Jeff Schuyler
Applied Geomechanics Inc.
1336 Brommer St.
Santa Cruz CA 95062
applied@geomechanics.com
408-462-2801

Ms. Mary Severson
Synchrotron Radiation Center
University of Wisconsin
3731 Schneider Drive
Stoughton WI 53589
severson@src.wisc.edu
608 877-2140

Dr. Sarvjit Shastri
Advanced Photon Source
Bldg. 431 Rm. A009
Argonne IL 60439
shastri@aps.anl.gov
630-252-0129

Dr. Qun Shen
CHESS
Cornell University
280 Wilson Lab.
Ithaca NY 14853
qs11@cornell.edu
(607) 255-0923

Mr. Steve Shim
Russian Scientific Products
K-TEK International, Inc
7000 SW Hampton Street, Ste 101
Portland OR 97223
503 624 0315

Dr. Deming Shu
Experimental Facilities Division
Argonne National Laboratory
9700 S. Cass Avenue, Bldg. 401/C1255
Argonne IL 60439
shud@aps.anl.gov
630-252-4686

Mr. Rick Smith
Osmic, Inc.
1788 Northwood Dr.
Troy MI 48084
rick@osmic.com
810-362-1290

Martin Smith
Argonne National Laboratory
9700 S. Cass Ave.
Argonne IL 90439
mls@aps.anl.gov
630-252-8881

Mr. Karl Smolenski
CHESS
Cornell University
280 Wilson Lab.
Ithaca NY 14853
kws4@cornell.edu
(607) 255-7163

Mr. Eric Spauldin
CHESS
Cornell University
Wilson Lab.
Ithaca NY 14853
(607) 255-7163

Dr. Caroline Stahle
Laboratory for High Energy Astrophysics
NASA / Goddard Space Flight Center
Code 662
Greenbelt MD 20771
stahle@tigers.gsfc.nasa.gov
301-286-2469

Mr. Jim Stathis
Newport Corporation
1791 Deere Ave.
Irvine CA 92606
ppaetzold@newport.com
603-891-2353

Prof. Robert Suter
Physics
Carnegie Mellon University
5000 Forbes Avenue
Pittsburgh PA 15208
suter@andrew.cmu.edu
412-268-2982

Ms. Dianne Swart
Kurt J. Lesker Co.
1515 Worthington Ave.
Clairton PA 15025-2700
sales@lesker.com
800-245-1656

Dr. D. Marion Szebenyi
MacCHESS
Wilson Laboratory
Ithaca NY 14853
szebenyi@quarterdec.chess.cornell.edu
607-255-7163

Dr. Charles Tarrio
National Institute of Standards and Technology
Physics Bldg. Room A253
Gaithersburg MD 20899
ctarrio@nist.gov
(301) 975-3737

Dr. Daniel Thiel
MacCHESS
Wilson Laboratory
Ithaca NY 14853
djt7@cornell.edu
607-255-7163

Dr. Jon Tischler
Bldg. 438D, Sector 33
ORNL
9700 S. Cass Ave.
Argonne IL 60439-4863
TischlerJZ@ornl.gov
630-252-0861

Dr. Thomas Toellner
XFD
Argonne National Laboratory
9700 S. Cass Ave.
Argonne IL 60439
toellner@aps.anl.gov
(630)252-0166

Mr. Patricia Tomkins
Micro Photonics, Inc.
PO Box 3129
Allentown PA 18106
surftest@aol.com
610-366-7103

Mr. Thomas Tonnessen
Engineering
Boeing North American
2511C Broadbent Pkwy
Albuquerque NM 87107
tom.w.tonnessen@boeing.com
505-345-2660x615

Dr. Emil Trakhtenberg
Argonne National Laboratory
9700 S. Cass Ave.
Argonne IL 60439
emil@aps.anl.gov
630-252-9400

Dr. Robert Vest
N.I.S.T.
Far Ultraviolet Physics Group
Bldg. 245, Rm. B119
Gaithersburg MD 20899
robert.vest@nist.gov
(301)975-3992

Dr. Yuli Vladimirsky
Center for X-ray Lithography
College of Engineering
University of Wisconsin-Madison
3731 Schneider Drive
Stoughton WI 53589-3097
yuli@xraylith.wisc.edu
608-877-2434

Dr. Jin Wang
Experiment Facilities Division
Argonne National Laboratory
9700 S. Cass Ave.
Argonne IL 60439
wangj@aps.anl.gov
(630) 252 - 9125

Mr. Weiru Wang
Cornell University
209 Biotech Building
Ithaca NY 14853
ww17@cornell.edu
607-255-2174

Dr. William Warburton
X-ray Instrumentation Associates
2513 Charleston Rd. STE 207
Mountain View CA 94043-1607
bill@xai.com
415-903-9980

Dr. Charles Weeks
Hauptman-Woodward Medical
Research Institute
73 High Street
Buffalo NY 14203-1196
weeks@hwi.buffalo.edu
716-856-9600

Mr. Jeffrey White
CHESS
Cornell University
181 Wilson Lab.
Ithaca NY 14853
jaw7@cornell.edu
(607) 255-0913

Mr. Arthur Woll
Dept. Of Physics
Cornell University
Ithaca NY 14853
woll@msc.cornell.edu
607-255-0819

Mr. James Wood
Osmic, Inc.
1788 Northwood Dr.
Troy MI 48084
jwood@osmic.com
810-362-1290

Ms. Shenglan Xu
APS/XFD 401/B3160
Argonne National Laboratory
9700 S. Cass Ave.
Argonne IL 60439
xus@aps.anl.gov
630-252-4543

Dr. Wenbing Yun
Experimental Facilities Division
Argonne National Laboratory
9700 S. Cass Avenue, Bldg. 401/B3161
Argonne IL 60439
yun@aps.anl.gov
630-255-5320

Dr. Zhong Zhong
NSLS
Brookhaven National Lab.
Bldg 725D, BNL
Upton NY 11973
zhong@bnlls1.nsls.bnl.gov
(516)344-4744

List of Attending Vendors

Advanced Research Systems

Ravi Bains, Beth Kohler, Pete Blanchet
1942 Riverbend Rd.
Allentown, PA 18103
610-439-8022
610-439-1184 FAX

Advanced Research Systems will display new products and state of the art Closed Cycle Systems and Flow Cryostats for sample cooling to reach temperatures as low as 1.7K. On display will be the Displex 201 for installation on the Huber 511 Four Circle Diffractometer.

A complete line of Temperature Controllers and Cryogenic Sensors will also be on display. New products include the Genesis Controller and Ruthenium Oxide sensors. Information on cryogenic dewars, Transfer Lines, vacuum systems and components will also be available.

Applied Geomechanics Inc.

G. Holzhausen and J. Schuyler
1336 Brommer St.
Santa Cruz CA 95062
408-462-2801
408-462-4418 FAX
E-mail:applied@geomechanics.com
URL: http://www.geomechanics.com

Founded in 1982, Applied Geomechanics now manufactures the world's broadest and most sensitive line of gravity-referenced angle measurement tiltmeters and inclinometers. Because they are gravity-referenced, you can install them anywhere. There is no need for shafts, linkages or other complicated and expensive fixturing. Whether you want to measure degrees, arcseconds, microradians, or nanoradians of angular rotation, Applied Geomechanics has the right product for your application.

Area Detector Systems Corporation

Chris Nielsen
12550 Stowe Drive
Poway, CA 92064
E-mail:cn@chem.ucsd.edu
URL: http://WWW.adsc_Xray.com

CCD detectors for x-ray applications.

Blake Industries, Inc.

David G. Rognlie, Kenneth V. Nelson
660 Jerusalem Road
Scotch Plains, NJ 07076
908-233-7240
908-233-1354 fax
E-mail:blake4xray@worldnet.att.net

X-ray diffraction equipment and accessories for Synchrotron and Laboratory use including Huber Rotary Tables, Goniometer Heads, Beamline Instrumentation and Accessories.

Boeing North American, Rocketdyne Albuquerque Operations

Thomas W. Tonnessen andRussel H. Bonn
2511C Broadbent Pkwy
Albuquerque, NM 87107
505-345-2660 X615
505-345-2589 FAX
E-mail:tom.w.tonnessen@boeing.com

Cooled and uncooled optics, specializing in synchrotron optics.

Brush Wellman - Electrofusion Products

Christopher Dorn
44036 South Grimmmer Blvd.
Fremont, CA 94538
510-661-9747
FAX: 510-623-7600
E-mail:Electrofusion@BrushWellman.com

Anything and everything Beryllium. Windows, domes and specialized components to your specifications.

Johnsen Ultravac, Inc.

Hans DeJong
3470 Mainway Ave.
Burlington Ontario
800-268-4980
905-335-3506 FAX
E-mail: juvinfo@ultrahivac.com
URL: http://www.ultrahivac.com

Monochromators, beamline components, UHV vacuum chambers, XYZ stages, linear motion devices, energy electron analyzer (EELS, AES, ARP). Vacuum systems.

K-Tek International, Inc.

Charles Howard, Steve Shim
7000 SW Hampton Street #101
Portland, OR
503-624-0315
Fax: 503-624-0735
E-mail:science@ktekintl.com

Beryllium windows, domes, and custom fabricated items. K-Tek International, Inc, is the international representative for russian institutes and their privatized companies. Fabricated equipment for x-ray and synchrotron radiation applications include beryllium vcuum barriers, instrumentation, crystals, and scanning probe microscopes (SPM) for surface science.

Kurt J. Lesker Co.

Dianne Swart and Nick Piscione
1515 Worthington Ave.
Clairton PA 15025-2700
800-245-1656
412-233-4275 FAX
E-mail:sales@lesker.com
URL: http://www.lesker.com

Kurt J. Lesker Company will be displaying: surface science components; a low cost XYZ manipulator; the "mechanical hand"; linear drives; and UHV valves and hardware.
"...vacuum science is our business."

Micro Photonics, Inc.

George Ferrio, Patricia Tomkins, Adrian Lyons
PO BOX 3129
Allentown, PA 18106
610-366-7103
610-366-7105 fax
E-mail:surftest@aol.com
URL: http://www.microphotonics.com/

X-ray Imaging Cameras

Newport Corporation

Peter Paetzold, Jim Stathis, and Scott Giancola
1791 Deere Avenue
Irvine, CA 92606
603-891-2353
603-888-7681 FAX
E-mail:ppaetzold@newport.com

Newport designs and manufactures a full line of precision general purpose multiaxis goniometers/diffractometers and accessories. The product line includes Kappa Geometry systems, Eulerian cradles with either open or full circle Chi designs. A wide selection of rotation devices are offered as well as a full line of slits and metrology equipment for measurement of sphere of confusion.

Osmic Inc.

Rick Smith
1788 Northwood Dr.
Troy, MI 48084
810-362-1290
fax: 810-362-4043
E-mail:rick@osmic.com
URL: http://www.osmic.com

Osmic manufactures Ovonyx(TM) multilayer optics for use in x-ray applications such as spectroscopy, diffraction, lithography, astronomy, plasma diagnostics, and synchrotron research. Ovonyx(TM) synchrotron optics are used internationally as supermirrors, monochromators, polarizers, and focusing mirrors. These coated optics provide high reflectivity for selected x-ray energies between 50 eV and 100 keV and have excellent durability and stability.

Oxford Instruments

Nagel Bolding, Dean Morris
130A Baker Ave.
Concord MA 01742
508-369-9933
508-369-6616 fax
E-mail:penn@oxford.usa.com

URL: http://www.oxinst.com/

Oxford Instruments, Accelerator Technology Group, designs, fabricates, and installs beamline equipment. Our range of products includes custom designed monochromator systems, mirror systems (including optics), photon shutters, beam position monitors, masks, beryllium windows, Bremsstrahlung stops, slits, integral shutters, filters, liquid nitrogen coolers, liquid gallium pumping systems, lead shielding transfer lines, supports and stands.

MAR-USA, Inc.

Jules Hendrix
1840 Oak Ave.
Evanston IL 60201 USA
info@mar-usa.com
847-869-1548

The new fast MAR345 Image Plate detector and the MAR CCD Systems.

Physical Sciences Laboratory, University of Wisconson-Madison

Don Holly, Bill Mason
3725 Schneider Drive
Stoughton, WI 53589-3034
608-877-2251
608-877-2201 fax
E-mail:psl@psl.wisc.edu
URL: http://www.psl.wisc.edu/

PSL designs and manufactures a variety of synchrotron instrumentation, including monochromators (SGM,PGM,DCM) for soft and hard x-ray energies, mirror chambers, slits, and other beamline components. PSL also builds many other types of instrumentation and can provide advanced engineering design, fabrication, and measurement services.

Xray Instrumentation Associates

William Warburton, Mark E. Daly
2513 Charleston Rd.
Suite 207
Mountain View, CA 94043
415-903-9980
415-903-9887 FAX
E-mail:bill@xia.com
URL: http://www.xai.com

X-ray Instrumentation Associates develops and sells advanced components for x-ray experimentation. We currently offer a differential pump which suppresses the straight-thru beam, an in-hutch shutter/filter set with NIM control unit, a very high speed digital x-ray processing module for Ge or Si(Li) array detectors, and a compact, inexpensive serially controlled stepper motor pack for Huber slits.

Author Index

A

Alp, E. E., 88, 179
Antonuk, L. E., 243
Arp, U., 48
Assoufid, L., 89

B

Barg, B., 15
Barraza, J., 173, 179
Bedzyk, M. J., 10
Bell, M. I., 117
Bennett, B. R., 117
Benson, C., 179
Berman, L. E., 71
Bilderback, D. H., 147
Billinge, S. J. L., 209
Bilski, P., 197
Bissen, M., 17, 135
Black, E., 16
Bosch, R. A., 35
Brewe, D., 15
Brock, J. D., 218
Brown, F., 15
Bucksbaum, P. H., 204
Bunker, G. B., 16
Burley, S. K., 71

C

Cai, Z., 101, 158, 161, 166
Caliebe, W. A., 6
Canfield, L. R., 234
Capel, M. S., 71
Carr, R., 29
Chang, J., 159, 179
Chang, Z., 204
Chapman, L. D., 95, 106
Chatterji, S., 197
Cheung, L. K., 233
Chubar, O. V., 35
Clarke, R., 196
Cross, J. O., 117
Crozier, E. D., 15

D

Deacon, A., 187
DeCarlo, F., 159, 165
Dejus, R. J., 55
Denbeaux, G., 56, 57, 58
Den Hartog, P. K., 219
de Souza, G. G. B., 22
Dierker, S. B., 71

Dilmanian, F. A., 106
Ding, H., 173
Dmowski, W., 209
Doing, P., 66
Dufresne, E., 71

E

Ealick, S. E., 187
Eberhardt, A. S., 209
Egami, T., 209
Eisert, D., 135
El-Mohri, Y., 243

F

Falco, C. M., 214
Falcone, R. W., 204
Fassò, A., 197
Feng, Y., 158
Fernandez, P. B., 89
Finkelstein, K. D., 80
Fischetti, R., 16
Fisher, M. V., 17, 130, 135
Fontes, E., 147, 192
Fornek, J., 178
Fraser, G., 48

G

Gaigalas, A. K., 224
Gluskin, E., 4, 101
Goldhaber, G., 241
Gordon, R. A., 15
Graber, T., 89
Grenko, J., 161
Groom, D. E., 241
Gückel, H., 156

H

Haeffner, D. R., 230
Hartog, P. D., 43
Hastings, J. B., 6
Headrick, R. L., 195, 218
Heald, S. M., 15
Heimann, P. A., 204
Hession, P., 88
Hight Walker, A. R., 48
Höchst, H., 17
Holland, S. E., 241

Holly, D. J., 124
Hower, N., 57
Hu, M. Y., 88
Huang, X., 106

I

Ilinski, P., 49, 101, 161
Ipe, N., 197
Irick, S. C., 118
Irving, T., 16
Isaacs, E., 161
Ivanov, I., 106

J

Jee, K. W., 243
Jeromin, L. S., 233
Jiang, D. T., 15
Job, P. K., 219
Johnson, E. D., 156
Johnson, L. E., 56, 57, 58
Judd, E., 204

K

Kao, C.-C., 6
Kapteyn, H., 204
Kase, K., 197
Keane, D. T., 10
Kim, K. H., 15
Klein, J. L., 156
Krasnicki, S., 89
Krisch, M., 6
Kubala, T., 130, 135
Kuhn, K. J., 117
Kupperman, D. S., 166
Kuzay, T. M., 173, 179
Kycia, S., 209, 218

L

Lai, B., 101, 159, 165, 179
LaIuppa, J. M., 187
Lang, J. C., 49
Larson, B. C., 80, 140
Larsson, J., 204
Le Duc, G., 95
Lee, D. L., 233
Lee, H.-R., 101, 161, 166
Lee, R. W., 204
Lee, W.-K., 89
Legnini, D., 101
Lehmann, K. K., 48
Lindenberg, A., 204
Lucatorto, T. B., 48, 214
Lyman, P. F., 10

M

Maciel, J. B., 22
Madden, R. P., 3
Madey, J. M. J., 56, 57, 58
Makarov, O. A., 43
Mancini, D. C., 159, 165
Maolinbay, M., 243
Mason, W. P., 124
McKinney, W. R., 118
McNulty, I., 179
Middleton, F. H., 124
Mills, D. M., 89, 230
Milne, J. C., 156
Mochrie, S. G. J., 71
Montanez, P., 6
Moog, E. R., 43, 219
Morikawa, E., 22
Moses, W. W., 241
Murnane, M., 204

N

Niaura, G., 224

O

Olko, P., 197
Ouimette, D., 245
Oversluizen, T., 6

P

Padmore, H. A., 204
Palecki, E. F., 233
Peck, S., 192
Pennypacker, C. R., 241
Permutter, S., 241

R

Ramanathan, M., 173
Randall, K., 179
Reininger, R., 17, 130, 135
Ren, B., 106
Rodricks, B., 233
Rodrigues, W., 101, 161, 166
Rogers, C. S., 61, 89
Rogers, G., 17, 135
Rong, X., 243
Rosenbaum, G., 5, 178, 186
Rossington, C. S., 241
Ruzgas, T., 224

S

Sailor, T., 124
Satyam, P., 158
Schuck, P. J., 204
Seefred, R., 197
Semones, E. J., 219
Severson, M., 130, 135
Shastri, S. D., 230
Shen, Q., 66
Shu, D., 159, 173, 179
Shu, F., 71
Siddons, D. P., 156
Siewerdsen, J. H., 243
Sinha, S. K., 158
Slaughter, J. M., 214
Smith, M. L., 43
Smith, R. E., 124
Smolenski, K. W., 66
Soares, C., 197
Sood, A. K., 158
Srajer, G., 49, 179
Stepanov, S., 16
Stern, E. A., 15
Stover, R. J., 241
Straub, K. D., 56, 57, 58
Sturhahn, W., 88, 179
Sutter, J., 88
Sweet, R. M., 71
Szebenyi, D. M. E., 187

T

Tarrio, C., 214
Thiel, D. J., 187
Thomlinson, W. C., 95, 106
Tischler, J. Z., 80, 140
Toellner, T. S., 88
Trakhtenberg, E., 101
Tsui, O. K. C., 71

V

Venkataraman, C. T., 49
Verma, M., 243
Vest, R. E., 234
Vladimirsky, Y., 157

W

Wang, J., 158
Wang, N. W., 241
Wang, S., 16
Wark, J. S., 204
Watts, R. N., 214
Wei, M., 241
Westbrook, E. M., 5, 186
White, J., 192
Woll, A. R., 218
Wood, W., 135
Wu, X. Y., 106, 158

X

Xie, X., 71
Xu, S., 101
Xu, Z., 179

Y

Yin, Z., 71
Yun, W., 101, 158, 159, 161, 166, 179

Z

Zhang, K., 16
Zhao, Q., 243
Zhong, Z., 95, 106
Zschack, P., 140

AIP Conference Proceedings

	Title	L.C. Number	ISBN
No. 230	Nonlinear Dynamics and Particle Acceleration (Tsukuba, Japan 1990)	91-55348	0-88318-824-4
No. 231	Boron-Rich Solids (Albuquerque, NM 1990)	91-53024	0-88318-793-4
No. 232	Gamma-Ray Line Astrophysics (Paris-Saclay, France 1990)	91-55492	0-88318-875-9
No. 233	Atomic Physics 12 (Ann Arbor, MI 1990)	91-55595	088318-811-2
No. 234	Amorphous Silicon Materials and Solar Cells (Denver, CO 1991)	91-55575	088318-831-7
No. 235	Physics and Chemistry of MCT and Novel IR Detector Materials (San Francisco, CA 1990)	91-55493	0-88318-931-3
No. 236	Vacuum Design of Synchrotron Light Sources (Argonne, IL 1990)	91-55527	0-88318-873-2
No. 237	Kent M. Terwilliger Memorial Symposium (Ann Arbor, MI 1989)	91-55576	0-88318-788-4
No. 238	Capture Gamma-Ray Spectroscopy (Pacific Grove, CA 1990)	91-57923	0-88318-830-9
No. 239	Advances in Biomolecular Simulations (Obernai, France 1991)	91-58106	0-88318-940-2
No. 240	Joint Soviet-American Workshop on the Physics of Semiconductor Lasers (Leningrad, USSR 1991)	91-58537	0-88318-936-4
No. 241	Scanned Probe Microscopy (Santa Barbara, CA 1991)	91-76758	0-88318-816-3
No. 242	Strong, Weak, and Electromagnetic Interactions in Nuclei, Atoms, and Astrophysics: A Workshop in Honor of Stewart D. Bloom's Retirement (Livermore, CA 1991)	91-76876	0-88318-943-7
No. 243	Intersections Between Particle and Nuclear Physics (Tucson, AZ 1991)	91-77580	0-88318-950-X
No. 244	Radio Frequency Power in Plasmas (Charleston, SC 1991)	91-77853	0-88318-937-2
No. 245	Basic Space Science (Bangalore, India 1991)	91-78379	0-88318-951-8
No. 246	Space Nuclear Power Systems (Albuquerque, NM 1992)	91-58793	1-56396-027-3 1-56396-026-5 (pbk.)
No. 247	Global Warming: Physics and Facts (Washington, DC 1991)	91-78423	0-88318-932-1
No. 248	Computer-Aided Statistical Physics (Taipei, Taiwan 1991)	91-78378	0-88318-942-9
No. 249	The Physics of Particle Accelerators (Upton, NY 1989, 1990)	92-52843	0-88318-789-2
No. 250	Towards a Unified Picture of Nuclear Dynamics (Nikko, Japan 1991)	92-70143	0-88318-951-8

Title	L.C. Number	ISBN
No. 251 Superconductivity and its Applications (Buffalo, NY 1991)	92-52726	1-56396-016-8
No. 252 Accelerator Instrumentation (Newport News, VA 1991)	92-70356	0-88318-934-8
No. 253 High-Brightness Beams for Advanced Accelerator Applications (College Park, MD 1991)	92-52705	0-88318-947-X
No. 254 Testing the AGN Paradigm (College Park, MD 1991)	92-52780	1-56396-009-5
No. 255 Advanced Beam Dynamics Workshop on Effects of Errors in Accelerators, Their Diagnosis and Corrections (Corpus Christi, TX 1991)	92-52842	1-56396-006-0
No. 256 Slow Dynamics in Condensed Matter (Fukuoka, Japan 1991)	92-53120	0-88318-938-0
No. 257 Atomic Processes in Plasmas (Portland, ME 1991)	91-08105	0-88318-939-9
No. 258 Synchrotron Radiation and Dynamic Phenomena (Grenoble, France 1991)	92-53790	1-56396-008-7
No. 259 Future Directions in Nuclear Physics with 4p Gamma Detection Systems of the New Generation (Strasbourg, France 1991)	92-53222	0-88318-952-6
No. 260 Computational Quantum Physics (Nashville, TN 1991)	92-71777	0-88318-933-X
No. 261 Rare and Exclusive B&K Decays and Novel Flavor Factories (Santa Monica, CA 1991)	92-71873	1-56396-055-9
No. 262 Molecular Electronics—Science and Technology (St. Thomas, Virgin Islands 1991)	92-72210	1-56396-041-9
No. 263 Stress-Induced Phenomena in Metallization: First International Workshop (Ithaca, NY 1991)	92-72292	1-56396-082-6
No. 264 Particle Acceleration in Cosmic Plasmas (Newark, DE 1991)	92-73316	0-88318-948-8
No. 265 Gamma-Ray Bursts (Huntsville, AL 1991)	92-73456	1-56396-018-4
No. 266 Group Theory in Physics (Cocoyoc, Morelos, Mexico 1991)	92-73457	1-56396-101-6
No. 267 Electromechanical Coupling of the Solar Atmosphere (Capri, Italy 1991)	92-82717	1-56396-110-5
No. 268 Photovoltaic Advanced Research & Development Project (Denver, CO 1992)	92-74159	1-56396-056-7
No. 269 CEBAF 1992 Summer Workshop (Newport News, VA 1992)	92-75403	1-56396-067-2
No. 270 Time Reversal—The Arthur Rich Memorial Symposium (Ann Arbor, MI 1991)	92-83852	1-56396-105-9

	Title	L.C. Number	ISBN
No. 271	Tenth Symposium Space Nuclear Power and Propulsion (Vols. I–III) (Albuquerque, NM 1993)	92-75162	1-56396-137-7 (set)
No. 272	Proceedings of the XXVI International Conference on High Energy Physics (Vols. I and II) (Dallas, TX 1992)	93-70412	1-56396-127-X (set)
No. 273	Superconductivity and Its Applications (Buffalo, NY 1992)	93-70502	1-56396-189-X
No. 274	VIth International Conference on the Physics of Highly Charged Ions (Manhattan, KS 1992)	93-70577	1-56396-102-4
No. 275	Atomic Physics 13 (Munich, Germany 1992)	93-70826	1-56396-057-5
No. 276	Very High Energy Cosmic-Ray Interactions: VIIth International Symposium (Ann Arbor, MI 1992)	93-71342	1-56396-038-9
No. 277	The World at Risk: Natural Hazards and Climate Change (Cambridge, MA 1992)	93-71333	1-56396-066-4
No. 278	Back to the Galaxy (College Park, MD 1992)	93-71543	1-56396-227-6
No. 279	Advanced Accelerator Concepts (Port Jefferson, NY 1992)	93-71773	1-56396-191-1
No. 280	Compton Gamma-Ray Observatory (St. Louis, MO 1992)	93-71830	1-56396-104-0
No. 281	Accelerator Instrumentation Fourth Annual Workshop (Berkeley, CA 1992)	93-072110	1-56396-190-3
No. 282	Quantum 1/f Noise & Other Low Frequency Fluctuations in Electronic Devices (St. Louis, MO 1992)	93-072366	1-56396-252-7
No. 283	Earth and Space Science Information Systems (Pasadena, CA 1992)	93-072360	1-56396-094-X
No. 284	US-Japan Workshop on Ion Temperature Gradient-Driven Turbulent Transport (Austin, TX 1993)	93-72460	1-56396-221-7
No. 285	Noise in Physical Systems and 1/f Fluctuations (St. Louis, MO 1993)	93-72575	1-56396-270-5
No. 286	Ordering Disorder: Prospect and Retrospect in Condensed Matter Physics: Proceedings of the Indo-U.S. Workshop (Hyderabad, India 1993)	93-072549	1-56396-255-1
No. 287	Production and Neutralization of Negative Ions and Beams: Sixth International Symposium (Upton, NY 1992)	93-72821	1-56396-103-2
No. 288	Laser Ablation: Mechanismas and Applications-II: Second International Conference (Knoxville, TN 1993)	93-73040	1-56396-226-8

	Title	L.C. Number	ISBN
No. 289	Radio Frequency Power in Plasmas: Tenth Topical Conference (Boston, MA 1993)	93-72964	1-56396-264-0
No. 290	Laser Spectroscopy: XIth International Conference (Hot Springs, VA 1993)	93-73050	1-56396-262-4
No. 291	Prairie View Summer Science Academy (Prairie View, TX 1992)	93-73081	1-56396-133-4
No. 292	Stability of Particle Motion in Storage Rings (Upton, NY 1992)	93-73534	1-56396-225-X
No. 293	Polarized Ion Sources and Polarized Gas Targets (Madison, WI 1993)	93-74102	1-56396-220-9
No. 294	High-Energy Solar Phenomena: A New Era of Spacecraft Measurements (Waterville Valley, NH 1993)	93-74147	1-56396-291-8
No. 295	The Physics of Electronic and Atomic Collisions: XVIII International Conference (Aarhus, Denmark, 1993)	93-74103	1-56396-290-X
No. 296	The Chaos Paradigm: Developments an Applications in Engineering and Science (Mystic, CT 1993)	93-74146	1-56396-254-3
No. 297	Computational Accelerator Physics (Los Alamos, NM 1993)	93-74205	1-56396-222-5
No. 298	Ultrafast Reaction Dynamics and Solvent Effects (Royaumont, France 1993)	93-074354	1-56396-280-2
No. 299	Dense Z-Pinches: Third International Conference (London, 1993)	93-074569	1-56396-297-7
No. 300	Discovery of Weak Neutral Currents: The Weak Interaction Before and After (Santa Monica, CA 1993)	94-70515	1-56396-306-X
No. 301	Eleventh Symposium Space Nuclear Power and Propulsion (3 Vols.) (Albuquerque, NM 1994)	92-75162	1-56396-305-1 (set) 156396-301-9 (pbk. set)
No. 302	Lepton and Photon Interactions/ XVI International Symposium (Ithaca, NY 1993)	94-70079	1-56396-106-7
No. 303	Slow Positron Beam Techniques for Solids and Surfaces Fifth International Workshop (Jackson Hole, WY 1992)	94-71036	1-56396-267-5
No. 304	The Second Compton Symposium (College Park, MD 1993)	94-70742	1-56396-261-6
No. 305	Stress-Induced Phenomena in Metallization Second International Workshop (Austin, TX 1993)	94-70650	1-56396-251-9
No. 306	12th NREL Photovoltaic Program Review (Denver, CO 1993)	94-70748	1-56396-315-9

Title	L.C. Number	ISBN
No. 307 Gamma-Ray Bursts Second Workshop (Huntsville, AL 1993)	94-71317	1-56396-336-1
No. 308 The Evolution of X-Ray Binaries (College Park, MD 1993)	94-76853	1-56396-329-9
No. 309 High-Pressure Science and Technology—1993 (Colorado Springs, CO 1993)	93-72821	1-56396-219-5 (set)
No. 310 Analysis of Interplanetary Dust (Houston, TX 1993)	94-71292	1-56396-341-8
No. 311 Physics of High Energy Particles in Toroidal Systems (Irvine, CA 1993)	94-72098	1-56396-364-7
No. 312 Molecules and Grains in Space (Mont Sainte-Odile, France 1993)	94-72615	1-56396-355-8
No. 313 The Soft X-Ray Cosmos ROSAT Science Symposium (College Park, MD 1993)	94-72499	1-56396-327-2
No. 314 Advances in Plasma Physics Thomas H. Stix Symposium (Princeton, NJ 1992)	94-72721	1-56396-372-8
No. 315 Orbit Correction and Analysis in Circular Accelerators (Upton, NY 1993)	94-72257	1-56396-373-6
No. 316 Thirteenth International Conference on Thermoelectrics (Kansas City, Missouri 1994)	95-75634	1-56396-444-9
No. 317 Fifth Mexican School of Particles and Fields (Guanajuato, Mexico 1992)	94-72720	1-56396-378-7
No. 318 Laser Interaction and Related Plasma Phenomena 11th International Workshop (Monterey, CA 1993)	94-78097	1-56396-324-8
No. 319 Beam Instrumentation Workshop (Santa Fe, NM 1993)	94-78279	1-56396-389-2
No. 320 Basic Space Science (Lagos, Nigeria 1993)	94-79350	1-56396-328-0
No. 321 The First NREL Conference on Thermophotovoltaic Generation of Electricity (Copper Mountain, CO 1994)	94-72792	1-56396-353-1
No. 322 Atomic Processes in Plasmas Ninth APS Topical Conference (San Antonio, TX)	94-72923	1-56396-411-2
No. 323 Atomic Physics 14 Fourteenth International Conference on Atomic Physics (Boulder, CO 1994)	94-73219	1-56396-348-5
No. 324 Twelfth Symposium on Space Nuclear Power and Propulsion (Albuquerque, NM 1995)	94-73603	1-56396-427-9
No. 325 Conference on NASA Centers for Commercial Development of Space (Albuquerque, NM 1995)	94-73604	1-56396-431-7

Title	L.C. Number	ISBN
No. 326 Accelerator Physics at the Superconducting Super Collider (Dallas, TX 1992-1993)	94-73609	1-56396-354-X
No. 327 Nuclei in the Cosmos III Third International Symposium on Nuclear Astrophysics (Assergi, Italy 1994)	95-75492	1-56396-436-8
No. 328 Spectral Line Shapes, Volume 8 12th ICSLS (Toronto, Canada 1994)	94-74309	1-56396-326-4
No. 329 Resonance Ionization Spectroscopy 1994 Seventh International Symposium (Bernkastel-Kues, Germany 1994)	95-75077	1-56396-437-6
No. 330 E.C.C.C. 1 Computational Chemistry F.E.C.S. Conference (Nancy, France 1994)	95-75843	1-56396-457-0
No. 331 Non-Neutral Plasma Physics II (Berkeley, CA 1994)	95-79630	1-56396-441-4
No. 332 X-Ray Lasers 1994 Fourth International Colloquium (Williamsburg, VA 1994)	95-76067	1-56396-375-2
No. 333 Beam Instrumentation Workshop (Vancouver, B. C., Canada 1994)	95-79635	1-56396-352-3
No. 334 Few-Body Problems in Physics (Williamsburg, VA 1994)	95-76481	1-56396-325-6
No. 335 Advanced Accelerator Concepts (Fontana, WI 1994)	95-78225	1-56396-476-7 (set) 1-56396-474-0 (Book) 1-56396-475-9 (CD-Rom)
No. 336 Dark Matter (College Park, MD 1994)	95-76538	1-56396-438-4
No. 337 Pulsed RF Sources for Linear Colliders (Montauk, NY 1994)	95-76814	1-56396-408-2
No. 338 Intersections Between Particle and Nuclear Physics 5th Conference (St. Petersburg, FL 1994)	95-77076	1-56396-335-3
No. 339 Polarization Phenomena in Nuclear Physics Eighth International Symposium (Bloomington, IN 1994)	95-77216	1-56396-482-1
No. 340 Strangeness in Hadronic Matter (Tucson, AZ 1995)	95-77477	1-56396-489-9
No. 341 Volatiles in the Earth and Solar System (Pasadena, CA 1994)	95-77911	1-56396-409-0
No. 342 CAM -94 Physics Meeting (Cacun, Mexico 1994)	95-77851	1-56396-491-0
No. 343 High Energy Spin Physics Eleventh International Symposium (Bloomington, IN 1994)	95-78431	1-56396-374-4

	Title	L.C. Number	ISBN
No. 344	Nonlinear Dynamics in Particle Accelerators: Theory and Experiments (Arcidosso, Italy 1994)	95-78135	1-56396-446-5
No. 345	International Conference on Plasma Physics ICPP 1994 (Foz do Iguaçu, Brazil 1994)	95-78438	1-56396-496-1
No. 346	International Conference on Accelerator-Driven Transmutation Technologies and Applications (Las Vegas, NV 1994)	95-78691	1-56396-505-4
No. 347	Atomic Collisions: A Symposium in Honor of Christopher Bottcher (1945-1993) (Oak Ridge, TN 1994)	95-78689	1-56396-322-1
No. 348	Unveiling the Cosmic Infrared Background (College Park, MD, 1995)	95-83477	1-56396-508-9
No. 349	Workshop on the Tau/Charm Factory (Argonne, IL, 1995)	95-81467	1-56396-523-2
No. 350	International Symposium on Vector Boson Self-Interactions (Los Angeles, CA 1995)	95-79865	1-56396-520-8
No. 351	The Physics of Beams Andrew Sessler Symposium (Los Angeles, CA 1993)	95-80479	1-56396-376-0
No. 352	Physics Potential and Development of $m^+ m^-$ Colliders: Second Workshop (Sausalito, CA 1994)	95-81413	1-56396-506-2
No. 353	13th NREL Photovoltaic Program Review (Lakewood, CO 1995)	95-80662	1-56396-510-0
No. 354	Organic Coatings (Paris, France, 1995)	96-83019	1-56396-535-6
No. 355	Eleventh Topical Conference on Radio Frequency Power in Plasmas (Palm Springs, CA 1995)	95-80867	1-56396-536-4
No. 356	The Future of Accelerator Physics (Austin, TX 1994)	96-83292	1-56396-541-0
No. 357	10th Topical Workshop on Proton-Antiproton Collider Physics (Batavia, IL 1995)	95-83078	1-56396-543-7
No. 358	The Second NREL Conference on Thermophotovoltaic Generation of Electricity	95-83335	1-56396-509-7
No. 359	Workshops and Particles and Fields and Phenomenology of Fundamental Interactions (Puebla, Mexico 1995)	96-85996	1-56396-548-8
No. 360	The Physics of Electronic and Atomic Collisions XIX International Conference (Whistler, Canada, 1995)	95-83671	1-56396-440-6
No. 361	Space Technology and Applications International Forum (Albuquerque, NM 1996)	95-83440	1-56396-568-2
No. 362	Two-Center Effects in Ion-Atom Collisions (Lincoln, NE 1994)	96-83379	1-56396-342-6

	Title	L.C. Number	ISBN
No. 363	Phenomena in Ionized Gases XXII ICPIG (Hoboken, NJ, 1995)	96-83294	1-56396-550-X
No. 364	Fast Elementary Processes in Chemical and Biological Systems (Villeneuve d'Ascq, France, 1995)	96-83624	1-56396-564-X
No. 365	Latin-American School of Physics XXX ELAF Group Theory and Its Applications (México City, México, 1995)	96-83489	1-56396-567-4
No. 366	High Velocity Neutron Stars and Gamma-Ray Bursts (La Jolla, CA 1995)	96-84067	1-56396-593-3
No. 367	Micro Bunches Workshop (Upton, NY, 1995)	96-83482	1-56396-555-0
No. 368	Acoustic Particle Velocity Sensors: Design, Performance and Applications (Mystic, CT, 1995)	96-83548	1-56396-549-6
No. 369	Laser Interaction and Related Plasma Phenomena (Osaka, Japan 1995)	96-85009	1-56396-445-7
No. 370	Shock Compression of Condensed Matter-1995 (Seattle, WA 1995)	96-84595	1-56396-566-6
No. 371	Sixth Quantum 1/f Noise and Other Low Frequency Fluctuations in Electronic Devices Symposium (St. Louis, MO, 1994)	96-84200	1-56396-410-4
No. 372	Beam Dynamics and Technology Issues for + - Colliders 9th Advanced ICFA Beam Dynamics Workshop (Montauk, NY, 1995)	96-84189	1-56396-554-2
No. 373	Stress-Induced Phenomena in Metallization (Palo Alto, CA 1995)	96-84949	1-56396-439-2
No. 374	High Energy Solar Physics (Greenbelt, MD 1995)	96-84513	1-56396-542-9
No. 375	Chaotic, Fractal, and Nonlinear Signal Processing (Mystic, CT 1995)	96-85356	1-56396-443-0
No. 376	Chaos and the Changing Nature of Science and Medicine: An Introduction (Mobile, AL 1995)	96-85220	1-56396-442-2
No. 377	Space Charge Dominated Beams and Applications of High Brightness Beams (Bloomington, IN 1995)	96-85165	1-56396-625-7
No. 378	Surfaces, Vacuum, and Their Applications (Cancun, Mexico 1994)	96-85594	1-56396-418-X
No. 379	Physical Origin of Homochirality in Life (Santa Monica, CA 1995)	96-86631	1-56396-507-0
No. 380	Production and Neutralization of Negative Ions and Beams / Production and Application of Light Negative Ions (Upton, NY 1995)	96-86435	1-56396-565-8
No. 381	Atomic Processes in Plasmas (San Francisco, CA 1996)	96-86304	1-56396-552-6
No. 382	Solar Wind Eight (Dana Point, CA 1995)	96-86447	1-56396-551-8

	Title	L.C. Number	ISBN
No. 383	Workshop on the Earth's Trapped Particle Environment (Taos, NM 1994)	96-86619	1-56396-540-2
No. 384	Gamma-Ray Bursts (Huntsville, AL 1995)	96-79458	1-56396-685-9
No. 385	Robotic Exploration Close to the Sun: Scientific Basis (Marlboro, MA 1996)	96-79560	1-56396-618-2
No. 386	Spectral Line Shapes, Volume 9 13th ICSLS (Firenze, Italy 1996)		1-56396-656-5
No. 387	Space Technology and Applications International Forum (Albuquerque, NM 1997)	96-80254	1-56396-679-4 (Case set) 1-56396-691-3 (Paper set)
No. 388	Resonance Ionization Spectroscopy 1996 Eighth International Symposium (State College, PA 1996)	96-80324	1-56396-611-5
No. 389	X-Ray and Inner-Shell Processes 17th International Conference (Hamburg, Germany 1996)	96-80388	1-56396-563-1
No. 390	Beam Instrumentation Proceedings of the Seventh Workshop (Argonne, IL 1996)	97-70568	1-56396-612-3
No. 391	Computational Accelerator Physics (Williamsburg, VA 1996)	97-70181	1-56396-671-9
No. 392	Applications of Accelerators in Research and Industry: Proceedings of the Fourteenth International Conference (Denton, TX 1996)	97-71846	1-56396-652-2
No. 393	Star Formation Near and Far Seventh Astrophysics Conference (College Park, MD 1996)	97-71978	1-56396-678-6
No. 394	NREL/SNL Photovoltaics Program Review Proceedings of the 14th Conference—A Joint Meeting (Lakewood, CO 1996)	97-72645	1-56396-687-5
No. 395	Nonlinear and Collective Phenomena in Beam Physics (Arcidosso, Italy 1996)	97-72970	1-56396-668-9
No. 396	New Modes of Particle Acceleration—Techniques and Sources (Santa Barbara, CA 1996)	97-72977	1-56396-728-6
No. 397	Future High Energy Colliders (Santa Barbara, CA 1997)	97-73333	1-56396-729-4
No. 398	Advanced Accelerator Colliders Seventh Workshop (Lake Tahoe, CA 1996)	97-72788	1-56396-697-2 (set) 1-56396-727-8 (cloth) 1-56396-726-X (CD-Rom)
No. 399	The Changing Role of Physics Departments: Proceedings of International Conference on Undergraduate Physics Education (College Park, MD 1996)	97-74866	1-56396-698-0

	Title	L.C. Number	ISBN
No. 400	High Energy Physics First Latin Symposium (Yucatan, México 1996)	97-73971	1-56396-686-7
No. 402	Astrophysical Implications of the Laboratory Study of Presolar Materials (St. Louis, MO 1996)	97-74679	1-56396-664-6
No. 401	Thermophotovoltaic Generation of Electricity Third NREL Conference (Colorado Springs, CO 1997)	97-74374	1-56396-734-0
No. 403	Radio Frequency Power in Plasmas 12th Topical Conference (Savannah, GA 1997)	97-74472	1-56396-709-X
No. 404	Future Generations Photovoltaic Technologies First NREL Conference (Denver, CO 1997)	97-74386	1-56396-704-9
No. 405	Beam Stability and Nonlinear Dynamics (Santa Barbara, CA 1996)	97-74676	1-56396-731-6
No. 406	Laser Interaction and Related Plasma Phenomena 13th International Conference (Monterey, CA 1997)	97-76763	1-56396-696-4
No. 407	Deep Inelastic Scattering and QCD 5th International Workshop (Chicago, IL 1997)	97-74677	1-56396-716-2
No. 408	The Ultraviolet Universe at Low and High Redshift: Probing the Progress of Galaxy Evolution (College Park, MD 1997)	97-76762	1-56396-708-1
No. 409	Dense Z-Pinches 4th International Conference (Vancouver, Canada 1997)	97-76959	1-56396-610-7
No. 410	Proceedings of the 4th Compton Symposium (Williamsburg, VA 1997)	97-77179	1-56396-659-X (set)
No. 411	Applied Non-Linear Dynamics Near the Millenium (San Diego, CA 1997)	97-77035	1-56396-736-7
No. 412	Intersections Between Particle and Nuclear Physics 6th Conference (Big Sky, MT 1997)	97-0564	1-56396-712-X
No. 413	Towards X-Ray Free Electron Lasers Workshop on Single Pass, High Gain FELs Starting from Noise, Aiming at Coherent X-Rays (Garda Lake, Italy 1997)	97-06161	1-56396-744-8
No. 414	Two-Dimensional Turbulence in Plasmas and Fluids Research Workshop (Canberra, Australia 1997)	97-06162	1-56396-764-2
No. 415	Beyond the Standard Model V Fifth Conference (Balholm, Norway 1997)	97-77246	1-56396-735-9
No. 416	Similarities and Differences Between Atomic Nuclei and Clusters: Toward a Unified Development of Cluster Science (Tsukuba, Japan 1997)		1-56396-714-6
No. 417	Synchrotron Radiation Instrumentation Tenth US National Conference (Ithaca, NY 1997)	97-77402	1-56396-742-1